# Methods in Cell Biology

**VOLUME 43**

Protein Expression in Animal Cells

**Series Editors**

Leslie Wilson
Department of Biological Sciences
University of California, Santa Barbara
Santa Barbara, California

Paul Matsudaira
Whitehead Institute for Biomedical Research and
Department of Biology
Massachusetts Institute of Technology
Cambridge, Massachusetts

# Methods in Cell Biology

Prepared under the Auspices of the American Society for Cell Biology

## VOLUME 43

Protein Expression in Animal Cells


Edited by

**Michael G. Roth**

Department of Biochemistry
University of Texas Southwestern Medical Center
Dallas, Texas

**ACADEMIC PRESS**

San Diego New York Boston London Sydney Tokyo Toronto

*Cover photograph (paperback edition only)*: Intermediate passage cells (passage 30) 48 hr after infection with the AdCMV-GLUT2 recombinant adenovirus. From Ferber *et al.* (1994). Reprinted with permission from *The Journal of Biological Chemistry*, Volume 269, p. 11526. Photograph kindly provided by Christopher B. Newgard and Richard Noel.

This book is printed on acid-free paper. ♾



Academic Press, Inc.
A Division of Harcourt Brace & Company
525 B Street, Suite 1900, San Diego, California 92101-4495

*United Kingdom Edition published by*
Academic Press Limited
24-28 Oval Road, London NW1 7DX

International Standard Serial Number: 0091-679X

International Standard Book Number: 0-12-564144-3 (Hardcover)

International Standard Book Number: 0-12-598560-6 (Paperback)

PRINTED IN THE UNITED STATES OF AMERICA
94 95 96 97 98 99 EB 9 8 7 6 5 4 3 2 1

# CONTENTS

# CONTRIBUTORS

*Numbers in parentheses indicate the pages on which the authors' contributions begin.*

**Tausif Alam** (161), Departments of Biochemistry and Internal Medicine, and Gifford Laboratories for Diabetes Research, University of Texas Southwestern Medical Center, Dallas, Texas 75235

**Thomas C. Becker** (161), Departments of Biochemistry and Internal Medicine, and Gifford Laboratories for Diabetes Research, University of Texas Southwestern Medical Center, Dallas, Texas 75235

**Colleen B. Brewer** (233), Department of Biochemistry, University of Texas Southwestern Medical Center, Dallas, Texas 75235

**Jane C. Burns** (99), Department of Pediatrics, Center for Molecular Genetics, University of California, San Diego, La Jolla, California 92093

**L. Carnell** (263), Department of Molecular and Cell Biology, University of California, Berkeley, Berkeley, California 94720

**R. A. Chavez** (263), Department of Molecular and Cell Biology, University of California, Berkeley, Berkeley, California 94720

**Y.-T. Chen** (263), Department of Molecular and Cell Biology, University of California, Berkeley, Berkeley, California 94720

**Ward S. Coats** (161), Departments of Biochemistry and Internal Medicine, and Gifford Laboratories for Diabetes Research, University of Texas Southwestern Medical Center, Dallas, Texas 75235

**Richard W. Compans** (3), Department of Microbiology and Immunology, Emory University School of Medicine, Altanta, Georgia 30322

**David G. Cook** (289), Department of Pathology and Laboratory Medicine, University of Pennsylvania Medical Center, Philadelphia, Pennsylvania 19104

**Robert W. Doms** (289), Department of Pathology and Laboratory Medicine, University of Pennsylvania Medical Center, Philadelphia, Pennsylvania 19104

**Paul Dupree** (43), European Molecular Biology Laboratory, 69012 Heidelberg, Germany

**Nicolas Fasel** (247), Institute of Biochemistry, University of Lausanne, CH-1066 Epalinges, Switzerland

**Theodore Friedmann** (99, 211), Department of Pediatrics, Center for Molecular Genetics, University of California, San Diego, La Jolla, California 92093

**Henrik Garoff** (43), Department of Molecular Biology, Karolinska Institute, 14157 Huddinge, Sweden

**Robert D. Gerard** (161), Departments of Biochemistry and Internal Medicine, and Gifford Laboratories for Diabetes Research, University of Texas Southwestern Medical Center, Dallas, Texas 75235

**Anna M. Gómez-Foix** (161), Departments of Biochemistry and Internal Medicine, and Gifford Laboratories for Diabetes Research, University of Texas Southwestern Medical Center, Dallas, Texas 75235

**Joachim Herz** (305), Department of Molecular Genetics, University of Texas Southwestern Medical Center, Dallas, Texas 75235

**Robert P. Hirt** (247), Institute of Biochemistry, University of Lausanne, CH-1066 Epalinges, Switzerland

**Dora Y. Ho** (191), Department of Biological Sciences, Stanford University, Stanford, California 94305

**David E. James** (55), Centre for Molecular Biology and Biotechnology, University of Queensland, Brisbane, Queensland 4072, Australia

**Paul A. Johnson** (211), Department of Pediatrics, Center for Molecular Genetics, University of California, San Diego, La Jolla, California 92093

**Stephen A. Johnston** (353), Departments of Internal Medicine and Biochemistry, University of Texas Southwestern Medical Center, Dallas, Texas 75235

**Jean-Pierre Kraehenbuhl** (247), Institute of Biochemistry, University of Lausanne, and Swiss Institute for Experimental Cancer Research, CH-1066 Epalinges, Switzerland

**Virginia M. -Y. Lee** (289), Department of Pathology and Laboratory Medicine, University of Pennsylvania Medical Center, Philadelphia, Pennsylvania 19104

**Guangpu Li** (55), Department of Cell Biology and Physiology, Washington University, St. Louis, Missouri 63110

**Peter Liljeström** (43), Department of Molecular Biology, Karolinska Institute, 14157 Huddinge, Sweden

**Jean-François Louvion** (335), Département de Biologie Cellulaire, Université de Genève, CH-1211 Genève 4, Switzerland

**Caroline E. Machamer** (137), Department of Cell Biology and Anatomy, Johns Hopkins University School of Medicine, Baltimore, Maryland 21205

**Tiziana Mattioni** (335), Département de Biologie Cellulaire, Université de Genève, CH-1211 Genève 4, Switzerland

**Hsiao-Ping Moore** (263), Department of Molecular and Cell Biology, University of California, Berkeley, Berkeley, California 94720

**Hussein Y. Naim** (113), Department of Biochemistry, University of Texas Southwestern Medical Center, Dallas, Texas 75235

**Christopher B. Newgard** (161), Departments of Biochemistry and Internal Medicine, and Gifford Laboratories for Diabetes Research, University of Texas Southwestern Medical Center, Dallas, Texas 75235

**Richard J. Noel** (161), Departments of Biochemistry and Internal Medicine, and Gifford Laboratories for Diabetes Research, University of Texas Southwestern Medical Center, Dallas, Texas 75235

**Greg Odorizzi** (79), Department of Cancer Biology, The Salk Institute, San Diego, California 92186

**Vesa M. Olkkonen** (43), European Molecular Biology Laboratory, 69012 Heidelberg, Germany

**Didier Picard** (335), Départment de Biologie Cellulaire, Université de Génève, CH-1211 Genève 4, Switzerland

**Robert C. Piper** (55), Institute of Molecular Biology, University of Oregon, Eugene, Oregon 97403

**Paul C. Roberts** (3), Department of Microbiology and Immunology, Emory University School of Medicine, Altanta, Georgia 30322

**Michael G. Roth** (113), Department of Biochemistry, University of Texas Southwestern Medical Center, Dallas, Texas 75235

**W. K. Schmidt** (263), Department of Molecular and Cell Biology, University of California, Berkeley, Berkeley, California 94720

**Kai Simons** (43), European Molecular Biology Laboratory, 69012 Heidelberg, Germany

**Jan W. Slot** (55), Department of Cell Biology, University of Utrecht, 3508 TC Utrecht, The Netherlands

**Phillip D. Stahl** (55), Department of Cell Biology and Physiology, Washington University, St. Louis, Missouri 63110

**De-chu Tang** (353), Simmons Cancer Center, and Department of Internal Medicine, University of Texas Southwestern Medical Center, Dallas, Texas 75235

**Ian S. Trowbridge** (79), Department of Cancer Biology, The Salk Institute, San Diego, California 92186

**Ora A. Weisz** (137), Department of Cell Biology and Anatomy, Johns Hopkins University School of Medicine, Baltimore, Maryland 21205

**Thomas E. Willnow** (305), Department of Molecular Genetics, University of Texas Southwestern Medical Center, Dallas, Texas 75235

**Jiing-Kuan Yee** (99), Department of Pediatrics, City of Hope, Duarte, California 91010

# PREFACE

The purpose of this volume is to provide information that will allow a biologist investigating cellular function to choose the optimum system for expressing exogenous proteins in avian and mammalian cells. Such studies often require that a protein be expressed in a particular cell type, in amounts suitable for a desired detection technique, and under conditions that minimize alterations of the cellular mechanisms under study. The requirements of this sort of experiment are quite demanding compared to experiments in which a protein is produced for the purposes of purification (where the goal is large amounts of functional protein), for studies of transcription, or for expression cloning (where extremely low expression levels can often be tolerated). Not only is there no universal expression vector capable of satisfying the conditions for investigations of cellular function, it is rare that all three of the requirements for such studies can be equally satisfied. Inevitably, compromises must be made between the choice of expression vector, cell type, and expression system.

The current list of expression vectors available for experiments in cell biology is extensive and includes naturally occurring viruses, recombinant viruses, and recombinant plasmids. Each class of vector can be used for transient or long-term expression and each has its limitations and strengths. This volume describes the structure, production, host range, and use for selected examples of each type of vector. Since the current direction of advances in this sort of experiment is towards expression of proteins in differentiated cells, both in culture and in animals, chapters have been included that describe techniques for accomplishing this goal. Several chapters describe the conditions for expression and analysis of proteins in cells of certain types that are currently of great interest. Examples of these are cell lines that can differentiate fully into neurons, epithelial cells, or cells capable of regulated secretion. A chapter is included that describes the use of homologous recombination to create cultures of embryonic fibroblasts derived from homozygous mice with gene replacements or knockouts. Such cells can be used for experiments that are difficult or impossible to perform in the intact animal.

Certain types of experiment impose additional difficulties. Investigations of cellular housekeeping functions frequently require detection of an exogenous protein against a background of abundant, highly conserved, endogenous protein. Since antibodies specific for many mammalian proteins show limited or no cross-reactivity with avian proteins, one solution to this problem is the use of recombinant Rous sarcoma viruses to express mammalian proteins in chicken or quail cells. Several chapters discuss techniques for introducing DNA into cells that are difficult to transfect, and several different approaches for regulating expression of a protein are included. Many chapters present the results of actual experiments and describe in some detail protocols for detecting a protein once it is expressed. The intent is to provide the reader with protocols proven to work with a particular vector in a certain cell type.

Rather than serve as a "cookbook" of approaches for getting DNA into cells, this volume is intended to provide a cell biologist or biochemist with sufficient information to choose the best expression system for a particular purpose. Expression of proteins from cDNAs is now a commonly applied technology, but my experience advising colleagues in biochemistry, cell biology, and pharmacology departments is that a great deal of time is lost by laboratories beginning expression experiments due to inappropriate choices of vectors or host systems. Most commonly this is due to the fact that the limitations for use of most available vectors are not published. In addition, there is a tendency to choose to expedite the process, usually trivial, of subcloning a cDNA into a vector, rather than pay the price of obtaining and engineering the best vector for the purpose. However, the gain in time obtained in the earliest stages of expression experiments is usually repaid with high interest later on when expression levels are too high, too low, or too variable, or cells are too sick to produce interpretable data.

The development of technology for the expression of recombinant proteins is rapid, and I wish to thank the authors of this book for adhering faithfully to a compressed production schedule. In the midst of their busy professional lives, they have taken the time to write down protocols in a way that can be used by a graduate student starting out in a lab. In addition, since it is rare that a laboratory has experience in more than a few expression systems, it is our intent that even relatively expert laboratories will find this book useful.

**Michael G. Roth**

# PART I

## The Use of Naturally Occurring and Recombinant Viruses

# CHAPTER 1

# Viruses as Model Systems in Cell Biology

**Richard W. Compans and Paul C. Roberts**

Department of Microbiology and Immunology
Emory University School of Medicine
Atlanta, Georgia 30322



## I. Introduction

Molecular and cell biological studies of animal viruses, and their replication processes in infected cells, have been at the forefront of research providing new insights into fundamental mechanisms of cell biology. Among the important advantages that viruses provide in such studies is their structural and genetic simplicity. Replication of many viruses intimately involves various cellular processes including DNA replication, transcription, translation, secondary modifications of proteins, and protein targeting processes. The nucleotide sequences of many viral genomes have been determined, and the complete three-dimensional structures of some viruses are known. The nucleotide sequences of viral genomes have revealed alternative coding strategies, and have led to the discovery of distinctive structural features of viral nucleic acids including terminal repeats, inverted terminal repeats, and inversion of genome segments. Insight into chromatin structure has been provided by studies of the nucleoprotein organization of genomes in small DNA viruses, consisting of histones bound to the viral DNA. The replication of viral genomes has also provided *in vivo* and *in vitro* model systems for studies of viral and cellular DNA replication (Challberg and Kelly, 1989). Viruses with limited coding capacities, such as parvo- and papovaviruses, replicate in the nucleus using host cell machinery. The development of *in vitro* systems for replicating the genomes of these viruses (Li and Kelly, 1984; N. Muzyczka, personal communication) is leading to a greater understanding of the biochemistry of eukaryotic DNA replication. The larger DNA viruses, including adenoviruses, herpesviruses, and poxviruses, encode many of the enzymes needed for their replication. The biochemistry of adenovirus replication has been investigated extensively in *in vitro* systems; the essential viral products required for the replication of herpes simplex virus and for Epstein–Barr virus have been genetically defined (Challberg and Kelly, 1979; Wu, 1988; Fixman, 1992).

Studies of transcription of viral genomes have provided many of the initial insights into the structure and biosynthesis of mRNA, including the finding of polyadenylation of the 3′ termini of mRNAs (Kates and Beeson, 1970; Philipson *et al.,* 1971), the identification of a 5′ cap structure (Shatkin, 1976), the discovery of RNA splicing (Aloni *et al.,* 1977; Berget *et al.,* 1977; Chow *et al.,* 1977; Klessig, 1977), and the identification and characterization of control elements including enhancer sequences (Khoury and Gruss, 1983; Nevins, 1983). The availability of viral genomes as templates has provided an important resource for identification and characterization of transcription factors, as well as for development of *in vitro* systems for transcription studies (Wu, 1978; Weil *et al.,* 1979; Manley *et al.,* 1980). Studies with paramyxoviruses have provided evidence for a new form of modification of mRNAs termed RNA editing, in which one or two non-template-coded nucleotides are introduced during mRNA synthesis, enabling access to an alternative reading frame (Thomas *et al.,* 1988; Vidal *et al.,* 1990).

In many types of virus-infected cells, cellular macromolecular synthesis is markedly inhibited and is replaced by the synthesis of viral macromolecules (Schneider and Shenk, 1987). Large quantities of viral components are produced, which has greatly simplified the analysis of the synthesis, modification, transport, and assembly of specific viral nucleic acids and proteins, and has led to their utilization as probes for many aspects of macromolecular biosynthesis and assembly.

The viral assembly process occurs at specific sites within the cell, that differ depending on the virus family (Table I). Various nonenveloped RNA viruses (e.g., polio, reo) are assembled within the cytoplasm, whereas nonenveloped DNA viruses (parvo, papova, adeno) are assembled in the nucleus. Other viruses possess lipid-containing envelopes that are acquired by budding at a

**Table I**
**Sites of Assembly of Animal Viruses**

| Virus family | Site of assembly | Examples |
|---|---|---|
| Enveloped viruses | | |
| RNA | | |
| Alphaviruses | Plasma membrane | Semliki forest, Sindbis |
| Arenavirus | Plasma membrane | Lymphocytic choriomeningitis |
| Bunyavirus | Golgi complex | Punta Toro, Uukuniemi |
| Coronavirus | Golgi complex | Mouse hepatitis |
| Flavivirus | ER/Golgi complex | West Nile, yellow fever |
| Orthomyxovirus | Plasma membrane | Influenza |
| Paramyxovirus | Plasma membrane | Sendai, SV5 |
| Retrovirus | Plasma membrane | HIV, Rous sarcoma |
| Rhabdovirus | Plasma membrane | Vesicular stomatitis |
| DNA | | |
| Herpesvirus | Nuclear envelope | Herpes simplex 1 |
| Nonenveloped viruses | | |
| DNA | | |
| Adenovirus | Nucleus | Adeno |
| Papovavirus | Nucleus | SV40, polyoma |
| Parvovirus | Nucleus | Canine parvovirus, minute virus of mice |
| RNA | | |
| Picornavirus | Cytoplasm | Polio |
| Reovirus | Cytoplasm | Reovirus, rotavirus |

cellular membrane. The site of budding differs depending on the virus family, and may occur at the plasma membrane (arena, influenza, paramyxo, retro, and rhabdoviruses), the Golgi complex (bunya, coronavirus), the rough endoplasmic reticulum (rotavirus), or the inner nuclear envelope (herpesvirus). Such viruses encode one or more membrane glycoproteins; the cellular site of viral glycoprotein accumulation correlates with the site of virus budding. Thus, the protein components of such viruses have provided excellent systems for studies of the mechanisms of targeting proteins to distinct cellular locations.

In this chapter, we have indicated some of the contributions that studies with viruses have made to current concepts in cell biology. We also describe methods for growth, assay, and purification of viruses and infection of cells by several viruses that have been widely utilized for studies of cellular processes, and that are not described elsewhere in this volume.

## II. Cell Biology of Virus Infection

### A. Overview of the Viral Replication Cycle

Most investigations of virus replication at the cellular level are carried out using animal cells in culture. For the events in individual cells to occur with a high level of synchrony, single cycle growth conditions are used. Cells are infected using a high multiplicity of infectious virus particles (usually at least 5–10 infectious particles per cell) in a low volume of medium to enhance the efficiency of virus adsorption to cell surfaces. After the adsorption period, the residual inoculum is removed and replaced with an appropriate culture medium. During further incubation, each individual cell in the culture is at a similar temporal stage in the viral replication process. Therefore, experimental procedures carried out on the entire culture reflect the replicative events occurring within an individual cell. The length of a single cycle of virus growth can range from a few hours to several days, depending on the virus type.

Although there is a great diversity in the genome structures and replicative strategies used by animal viruses, the following steps are common features of all viral replication cycles.

1. Adsorption: The virus attaches to specific receptors on the cell surface by means of a specific viral surface component. For enveloped viruses, one of the surface glycoproteins serves as the viral attachment protein. Some viruses attach to specific protein receptors that may only be expressed by cells of some species and on some cell types, for example, human immunodeficiency virus (HIV) infection of CD4+ cells. Other viruses with a broad host range may attach to a cell surface constituent that is widely distributed on cell surfaces. An example of this is influenza virus, which binds to sialic acid residues.

2. Penetration/Uncoating: The viral genome must enter the cell in a suitable form and reach an appropriate site in the cell for its expression and replication. Several alternative mechanisms are used by different viruses to accomplish this goal. Membrane fusion is used as the entry mechanism by enveloped viruses, and is discussed further in a subsequent section. Nonenveloped viruses generally penetrate cells by endocytosis, but the precise mechanism by which the genome is released from the virus particle and reaches the appropriate cellular site (cytoplasm or nucleus) to initiate replication is not well understood. For many viruses, essential enzymes are associated with the viral nucleic acid in the form of a nucleoprotein complex, and remain associated with the genome to initiate the biosynthesis of virus-specific macromolecules. Examples include the reverse transcriptase of retroviruses and RNA-dependent RNA polymerases of negative strand RNA viruses such as influenza virus and vesicular stomatitis virus (VSV).

3. Biosynthesis: All viruses must accomplish the same two essential biosynthetic processes: synthesis of virus-specific proteins and replication of the viral genome. The precise mechanisms used to accomplish this vary greatly among virus families, depending on the structure and coding arrangements of their genomes [whether they are DNA or RNA, single or double-stranded, message sense (positive stranded) or antisense (negative stranded), etc.]. The replication processes for some viruses are briefly described in this chapter and in other chapters of this volume.

4. Assembly: Following the production of newly synthesized viral genomes and viral structural proteins, assembly of progeny virions occurs. The precise cellular site used for virus assembly depends on targeting signals contained in viral proteins, as discussed in detail in a subsequent section.

5. Release: The final step in replication involves the release of progeny virions from the cell. For viruses that are assembled by budding at the plasma membrane, assembly and release occur simultaneously. In contrast, for viruses that are assembled intracellularly, release occurs by alternative mechanisms that include cell lysis or vesicular transport. For some nonenveloped viruses that are released from cells without apparent cell lysis, the mechanism of release remains uncertain.

## B. Virus Receptors and Virus Entry

The process of infection of a susceptible cell is initiated by virus attachment to specific receptors on the cell surface. The presence of specific receptors is an important determinant of cellular susceptibility to virus infection. Several virus receptors have now been identified; the details of their molecular interactions with viral surface proteins are being actively investigated. The best characterized of these receptors is the CD4 molecule, which serves as a receptor for HIV. Other cellular factors may also be involved as determinants of cell tropism at a postreceptor stage. Following virus entry, the replication process of some viruses is initiated in the cytoplasm, whereas for other viruses the incoming genome is transported to the nucleus. Evidence has been obtained that specific internal proteins of HIV are associated with viral nucleoproteins during their transport to the nucleus (Bukrinsky *et al.*, 1993). The transport of such viral nucleoproteins may serve as a model for cellular transport processes. Viruses that enter cells by endocytosis serve as convenient markers for tracing cellular endocytic pathways, and have been used to recognize distinct classes of endocytic membranes following infection (Marsh *et al.*, 1986; Schmid *et al.*, 1989).

Many glycoproteins of enveloped viruses possess membrane fusion activity. Virus-induced membrane fusion plays an essential role in the early stage of replication, since the incoming viral envelope fuses with a cellular membrane and thereby enables the viral genome to enter into the cytoplasm. For some viruses, fusion can occur at neutral pH at the cell surface, whereas other viruses first enter cells through endocytosis; the acidic pH within the endosomal compartment results in a conformational change that activates the viral membrane fusion activity (White *et al.*, 1983). Viral membrane fusion serves as a model for investigation of cellular fusion processes occurring at the cell surface. One example of a cellular fusion protein that resembles viral fusion proteins in some respects has been described (White, 1992).

Many viral glycoproteins, such as the paramyxovirus F protein and influenza hemagglutinin (HA) protein, are synthesized as precursors that are activated by a proteolytic cleavage event (Homma and Ohuchi, 1973; Scheid and Choppin,

1974; Klenk *et al.*, 1975; Lazarowitz and Choppin, 1975). These proteins possess a hydrophobic stretch of amino acids that is exposed at the N-terminal end of the cleavage site. Mutagenesis studies (Daniels *et al.*, 1985) have revealed that this stretch of amino acids is important for membrane fusion activity; thus it has been designated the "fusion peptide." In influenza virus, a dramatic change in the three-dimensional structure of HA occurs on exposure to low pH; this change exposes the hydrophobic fusion peptide (Carr and Kim, 1993). Other viral fusion proteins that do not undergo cleavage activation also possess functional regions involved in fusion activity (White *et al.*, 1983). Ongoing studies of viral fusion proteins are providing new information on the structural requirements for fusion activity.

The discovery that influenza virus encodes a transmembrane protein (the M2 protein) that exhibits ion channel activity (Pinto *et al.*, 1992) has stimulated great interest in the role of such proteins during the early steps of virus infection. The antiviral drug amantadine, which inhibits influenza virus penetration (Kato and Eggers, 1969; Skehel *et al.*, 1978), exerts its inhibitory effect on the M2 protein (Hay *et al.*, 1985). Amantadine blocks the M2 ion channel activity (Pinto *et al.*, 1992; Wang *et al.*, 1993), indicating that this activity functions during the early stages of virus infection. Evidence has been obtained that dissociation of the viral matrix protein (M1) from the nucleocapsids is inhibited in the presence of amantadine (Bukrinskaya *et al.*, 1982; Martin and Helenius, 1991), indicating that the transport of $H^+$ ions by the ion channel is involved in promoting this dissociation. An additional function of M2 in some virus strains is the regulation of Golgi complex pH, preventing the intracellular activation of the fusion activity of the influenza HA protein (Ciampor *et al.*, 1992; Sugrue *et al.*, 1990). The identification of such a viral ion channel protein provides a convenient system for detailed studies of its essential structural features.

### C. Protein Synthesis and Modification

Viral proteins have been employed to elucidate many features of the cellular processes involved in synthesis of proteins and their subsequent modification. Viral genomes utilize alternative coding strategies to produce their protein products. Some viruses encode their polypeptide products in the form of a polyprotein that undergoes subsequent proteolytic processing (Summers and Maizel, 1968), whereas in other viruses individual mRNAs encode distinct gene products. Some viral mRNAs encode multiple proteins in distinct reading frames. Among the most striking examples is the P/C gene of paramyxoviruses, which encodes as many as nine distinct polypeptide products in overlapping reading frames that are accessed by alternative initiation codons as well as by RNA editing (Kolakofsky *et al.*, 1991).

Key information about the role of N-terminal signal peptides in targeting membrane proteins to the rough ER was obtained using viral glycoproteins as model systems. Translocation of the VSV-G protein into the rough ER was shown to be directed by a hydrophobic N-terminal signal sequence that is subsequently cleaved (Rothman and Lodish, 1977; Toneguzzo and Ghosh, 1978; Lingappa *et al.*, 1978). The mature protein remains anchored to the membrane by a hydrophobic stretch of 20 amino acids near the C terminus (Rose and Bergmann, 1982).

Viral glycoproteins have also been utilized to study the process of protein glycosylation as well as the functional role of carbohydrate side chains. The transfer of high mannose oligosaccharides to the polypeptide chain and their subsequent modification during intracellular transport have been studied extensively with viral glycoproteins such as influenza HA protein (Compans, 1973; Klenk *et al.*, 1974; Nakamura and Compans, 1979) and VSV-G protein (Hubbard and Ivatt, 1981). The use of glycosylation inhibitors (Gibson *et al.*, 1979,1980) as well as site-directed mutagenesis studies of individual glycosylation sites (Machamer *et al.*, 1985; Ng *et al.*, 1990; Roberts *et al.*, 1993; Gallagher *et al.*, 1992) has led to the general conclusion that the presence of specific oligosaccharides is not required as a positive signal for intracellular transport of glycoproteins, but that in some cases glycosylation plays a role in protein folding as well as in the stability of some glycoproteins.

Another type of protein modification that has been investigated extensively is fatty acid esterification. The covalent attachment of a fatty acid (palmitate) to a glycoprotein was first demonstrated for VSV and Sindbis virus (Schmidt and Schlesinger, 1979; Schmidt *et al.*, 1979). The fatty acid residues are linked to cysteine residues via a thio ester linkage (Magee *et al.*, 1984); removal of a cysteine residue in the cytoplasmic domain by site-directed mutagenesis prevents fatty acid incorporation (Rose *et al.*, 1984). The acylation of the VSV-G protein is not required for intracellular transport; some viruses such as the New Jersey serotype of VSV lack the fatty acid modification entirely. Also, evidence has been obtained that acylation is not required for assembly of HA into influenza virus particles (Naim *et al.*, 1992; Simpson and Lamb, 1992). The precise role of fatty acid addition to viral glycoproteins remains to be determined.

N-Terminal myristylation has been demonstrated as a modification of certain viral proteins such as the matrix proteins of retroviruses (Henderson *et al.*, 1983) and the capsid proteins of some nonenveloped viruses (Chow *et al.*, 1987). In the case of retrovirus cores, this modification was found to be important for targeting core proteins to the plasma membrane where virus assembly occurs (Schultz and Rein, 1989). Among other modifications of viral proteins, phosphorylation is frequently observed; its possible regulatory role for the functions of proteins such as the SV40 T antigen is actively being investigated (Fanning and Knippers, 1992).

## D. Membrane Structure and Biogenesis

Enveloped viruses have been used extensively for studies of membrane structure and biogenesis. The viral envelope is acquired by a process of budding at a cellular membrane; during the budding process the envelope of the emerging virus particle is continuous with the cellular membrane where virus assembly takes place. The lipids of the viral envelope closely resemble the lipid composition of the membranes of the host cell (Klenk and Choppin, 1969). However, the proteins that are incorporated into the viral envelope are encoded by the viral genome. Thus, assembly of enveloped viruses involves the formation of a localized region on a cellular membrane, in which cellular membrane proteins are replaced by virus-encoded membrane proteins, that is recognized by the viral nucleocapsid leading to the release of an enveloped virus particle by a budding process. Therefore it is possible to obtain virus preparations containing the same protein components but different membrane lipids by growth of the same virus in two different cell types. Conversely, growth of two different enveloped viruses in the same cell type will yield viruses with similar membrane lipids but different envelope proteins. This ability to modulate viral lipids or proteins independently has been useful in studies of viral membrane biology (Lenard and Compans, 1974). Enveloped virus particles also represent useful systems for studies of physiological properties such as membrane permeability. The membranes of VSV were found to undergo swelling and shrinking in response to changes in osmotic conditions (Bittman *et al.*, 1976). Thus, it is possible to study the effects of specific changes in lipid or protein composition on permeability properties, since these changes can be readily introduced into viral membranes.

Because enveloped viruses contain a small number of virus-encoded proteins, and because preparing highly purified preparations of enveloped particles is comparatively simple, such virus particles constituted one of the systems of choice for early studies of the organization of lipids and proteins in biological membranes (Harrison *et al.*, 1971; Landsberger *et al.*, 1971). The protein components of viral envelopes were found to be asymmetrically distributed, with the external surfaces consisting of glycosylated proteins forming a layer of spike-like projections covering the surface of the virus particle (Cartwright *et al.*, 1970; Compans *et al.*, 1970; Schulze, 1970; Rifkin and Compans, 1971). In contrast, the internal proteins of enveloped viruses were found to be free of carbohydrate. Evidence that viral proteins span the lipid bilayer and interact with internal proteins has been obtained using several approaches (Lyles, 1979; Bowen and Lyles, 1981; Katz and Lodish, 1979). These general structural features were therefore similar to those of other membranes such as that of the erythrocyte, in which the spatial arrangements of proteins and lipids were being elucidated in parallel. More recently, the complete three-dimensional structures of the external domains of two viral glycoproteins, influenza HA and neuraminidase, have been determined (Wilson *et al.*, 1981; Varghese *et al.*, 1983). These and subsequent studies have provided great insight into structure–function relationships for viral membrane glycoproteins. Like cellular membrane glycoproteins, viral glycoproteins fall into several classes: type I proteins have a cleaved N-terminal signal peptide and a C-terminal hydrophobic anchor sequence, type II proteins are anchored to membranes by an uncleaved N-terminal signal–anchor sequence, and type III proteins have multiple membrane-spanning domains.

Many enveloped viruses with helical nucleocapsids possess a major nonglycosylated structural protein designated the matrix protein, which is thought to line the internal surface of the viral envelope. Such proteins may provide structural stability to the viral envelope, and play an essential role in the process of virus assembly. These viral matrix proteins may serve as models for cellular proteins that interact with the cytoplasmic surfaces of cellular membranes.

Cells infected with enveloped viruses that inhibit cellular biosynthesis have been used extensively to investigate the pathway followed by viral membrane glycoproteins from their sites of synthesis in the rough ER, through the Golgi complex, to the plasma membrane (e.g., Bergmann *et al.*, 1983; Matlin and Simons, 1983; Griffiths *et al.*, 1985). Mutants of viral glycoproteins that are conditionally defective in a stage of their intracellular transport have been used extensively to define vesicular transport intermediates (Rothman *et al.*, 1984). Important studies on protein folding and oligomerization as determinants of transport of membrane glycoproteins have also been carried out with viral glycoproteins (Gething *et al.*, 1986; Copeland *et al.*, 1986; Doms *et al.*, 1993).

The insertion of viral glycoproteins into cell surfaces and the maturation of viruses by budding at the plasma membrane provide a mechanism by which surfaces of cells can be modified under carefully controlled conditions. Viral antigens provide readily identifiable markers that can be recognized on cell surfaces by specific antibodies. The composition of cellular plasma membranes can undergo significant biochemical changes as a result of expression of virus-specific proteins such as the neuraminidases of influenza viruses or paramyxoviruses, which remove sialic acid residues from cellular as well as viral proteins and lipids (Klenk *et al.*, 1970).

## E. Protein Targeting

Viral assembly processes take place at distinct sites within the cell, depending on the virus family, and involve the targeting of viral proteins to specific locations. The proteins of nonenveloped DNA viruses such as adeno, papova-, and parvoviruses possess targeting signals directing them to the nucleus, where assembly of these viruses occurs. The first example of a sequence that specifies nuclear targeting was defined for a virus-encoded protein, the SV40 T antigen (Kalderon *et al.*, 1984a,b; Lanford and Butel, 1984).

The maturation of enveloped viruses occurs by budding at specific cellular membranes. Rotaviruses form by budding into the ER and encode two glycoproteins, VP7 and NCVP5, that have unusual structural features. VP7 possesses two in-frame initiation codons, each of which is followed by hydrophobic sequences (Both *et al.*, 1983). Evidence has been obtained that the second hydrophobic domain is responsible for anchoring VP7 in the ER membrane

(Poruchynsky *et al.*, 1985), whereas the first domain is thought to act as a signal sequence. The signal peptide of VP7 is apparently essential for its retention in the ER as an integral membrane protein (Stirzaker and Both, 1989). A different type of ER retention signal has been reported for a nonstructural membrane protein encoded by adenoviruses, in which a 6-amino-acid C-terminal sequence in the cytoplasmic tail was implicated in ER retention (Nilsson *et al.*, 1989).

Bunyaviruses are enveloped RNA viruses that are assembled by budding at the Golgi complex (Murphy *et al.*, 1973). They possess two glycoproteins designated G1 and G2 that are targeted to the Golgi complex (Matsuoka *et al.*, 1988; Pensiero *et al.*, 1988; Petterson *et al.*, 1988; Wasmoen *et al.*, 1988). For one such virus (Punta Toro virus) the G1 protein, when expressed in the absence of G2, also is targeted to the Golgi complex (Matsuoka *et al.*, 1993). In contrast, the G2 protein is transported to the cell surface when expressed in the absence of G1 (Chen *et al.*, 1991). Truncated G1 proteins with partial deletions in their cytoplasmic domains did not exhibit as clearly defined a pattern of accumulation in the Golgi as the native G1 protein, but appeared to be distributed throughout the ER and the Golgi complex. Proteins lacking most of the cytoplasmic domain, and in some cases part of the transmembrane domain sequences as well, were transported to the cell surfaces. Chimeric proteins constructed with the envelope protein of a murine leukemia virus, which is efficiently transported to the plasma membrane, were also examined; molecules that contained the G1 transmembrane and cytoplasmic domains were efficiently retained in the Golgi complex (Matsuoka *et al.*, 1994). Thus, the transmembrane domain, as well as a portion of the cytoplasmic domain adjacent to the transmembrane domain, is apparently crucial for Golgi retention of the G1 protein.

Coronaviruses are assembled by budding at membranes of the ER and/or Golgi complex. The mature virion contains two surface glycoproteins designated E1 and E2. E1 has three membrane-spanning hydrophobic stretches of amino acids, and is targeted to intracellular membranes where virus budding occurs (Machamer *et al.*, 1990). By analysis of deletion mutants of the avian coronavirus E1 protein, evidence was obtained that the first of these hydrophobic domains is involved in intracellular retention (Machamer and Rose, 1987). Chimeric proteins containing this hydrophobic domain formed large aggregates that were resistant to denaturation with SDS, and suggested that formation of such aggregates could be involved in the mechanism for retention of these proteins in the Golgi complex (Weisz *et al.*, 1993).

The transmembrane domains of the Punta Toro virus G1 protein and the first hydrophobic domain of the avian coronavirus E1 protein are both rich in polar amino acids (Ihara *et al.*, 1985; Swift and Machamer, 1991). A similar composition has been found for the membrane-spanning segments of cellular membrane proteins that are retained in the Golgi complex (Munro, 1991; Nilsson *et al.*, 1991), suggesting that interactions between such polar amino acids could mediate interactions among the proteins retained in the Golgi complex, and evidence has been obtained with the avian coronavirus that several of the polar residues are required for intracellular protein retention (Machamer *et al.*, 1993).

After their intracellular budding, coronaviruses as well as bunyaviruses appear to be released from cells by a vesicular transport process (Dubois-Dalcq *et al.*, 1984; Chen *et al.*, 1991). The murine coronavirus has been used as a marker for the constitutive secretory pathway, and was found to be sorted into a compartment that was distinct from condensed secretory proteins in murine pituitary cells (Tooze *et al.*, 1987).

Viruses of various other families are assembled at the plasma membranes. In polarized epithelial cells, junctional complexes between adjacent cells divide the plasma membrane into two distinct domains, the apical and the basolateral domains. Enveloped viruses of some families, for example, influenza and paramyxoviruses, are found to bud exclusively at apical plasma membrane domains, whereas viruses of other families such as VSV and C-type retroviruses bud exclusively at basolateral membrane domains (Rodriguez-Boulan and Sabatini, 1978; Roth *et al.*, 1983a). The site of virus assembly reflects the site of expression of the viral envelope proteins; viral glycoproteins expressed in the absence of other viral components are targeted to the same membrane at which virus assembly occurs (Roth *et al.*, 1983b; Jones *et al.*, 1985), so the site of expression of viral glycoproteins is likely to determine the site of virus assembly. Since nonglycosylated forms of viral glycoproteins that are produced in the presence of glycosylation inhibitors are correctly targeted to apical or basolateral plasma membranes (Roth *et al.*, 1979), the signals for protein targeting are apparently contained in specific amino acid sequences. Modification of the cytoplasmic tail of the influenza HA protein by introduction of a tyrosine residue resulted in redirection of the protein to basolateral membranes (Brewer and Roth, 1991); similar structural features of cytoplasmic domains have been implicated as basolateral targeting signals for several cellular glycoproteins (Rodriguez-Boulan and Powell, 1992). In several cases these structural features resemble signals for endocytosis.

When polarized epithelial cells are simultaneously infected with two different enveloped viruses that assemble at different plasma membrane domains, each virus continues to exhibit maturation at a restricted membrane domain. Thus, such doubly infected cells have provided a system in which the cellular site of sorting of apical versus basolateral plasma membrane proteins can be analyzed. Such studies have indicated that sorting occurs at a late stage in transport, probably as the proteins exit the Golgi complex (Rindler *et al.*, 1984).

Enveloped viruses have also been used to investigate the sorting of membrane glycoproteins in neuronal cells (Dotti and Simons, 1990). When cultured hippocampal neurons were infected with VSV, the VSV-G protein was found to be expressed on dendritic surfaces. In contrast, the influenza HA protein was preferentially expressed on surfaces of the axon. Based on these observations, it was proposed that the mechanism for sorting membrane glycoproteins in neuronal cells shares features with the mechanism observed in epithelial cells.

# III. Methods for Virus Growth, Assay, and Purification

## A. General Comments

When viruses are passaged under conditions of high multiplicity infection, particles containing truncated or aberrant genomes arise at high frequency. These particles are replication defective; that is, they require the presence of standard wild-type viruses for their replication. When cells are co-infected with such defective particles and standard virus, the yield of infectious virus is significantly reduced. Thus, the defective particles interfere with the replication of infectious virus and are designated defective interfering (DI) particles. To avoid problems caused by their interference with replication, it is important to avoid the presence of high levels of DI particles when preparing virus stocks. This is easily accomplished by using a diluted virus inoculum to prepare virus stocks; for example, a multiplicity of infection of ~0.1 infectious particles per cell. Under such conditions, DI particles and standard particles infect different cells and the DI particles are unable to replicate. Thus, as described in the following examples, virus stocks are usually prepared using a low multiplicity of infection and allowing multiple cycles of replication to occur.

An initial inoculum that is relatively free of DI particles can also be obtained by "plaque purification" of the virus, ensuring that the viral progeny are derived from a single parental genome. This procedure is described subsequently for influenza virus.

Nonenveloped viruses such as picornaviruses, papovaviruses, and adenoviruses are stable structures. The infectivity of stocks of these viruses can be preserved by storage at −20°C; it is also possible to store virus for several days at 4°C without a drastic loss of infectivity. In contrast, enveloped viruses are relatively fragile, and it is necessary to store virus stocks below −60°C to maintain their infectivity. These viruses are also highly susceptible to inactivation during freezing and thawing; to stabilize their infectivity, a protein solution (usually bovine serum albumin, BSA) is added to virus stocks prior to freezing. Also, to minimize inactivation of infectivity, stocks are frozen in multiple small aliquots. Once an aliquot is thawed, it is used and any remainder is discarded. If a 37°C water bath is used for thawing an aliquot of stock virus, the virus should be removed and placed on ice as soon as it is thawed, since the half-life of some enveloped viruses at 37°C is as short as 1 hr.

In the following section, we have provided a description of methods used in our laboratory for the growth, assay, and purification of viruses and single cycle infection of cells by three viruses that have been used widely in cell and molecular biology: influenza virus, poliovirus, and VSV. Influenza virus is the only lipid-containing virus for which the three-dimensional structure of its surface glycoproteins has been determined. Studies with influenza virus have contributed greatly to our knowledge of membrane structure, folding and intracellular transport of membrane glycoproteins, the role of glycosylation in protein function, and virus-induced membrane fusion. The surface glycoprotein (G) of VSV, another enveloped virus, has also been used extensively in studies of structure–function relationships in membrane glycoproteins, and in studies of protein trafficking and membrane biogenesis. Additionally, as described earlier, these two viruses are assembled and released at distinct plasma membrane domains in polarized epithelial cells: influenza virus is assembled at the apical plasma membrane whereas VSV maturation occurs at the basolateral surface. Therefore, cells infected with VSV or influenza virus have provided attractive systems in which to study the molecular and cellular aspects of protein sorting in polarized cells.

The initial demonstration that poliovirus could replicate in cultured cells (Enders *et al.*, 1949) was the key to enabling virology to advance to the molecular and cellular level, as well as to the development of poliovirus vaccines. Thus, poliovirus was the first virus for which most of the molecular events in the viral life cycle were elucidated; it has served as a prototype for studies of replication and assembly of small nonenveloped positive-stranded RNA viruses. Important advances made with polio or other closely related viruses include the first demonstration of RNA-dependent RNA polymerase activity (Baltimore *et al.*, 1963), the finding of base-paired double-stranded RNA that serves as an intermediate in replication (Montagnier and Sanders, 1963), and the demonstration of cleavage of large polyprotein precursors in viral protein synthesis (Summers and Maizel, 1968). The interaction of the virus with its cellular receptor has been studied in detail, as has the mechanism by which the virus alters host cell protein synthesis. In addition, studies in which poliovirus has been used as a vector for expression of foreign genes have been reported (Choi *et al.*, 1991; Percy *et al.*, 1992; Andino *et al.*, 1993; Porter *et al.*, 1993).

Methods for studies with a number of other animal viruses are described in other chapters in this volume, which focus on the use of these viruses as expression vectors.

## B. Influenza Virus

Influenza A viruses have a lipid-containing envelope enclosing a segmented RNA genome of negative polarity (i.e., vRNA is complementary to mRNA). The genome consists of eight single-stranded RNA molecules that are packaged as helical ribonucleoprotein (RNP) complexes consisting of the vRNA, the nucleocapsid structural protein (NP, nucleoprotein), and a few copies of the viral $P_1$, $P_2$, and $P_3$ (polymerase) proteins. The negative sense of the vRNA requires that a virus-encoded RNA-dependent RNA polymerase be packaged in mature virions to initiate transcription after infection. The nucleocapsids are enclosed in a viral envelope containing two major surface glycoproteins, the HA and neuraminidase proteins, as well as the M2 protein that possesses ion channel activity. A major nonglycosylated protein, the matrix (M1) protein, lines the inner surface of the viral envelope. *In vivo,* influenza virus preferentially infects cells of the upper respiratory tract, which possess sialic acid-containing receptors.

Infection is initiated by the binding of the major viral surface glycoprotein, HA, to sialic acid-containing receptors present on the surface of the host cell. This process is followed by virus entry into the cells by receptor-mediated endocytosis. Release of the RNPs into the cytoplasm is thought to occur at secondary endosomal compartments by activation of the low pH-induced fusion activity of the HA protein, resulting in fusion of the viral envelope with endosomal membranes. Following their release from the virions, the RNPs are transported to the nucleus, where transcription of mRNA is initiated by the accompanying viral RNA-dependent RNA polymerase. Two types of RNA transcripts are synthesized by the virus: mRNA and full length positive-strand RNA transcripts that are used as templates for replication of the negative-strand viral genome. Transcription of mRNA requires the presence of functional capped cellular mRNAs to serve as primers for the synthesis of viral mRNA (Krug *et al.,* 1989). The viral genome encodes 10 polypeptides, two of which are produced from spliced products of primary mRNA transcripts. Assembly of progeny virions occurs by budding at the plasma membrane. The neuraminidase (NA) plays a role in the final release of the virion from the cell surface (Palese *et al.,* 1974).

## 1. Preparation of Influenza Virus Stocks

### *a. Cell Culture*

Influenza virus stocks can be grown in either animal cell cultures or embryonated eggs. Madin–Darby bovine kidney (MDBK) cells have been found to produce high yields of infectious virus of the A/WSN strain, and very low levels of DI particles (Choppin, 1969).

1. MDBK cells can be obtained from the American Type Culture Collection (ATCC) (CCL #22; ATCC, Rockville, MD). The cells are passaged twice a week in Dulbecco's reinforced Eagle's medium (Bablanian *et al.,* 1965) supplemented with 10% calf serum (CS).

2. Cells are split 1:4 and maintained in a humidified 37°C incubator containing 5% $CO_2$. Routinely, we plate MDBK cells on 100-mm tissue culture dishes (No. 3100; Costar, Cambridge, MA) for large-scale preparation of virus stocks.

3. Once confluence is reached, cell monolayers are washed twice with warm (37°C) phosphate-buffered saline (PBS) to remove residual serum and then are overlaid with 1.5 ml 1:1000 dilution of virus stock. Virus stocks usually have a titer of at least $1–5 \times 10^8$ pfu/ml which, when diluted 1:1000, gives a titer

of 1–5 × $10^5$ pfu/ml. A confluent MDBK monolayer in a 100-mm dish contains approximately 1.5 × $10^7$ cells. Thus, infection is carried out at a low multiplicity (0.01–0.05 pfu/cell) to avoid problems caused by DI particles (see preceding discussion).

4. Once infected, the cell cultures are returned to the 37°C incubator for 2 hr. During this virus adsorption period, the plates are tilted periodically (every 15 min) to ensure that the virus is distributed uniformly over the culture.

5. After the virus adsorption period, the inoculum is removed and 7 ml Dulbecco's medium supplemented with 2% CS is added to each plate. In some instances, depending on the strain of influenza virus, it is necessary to incubate the infected cells in the presence of trypsin to ensure cleavage of the virus HA protein (Klenk *et al.*, 1975; Lazarowitz and Choppin, 1975). In this case, serum is not included and Dulbecco's medium is supplemented with 2.5 $\mu$g/ml sterile *p*-tosyl-phenylalanine chloromethyl ketone hydrochloride-treated trypsin (TPCK-trypsin; Sigma, St. Louis, MO).

6. The cells are incubated at 37°C for 2 days until rounding and detachment of cells is observed due to the cytopathogenicity of the virus. At this time, the virus yield can be checked rapidly by determining its HA titer (see subsequent section), which should be >512. If a lower titer is observed, the incubation can be continued for an additional day.

7. Virus is then harvested by collecting the medium. We typically transfer the medium into sterile conical 50-ml tubes and pellet the cell debris by centrifugation at 2000 rpm for 20 min.

8. After centrifugation, the supernatants are decanted into a sterile flask and 1/5 volume Eagle's medium with 5% BSA is added to give a final BSA concentration of 1%, which is important in stabilizing the virus during freezing/thawing.

9. After mixing, aliquots of the virus solution are then dispensed into sterile glass vials and quickly frozen using a dry ice/ethanol (ETOH) bath. It is essential that the vials used are not penetrable by $CO_2$ vapors during this step, because the resulting acid pH would inactivate virus infectivity.

10. Frozen stocks are stored at −80°C. It is important to maintain sterile conditions at each step of the preceding process.

***b. Growth of Virus in Embryonated Eggs***

Although many strains of influenza viruses have been adapted for growth in mammalian cell cultures, embryonated eggs remain important hosts for laboratory isolation and growth of some influenza viruses. Viruses are usually prepared by intra-allantoic inoculation in 10- to 11-day-old embryonated eggs. For a more detailed description of egg inoculation and harvesting of virus, consult any of several basic virology laboratory manuals (Department of the Army, 1964; Palmer *et al.*, 1975).

1. Embryonated hens' eggs are incubated in a humidified chamber with forced air flow at 37.5–38°C. Since the size of the allantoic cavity increases as the

embryo grows, 10- to 11-day-old hens' eggs are preferable for growth of established influenza virus strains.

2. Prior to inoculation, the eggs are candled in a darkened room and the positions of the embryo and air sac (blunt end of egg) are marked. The embryos must be viable, as evident by vascularization and movement of the embryo. Contaminated or dead embryos—evident by their black, brown, or green cast—should be discarded.

3. The shell is disinfected with iodine or 70% EtOH and a small hole is drilled 2–5 mm above the air sac. Seed virus from allantoic fluid is then injected into the allantoic cavity by gently inserting a syringe about 0.5 in deep under the airline with a 25-gauge needle into the hole aiming for the shell (needle held at a 45° angle). The seed virus is usually diluted by a factor of $10^3$–$10^4$; 0.1 ml is inoculated per egg.

4. The needle is removed and the hole sealed with wax, scotch tape, or household glue. Following inoculation, the eggs are returned to the incubator (blunt end up) for 2–3 days.

5. Eggs should be candled 16–20 hr postinfection to check for viability, since contamination and/or injection trauma can lead to premature death of the embryo. Such eggs should not be used in harvesting virus.

6. After 2–3 days at 35–36°C, infected eggs are transferred to 4°C overnight.

7. Virus is then harvested by collecting the allantoic fluid. The shell above the air sac is disinfected and a small portion is removed with sterilized forceps, taking care not to puncture the shell membrane. The shell membrane is then removed aseptically. Using sterile forceps the chorioallantoic membrane is gently punctured and the membranes are pushed aside allowing the allantoic fluid to empty into this cavity. The allantoic fluid (5–10 ml/egg) is then collected with a sterile pipet and transferred to sterile conical tubes.

8. As done for virus grown in cell culture, the virus-containing fluid is precleared by centrifugation, and aliquots are stored frozen at −80°C (see preceding procedure).

## 2. Plaque Assay

The plaque assay (Dulbecco and Vogt, 1954) is a quantitative assay for infectivity, which quantifies the number of infectious virions in a virus suspension. The principle behind the plaque assay is to infect cells with dilutions of virus and then culture them under agar or another support such as agarose so progeny virus is transmitted only from an infected cell to its immediate neighbors. The agar effectively prevents the diffusion of virus through the medium to other regions of the monolayer. After several rounds of infection, the virus will spread to produce a localized area of dead cells, referred to as a plaque, within the cell monolayer. Thus, if a cell monolayer is infected with a single infectious particle, a single plaque would arise from the one infected cell.

Scoring the number of plaques allows determination of the virus titer, which is then expressed in plaque-forming units (pfu):

$$\text{pfu/ml} = \text{number of plaques scored (mean from triplicate plates)} \times 2 \text{ (for 0.5 ml inoculum)} \times \text{reciprocal of dilution.}$$

Thus, if a monolayer is infected with 0.5 ml of a $10^{-8}$ dilution of the stock virus and yields 20 plaques, the titer would correspond to $4.0 \times 10^9$ pfu/ml.

For plaque assay of influenza virus stocks, Madin-Darby canine kidney (MDCK) cells cultured in 60-mm petri dishes are routinely used.

1. MDCK cells (CCL #34; ATCC) are passaged biweekly 1 : 4 in Dulbecco's reinforced Eagle's medium with 10% CS.

2. The number of plaques that can be counted to give an accurate result (nonoverlapping) is in the range of 30–50 per dish. The virus sample of unknown titer is diluted in 10-fold increments in Eagle's medium with 1% BSA. Prior to infection, confluent monolayers are washed twice with warm PBS and then overlaid with 0.2–0.5 ml appropriate virus dilutions.

3. Cells are then placed in an incubator at 37°C in an atmosphere of 5% $CO_2$ for 2 hr for virus adsorption. Plates are tilted every 10–15 min to ensure uniform distribution of the virus.

4. Following virus adsorption, the inoculum is removed and cell monolayers are washed twice with warm (37°C) PBS to remove unadsorbed virus.

5. The infected cells are then overlaid with 2 ml agar overlay medium consisting of a 1 : 1 mixture of 2× Dulbecco's medium and white agar (1.9% agar in $dH_2O$, autoclaved) supplemented with 2.5 $\mu$g/ml TPCK-trypsin. The agar overlay medium should be prepared in advance and kept at a suitable temperature. We typically melt the agar solution in a microwave oven and then place it in a 45°C water bath for 15–20 min before mixing with 2× Dulbecco's medium, which is at 37°C. The temperature of the overlay medium should be 40–42°C. The overlay will solidify if it cools below 40°C, and will damage the cells if it is much warmer than 42°C. Trypsin is added directly to the agar overlay media before overlaying because it is sensitive to heat.

6. The dishes are allowed to stand until the agar has solidified, and are then returned to the 37°C incubator in an inverted position, to prevent moisture from getting below the agar overlay, until plaques are evident. Visible plaque formation occurs in 2–3 days, but we recommend that you check for plaques every day. Plaques are readily visible by phase-contrast microscopy.

7. For counting, plaques are more readily visualized if the cell monolayer is stained with a vital dye such as neutral red. For staining with neutral red, the cells are overlaid 2–3 days postinfection with a 1.5-ml second agar overlay medium containing a 1:1 mixture of 2× growth medium and 1.9% agar/0.025% neutral red (in $dH_2O$, autoclaved). Since only living cells take up neutral red, plaques will appear as opaque, clear regions in a red background. Neutral red will be taken up by the living cells in approximately 3–5 hr at 37°C.

Note that trypsin is only needed with influenza virus strains in which the HA glycoprotein does not undergo intracellular cleavage. The cleavage of the HA protein is needed to activate its low pH-dependent membrane fusion activity, which is essential for virus penetration (Klenk *et al.*, 1975; Lazarowitz and Choppin, 1975). Thus, virulent avian subtypes (i.e., influenza A/FPV/Rostock/34 H7N1) do not require trypsin in the agar overlay because cleavage of the HA glycoprotein occurs intracellularly in the Golgi complex. In general, most human influenza virus strains and many avirulent avian strains do require trypsin (Tobita *et al.*, 1975). The A/WSN strain, however, forms plaques in chicken embryo fibroblasts in the absence of trypsin.

## 3. Hemagglutination Test

The hemagglutination test is a convenient rapid assay for influenza virus. Influenza virus adsorbs to cells via its major surface glycoprotein, HA, which binds to *N*-acetylneuraminic acid residues present on target cells. The hemagglutination test takes advantage of the binding activity of viral HA to neuraminic acid-containing receptors found on red blood cells. Virus binds to these receptors and causes clumping or agglutination of red blood cells in suspension. However, this assay will not measure virus infectivity because defective particles or virus fragments that are not infectious also contain functional hemagglutinin proteins.

1. Chicken red blood cells can be obtained commercially, or can be obtained by bleeding chickens directly.

2. For preparation of chicken red blood cells (cRBCs), place the cells in a 50-ml conical tube and add PBS without $CaCl_2$ and $MgCl_2$ (PBS-deficient) up to a volume of 45 ml.

3. Pellet the cells by centrifugation at 1500 rpm for 5 min and remove the supernatant by aspiration. Resuspend cells gently in 45 ml PBS-deficient, and repellet at 1500 rpm for 5 min.

4. Wash the cells again by the same procedure, and resuspend the cell pellet in 5 ml PBS-deficient; transfer them to a graduated 10- to 15-ml conical tube. Add PBS-deficient up to the neck of the centrifuge tube and pellet cells by centrifugation at 1500 rpm for 10 min. Determine the volume of the packed cRBCs and prepare a 10% suspension (e.g., 2 ml packed cells resuspended in 20 ml), which can be stored at 4°C for ~1 wk. Dilute this suspension 1 : 20 with PBS-deficient to prepare a 0.5% solution of cRBCs for the hemagglutination test.

5. The hemagglutination assay is carried out in 96-well microtiter plates. To each 12-well row of the microtiter plate that will be used, 100 $\mu$l 0.85% saline, pH 7.3–7.4, is added.

6. To the first well in each row, add 100 $\mu$l test virus (undiluted) or, as a negative control, add 100 $\mu$l Eagle's medium with 1% BSA. Using a multipi-

pettor, dilute the virus suspension serially from the first to the twelfth well by first mixing thoroughly, and then transferring 100 μl to the next adjacent well.

7. Discard the 100 μl derived from well 12.

8. To each well, add 100 μl of 0.5% solution of prewashed cRBCs, and allow the cells to settle for 60 min at room temperature.

9. After 60 min negative controls will exhibit small red buttons, whereas positive wells appear diffuse with no clear indication of buttons. The positive wells have a sufficient concentration of virus particles to induce agglutination of the red blood cells. The last well that clearly shows a diffuse pattern and contains no evidence of red blood cells settling to a button is the end point. This dilution is used in calculating the HA titer, which is the reciprocal of the dilution factor. The first well represents a 1 : 4 dilution, the second 1 : 8, the third 1 : 16, and so on. Thus, if the tenth well is scored as positive, the virus stock has an HA titer of 2048, since well 10 represents a 1:2048 dilution of the virus.

### 4. Influenza Virus Polypeptide Synthesis

Influenza A and B viruses possess eight genomic RNA segments that encode 10 viral polypeptides. Influenza virus will infect a number of mammalian cell lines such as MDBK, baby hamster kidney (BHK-21, CCL #10; ATCC), MDCK, or monkey kidney (African Green monkey kidney, CV-1, CCL #70; ATCC). The time course of viral polypeptide synthesis depends on the strain of virus and the host cell, as well as on the multiplicity of infection (Skehel, 1972; Meier-Ewert and Compans, 1974; Lamb and Choppin, 1976; Shapiro *et al.*, 1987). Thus, it is important to establish the growth properties and time course of replication of a given influenza virus strain in the cell type of interest before planning specific experiments.

In general, the first detectable viral polypeptides are NP and a nonstructural protein (NS1) derived from segment 8 of the viral genome. These two viral polypeptides are detectable as early as 0.5 hr postinfection in chicken embryo fibroblast (CEF) cells, and their rates of synthesis reach a maximum at about 4.5 hr postinfection. These proteins represent the most abundant proteins in the infected cells. The polymerase polypeptides (PA, PB1, PB2) encoded by RNA segments 1–3 are also observed early in infection, but their levels remain relatively low throughout the infection cycle. The M1 protein (derived from segment 7) is a late influenza virus protein, as are the viral glycoproteins, HA (vRNA segment 4), and NA (vRNA segment 5). Detectable levels are usually observed between 2.5 and 3.5 hr and steadily increase, reaching high rates of synthesis 4–5 hr postinfection. The other viral proteins (M2, segment 7, and NS2, segment 8) are present at intermediate levels.

In the experiment shown in Fig. 1, BHK-21 and MDCK cells were infected with 30 pfu/cell of A/WSN in Eagle's medium with 1% BSA and the virus was adsorbed for 1 hr. After virus adsorption, cells were washed twice with PBS

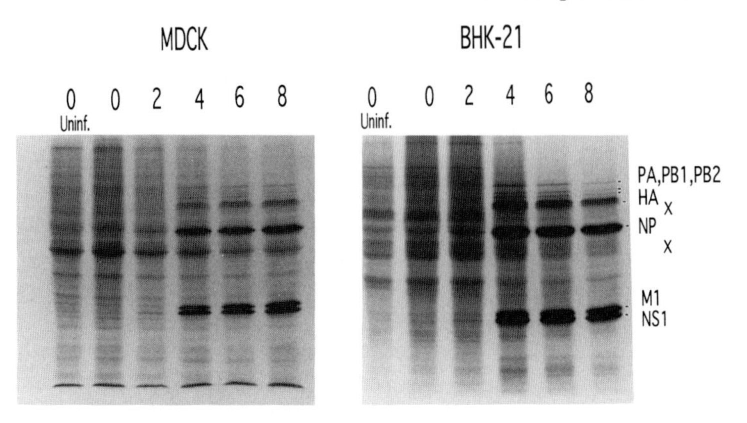


**Fig. 1** Time course of influenza virus polypeptide synthesis in two cell types. BHK-21 and MDCK cells infected with 30 pfu/cell of A/WSN were labeled with L-[$^{35}$S]methionine/cysteine for 15 min at the times postinfection shown. Cell lysates were directly analyzed by SDS–polyacrylamide gel electrophoresis (PAGE) and processed for fluorography without prior immunoprecipitation of viral proteins with antiviral antiserum.

and incubated in Dulbecco's medium with 2% CS until labeling at the desired times postinfection. At 0, 2, 4, 6, and 8 hr postinfection, a well of a 6-well plate of cells was washed twice with Dulbecco's modified Eagle's medium (DMEM) without methionine and cysteine and then incubated in DMEM containing 50 $\mu$Ci of L-[$^{35}$S]methionine/cysteine (Amersham, Arlington Heights, IL) for 15 min at 37°C. After labeling, the cells were immediately placed on ice, washed twice with cold PBS, and lysed in 300 $\mu$l SDS–polyacrylamide gel electrophoresis (PAGE) sample buffer (10% glycerol, 3% SDS, 200 m*M* Tris-HCl, pH 6.8, 0.004% bromophenol blue, 1 m*M* EDTA). The disrupted cells were collected from the plates, boiled for 5 min, sonicated for 20 sec to shear cellular DNA, and then stored at −20°C. Because influenza virus infection shuts off host cell protein synthesis, viral proteins can be directly analyzed by SDS–PAGE and processed for fluorography without prior immunoprecipitation of viral proteins with antiviral antiserum. For analysis of metabolically labeled viral proteins, aliquots of the cell lysates were separated on 13% polyacrylamide–SDS gels with an acrylamide:bis-acrylamide ratio of 130:1 (Lamb and Choppin, 1976); these conditions result in resolution of the M1 and NS1 proteins which comigrate under standard acrylamide:bis-acrylamide ratios. As can be observed in Fig. 1, influenza virus affects host cell protein synthesis rapidly (~1 hr postinfection) with maximal shutdown occurring approximately 2–4.5 hr postinfection.

### 5. Infection of Polarized MDCK Cells with Influenza Virus

Procedures for growth of MDCK cells on permeable supports are described in Section III,D,3,a for VSV. Because influenza virus is able to infect cells at their apical surfaces, the inoculation of filter-grown MDCK cells is carried out as described in Section III,B,4 for cells grown on plastic surfaces. Examples of the polarized release of influenza virus and VSV from MDCK cells are shown in Fig. 2.

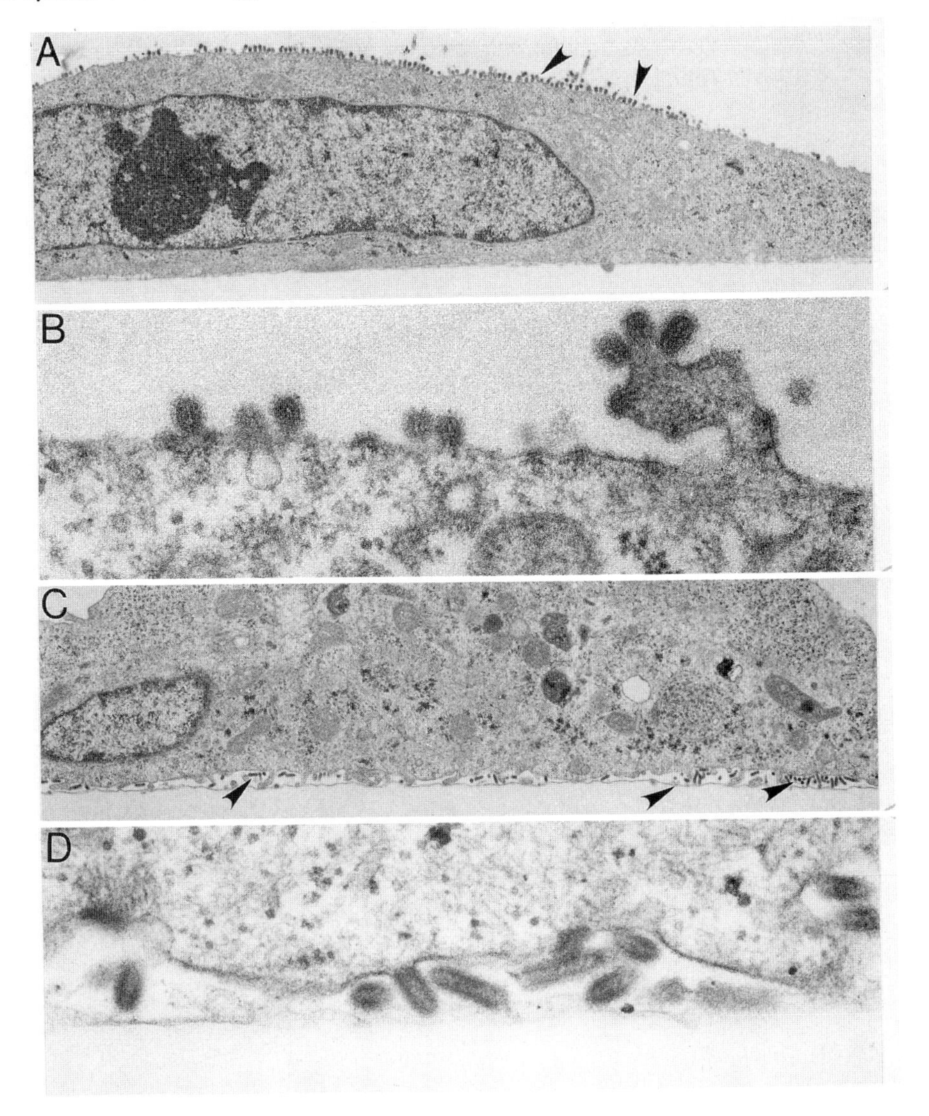


**Fig. 2** Assembly and release of influenza virus and VSV in polarized MDCK cells. (A) Influenza virus (arrows) budding at the apical surface of a cell (10,000×). (B) Higher magnification of particles in various stages of the budding process (100,000×). (C) VSV particles (arrows) at the basal surfaces of a cell (10,000×). (D) Higher magnification of budding and released particles (100,000×).

## 6. Purification of Virus

1. To prepare purified influenza virions, MDBK cells (100-mm dishes) are inoculated at a multiplicity of 1–10 pfu/cell. The cells are incubated in a humidi-

fied 37°C chamber under an atmosphere of 5% $CO_2$ for 2 hr with periodic tilting.

2. The virus inoculum is removed and 6 ml Dulbecco's reinforced Eagle's medium with 2% CS is added per plate. The medium from injected plates is collected 24–36 hr postinfection and is precleared to remove cell debris by centrifugation at 5,000 *g* for 20 min.

3a. Virus is then pelleted by centrifugation at 52,000 *g* for 60 min in a SW28 rotor.

3b. Alternatively, virus may be precipitated from supernatants of infected cells with polyethylene glycol (PEG) by adding PEG (6000, Sigma, St. Louis, MO) and NaCl to the precleared supernatant at final concentrations of 7.5% and 2.3% respectively. Once the crystals have dissolved, the virus–PEG solution is placed at 4°C for at least 1.5 hr. The resulting precipitate is pelleted at 1,000 *g* for 30 min.

4. The drained pellet from Step 3a is left overnight on ice in 0.25 ml medium, and is resuspended in Eagle's medium and layered onto a 10–40% continuous potassium tartrate gradient (w/w in PBS or $H_2O$). The pellet from Step 3b may be resuspended in Eagle's medium directly by pipetting gently until it is dissolved.

5. Following centrifugation at 23,000 rpm in a SW28 or similar rotor, the virus should resolve into a well-defined band that is then collected by dripping or by pasteur pipet. Some investigators prefer to purify virus by 15–60% sucrose or successive sucrose and potassium tartrate gradients. Under these conditions, the visible virus band is collected after sucrose gradient centrifugation at 22,000 rpm for 90 min and is diluted to give a sucrose concentration of 10%. The virus is then pelleted by centrifugation (52,000 *g* for 1 hr), resuspended in Eagle's medium, and separated by potassium tartrate gradient centrifugation.

6. Following the one-step or two-step centrifugation procedure, the virus is collected and dialyzed against buffer or medium.

## C. Poliovirus

Poliovirus is an enterovirus belonging to the Picornaviridae family of small RNA viruses. The poliovirion consists of an icosahedral protein capsid structure without a lipid envelope that contains a single-stranded RNA genome (positive polarity) that functions as messenger RNA. The naked RNA of poliovirus is infectious.

The mature poliovirus virion is composed of four capsid proteins designated VP1, VP2, VP3, and VP4 in order of decreasing molecular weight. The outer surface of the mature capsid is composed of 60 identical subunits, termed protomers or capsomers, each of which contains one copy of VP1, VP3, and VP2, arranged in icosohedral symmetry (Hogle *et al.*, 1985). VP4 is found internally and associates with the inner face of the capsid structures and with viral RNA. The viral RNA contains a virus-encoded protein VPg that is covalently attached to the 5′ terminus. VPg is essential for the initiation of RNA replication (Rueckert, 1990).

Poliovirus infection is restricted to primate hosts, and depends on the presence of functional receptors on the cell surfaces (Holland, 1961; Racaniello, 1988; Ren and Racaniello, 1992). However, nonsusceptible cell lines and tissues can be infected with naked poliovirus RNA (Holland *et al.*, 1959). The poliovirus receptor is well characterized and is a member of the immunoglobulin superfamily (Mendelsohn *et al.*, 1989). The functional poliovirus receptor (PVR) is a transmembrane glycoprotein of approximately 43–45 kD containing three immunoglobulin-like domains (I, II, and III, numbered consecutively from the N terminus). Mutational analysis has revealed that domain I of the PVR is primarily responsible for virus binding (Koike *et al.*, 1991; Freistadt and Racaniello, 1991). The precise cellular function of PVR is unknown.

Following virus binding, entry occurs by endocytosis. The subsequent uncoating process is not well understood. However, the incoming virion RNA is released into the cytoplasm where it functions as mRNA for initial viral protein synthesis (primary translation). Host cell protein synthesis is rapidly shut down between 1 and 2 hr postinfection as a result of disaggregation of cellular polyribosomes and their recruitment by viral RNA for synthesis of viral proteins. The genome of poliovirus encodes a large precursor polyprotein that undergoes subsequent cleavage, giving rise to the final viral protein species (Summers and Maizel, 1967; Leibowitz and Penman, 1971). Viral protein synthesis is readily detected beginning approximately 1.5 hr postinfection and reaches maximal levels at about 2.5 hr postinfection. Thereafter, protein synthesis levels off and begins to decline at 3.5 hr postinfection. Viral RNA synthesis is somewhat delayed after the onset of infection, after early viral protein synthesis has been initiated. Viral RNA synthesis is first detected between 1 and 2 hr postinfection and reaches a maximum at about 3 hr postinfection, declining considerably after 4 hr. Virus maturation is initiated by the association of vRNA with a newly synthesized viral proteins. Intracellular progeny virions are detectable between 3 and 4 hr postinfection. During virus maturation, the capsid proteins are assembled and precursor proteins are cleaved, resulting in the final virion coat proteins enclosing the vRNA genome. The assembly of infectious virions continues and large amounts accumulate in the cytoplasm until, at 6–8 hr postinfection, the progeny virions are released from the infected cell. A single cycle of virus growth is complete in about 8 hr (see Darnell and Levintow, 1960; Scharff *et al.*, 1963,1964; Baltimore *et al.*, 1966; Levintow, 1974).

## 1. Growth, Purification, and Plaque Assay of Poliovirus

### *a. Preparation of Virus Stocks and Virus Purification*

The S3 clonal derivative of the HeLa parent line (CCL #2.2; ATCC) is a suspension cell line that is particularly susceptible to poliovirus (Darnell and

Levintow, 1960). The parent HeLa cell line (CCL #2; ATCC) is well suited to growing poliovirus in cell monolayers.

1. HeLa cells are cultivated in 75-cm$^2$ tissue culture flasks in Dulbecco's reinforced Eagle's medium supplemented with 10% newborn calf serum (NCS) and can be passaged biweekly at a 1 : 5 dilution.

2. For preparation of virus stocks, freshly confluent monolayers (75-cm$^2$ flask, $\sim 3 \times 10^7$ cells) are washed twice with serum-free Dulbecco's medium and inoculated with 2 ml diluted virus stock ($\sim 1 \times 10^5$ pfu/ml) per flask. The cells are incubated for 1 hr at 37°C with occasional rocking of the plates.

3. Following the 1-hr adsorption period, the cells are overlaid with growth medium containing 10% NCS and returned to the incubator for 48 hr.

4. The cells are scraped and virus is harvested 48 hr postinfection by three cycles of freeze/thawing, followed by centrifugation of cell debris at 2000 rpm for 10 min at 4°C.

5. Supernatants from the centrifugation are stored at −20°C to −80°C. Virus titer is determined by plaque assay (see next procedure).

*b. Plaque Assay*

We routinely use HEp-2 cells for plaque assay of poliovirus.

1. HEp-2 cells are passaged biweekly (1 : 5) in DMEM supplemented with 10% CS. Freshly confluent HEp-2 cells cultivated in 6-well plates (~3 cm diameter; No. 3506; Costar, Cambridge, MA) are washed twice with warm PBS and inoculated with appropriate dilutions of poliovirus stock (calculated to give 20–50 plaques at one of the dilutions).

2. The cells are incubated at 37°C for 60 min with occasional rocking of the plates.

3. The cells are then washed once with PBS and overlaid with 2 ml 1 : 1 solution of 2× Dulbecco's and 1.9% agarose.

4a. At 3 days postinfection, the plaques are visualized by overlaying the agarose medium with a second layer containing 1% low-melting agarose with 0.025% neutral red.

4b. As an alternative, the cells can be stained with the vital dye MTT [3-(4,5-dimethylthiazol-2-yl)-2,5,-diphenyltetrazolium bromide; Sigma]. The MTT solution is pale yellow and, when taken up by living cells, is metabolized by cellular dehydrogenases into a purple precipitate. Thus plaques will appear pale yellow in a purple to black background. MTT is stored as a 5 mg/ml stock solution in PBS at 4°C. Add 2 $\mu$l stock solution to 2 ml 1% agarose (in $dH_2O$) and overlay cells. Following solidification of the second overlay, the cells are returned to the 37°C incubator for 30–120 min, at which time the color reaction should be complete.

*c. Virus Purification*

1. After preclearance of cellular debris by low-speed centrifugation at 10,000 rpm for 30 min, virus can be concentrated either by high-speed centrifuga-

tion (28,000 rpm, SW28 rotor, 2 hr) or by precipitation with PEG 6000 in the presence of NaCl.

2. In the latter case, a concentrated BSA solution is added to the precleared culture fluid to obtain a final concentration of 1% and PEG 6000 and NaCl are added to final concentrations of 10% and 2.3%, respectively. This suspension is stirred slowly overnight at 4°C. The virus is then pelleted at 3,000 rpm for 20 min at 4°C.

3. If only high-speed centrifugation of the cell culture suspension is used, the virus pellet is softened overnight in a small volume of PBS and resuspended, and carrier protein (BSA) is added to 1%. Large clumps are dispersed by sonification or homogenization, and insoluble aggregates are removed by low-speed centrifugation (3,000 rpm, 4°C).

4. Virus is subsequently purified by CsCl equilibrium gradient centrifugation. CsCl is added to 0.5 g/ml and the mild nonionic detergent NP-40 is added to 1% to the final volume of the virus suspension. Virus is then banded in CsCl by centrifugation at 146,000 *g* for 16 hr at 4°C.

5. Viral bands are harvested by fractionation. Note that empty capsids, DI particles, and dense particles may be evident in these preparations. However, because of their different buoyant densities in CsCl, they can be readily identified and separated from mature infectious virus particles, which band at a density of 1.35 g/ml in CsCl. Dense particles typically band at 1.44 g/ml in cesium chloride (Rowlands *et al.*, 1975; Yamaguchi-Koll *et al.*, 1975), whereas DI particles band at 1.31–1.32 g/ml and empty capsids at 1.29–1.30 g/ml (Cole *et al.*, 1971).

Alternatively, poliovirus can be purified on linear 15–45% sucrose gradients.

1. The sample in PBS is layered onto a 15–45% sucrose gradient (w/v in 10 m*M* Tris-HCl, pH 7.4, and 50 m*M* NaCl).

2. The samples are then centrifuged at 80,000 *g* for 4 hr at 4°C (25,000 rpm, Beckman ultracentrifuge, SW28 rotor).

3. Virus is collected by fractionation, the band diluted 5- to 10-fold with Tris-HCl, NaCl buffer, and pelleted by centrifugation at 100,000 *g* at 4°C.

4. The pellet is resuspended in PBS and stored frozen at −80°C. Mature infectious virus sediments in sucrose at 155S. (Fernandez-Tomas and Baltimore, 1973; Rowlands *et al.*, 1975).

## D. Vesicular Stomatitis Virus

Vesicular stomatitis virus (VSV) is the prototype rhabdovirus, containing a nonsegmented single-stranded RNA genome of negative sense enclosed in a host-cell-derived lipid envelope. VSV has a typical rod-shaped bullet-like morphology. VSV virions contain a functional virus-encoded RNA transcriptase to initiate transcription on infection. VSV is very cytopathogenic for the infected cell. Prior to cell death, VSV infection leads to the efficient shut-down of host RNA, DNA, and protein synthesis.

Although cellular receptors for VSV have not been identified conclusively, VSV does possess high affinity for phosphatidylserine (Schlegel *et al.,* 1983). VSV penetration occurs via receptor-mediated endocytosis followed by a pH-sensitive fusion event between viral and endosomal membranes (Matlin *et al.,* 1982a,b). Replication and assembly are solely cytoplasmic and can occur in enucleated cells. Transcription of the genome yields five monocistronic mRNAs, all of which contain a common leader sequence that is also present at the 3′ terminus of the viral genome. Replication of the negative-strand genome involves synthesis of full-length plus strands as templates.

The genome encodes five proteins, all of which are found associated with purified virions. The genomic organization of VSV is common to all rhabdoviruses with a small leader sequence at the extreme 3′ terminus of the genome followed by the coding regions for N, NS, M, G, and L proteins. The L (large) and NS proteins function as the viral RNA transcriptase and are associated with nucleocapsids consisting of genomic RNA and the abundant N (nucleocapsid) protein. The G protein is an integral membrane glycoprotein and forms the characteristic spike-like projections emanating from the surface of the virions, whereas the M (matrix) protein lines the inner surface of the virion lipid envelope and is in close association with the RNPs, as well as the surface G proteins.

## 1. Preparation and Plaque Assay of VSV Stocks

The procedure for preparation of VSV stocks is similar to that described for influenza virus (see preceding discussion) except for the choice of host cell. We routinely use continuous lines of baby hamster kidney (BHK-21F or BHK-21/c13) cells (McPherson and Stoker, 1962; Roth *et al.,* 1979; Roth and Compans, 1981) for growth and plaque assay of VSV.

1. BHK-21 cells are maintained at 37°C in reinforced Eagle's medium with 10% CS and 10% tryptose and phosphate broth in an atmosphere of 5% $CO_2$ (Holmes and Choppin, 1966). These cells can be passaged biweekly at 1 : 10 to 1 : 20 dilutions. For preparation of virus stocks, we use freshly confluent monolayers of BHK-21 cells cultivated in 100-mm dishes ($\sim 1 \times 10^7$ cells).

2. The monolayers are washed twice with warm PBS prior to inoculation with virus stocks. Cells infected at a multiplicity of infection of 0.01–0.1 pfu/cell typically yield high titer stocks, $>10^9$ pfu/ml. Infection at low multiplicity is important to reduce the presence of DI particles. The virus inoculum is diluted in Eagle's medium with 1% BSA to give $0.1–1 \times 10^6$ pfu/ml; 1.5 ml virus dilution is added to each monolayer.

3. After a 1-hr adsorption period at 37°C with occasional tilting of the plates, the virus inoculum is removed and 6 ml DMEM with 2% CS is added.

4. Virus is harvested from the medium after a 24-hr incubation at 37°C. The medium is precleared by low-speed centrifugation (2000 rpm, 10 min at 4°C) and BSA is added to 1%.

5. The virus suspensions are aliquoted and stored frozen at −80°C.

6. The virus titer is determined by plaque assay on BHK-21 cells. BHK-21 cells are seeded on 60-mm tissue cultures dishes and incubated at 37°C until confluence is reached.

7. Freshly confluent monolayers are then washed twice with PBS and incubated with 0.5 ml selected serial 10-fold dilutions of virus stock. The monolayers are returned to the incubator with occasional rocking for 60 min, and then washed with PBS to remove unadsorbed virus.

8. The cells are then overlaid with 4 ml DMEM containing 0.9% Noble Agar (Difco Laboratories, Detroit, MI). Plaques should be evident after a 24-hr incubation at 37°C.

9. Visualization of plaques is enhanced by incubation with an additional agar overlay containing the vital dye neutral red (2 ml 0.95% agar with 0.01% neutral red). For counting, dilutions that yield 10–50 well-defined plaques per dish are preferable.

It is sometimes necessary to plaque purify VSV to reduce formation of DI particles.

1. To plaque purify, individual plaques that are well defined and separated from other plaques are marked on the bottom of the dish and gently picked via suction with a sterile pasteur pipet. A sterile pipet is carefully inserted above the marked plaque until the pipet tip touches the bottom of the tissue culture plate. By gently removing the pipet, an agar plug containing virus and cells is aspirated into the pipet.

2. The pipet contents are then transferred to 0.5 ml Eagle's medium with 1% BSA and vortexed vigorously to free virus from the agar.

3. This suspension is centrifuged briefly and used to infect a 100-mm dish of freshly confluent BHK-21 cells.

4. After a 1-hr adsorption period, 6 ml Dulbecco's medium containing 2% CS is added and the cells are incubated for 48 hr at 37°C.

5. The culture medium is then collected and cellular debris is removed by low-speed centrifugation (2000 rpm, 10 min). Virus titer is determined by plaque assay. The process can be repeated an additional time to ensure a genetically cloned population; the resulting virus can be used as the inoculum to prepare larger virus stocks, as described already.

## 2. Purification of VSV

1. To prepare purified VSV, BHK-21 cells cultivated on 100-mm dishes are inoculated at a multiplicity of 5–10 pfu/cell.

2. After a 2-hr adsorption period at 37°C, the virus inoculum is removed and replaced with 7.5 ml Dulbecco's reinforced Eagle's medium with 2% CS. Cells are cultured overnight in a humidified $CO_2$ incubator.

3. VSV is released by budding from the plasma membrane, and virus is recovered from the extracellular medium. We recommend harvesting the virus 18–20 hr postinfection. At this time, the cells remain relatively intact and the extracellular medium is not extensively contaminated by cell debris. The culture medium is harvested and precleared by centrifugation at 2,000 *g* for 30 min.

4. Released virus is then precipitated with 7% PEG 6000 and 2.3% NaCl at 4°C for 10–16 hr (McSharry and Benzinger, 1970), followed by centrifugation at 1,000 *g* for 20 min.

5. The virus pellet is gently resuspended in Eagle's medium with 1% BSA, and further purified by isopycnic centrifugation in a linear 15–40% (w/w) potassium tartrate gradient at 90,000 *g* for 2.5 hr.

6. After centrifugation, the virus band is collected by side puncture with a needle attached to a syringe. In some instances, two viral bands may be discernible upon potassium tartrate gradient centrifugation (Klenk and Choppin, 1971). The lower band consists largely of cellular components with small amounts of virus particles, whereas the upper band contains infectious VSV particles. The upper band should be harvested.

7. The virus band is then dialyzed overnight at 4°C against buffer or medium for subsequent experiments.

Alternatively, virus can be purified by layering on a linear 10–40% sucrose gradient, followed by centrifugation at 50,000 *g* for 90 min. The virus band is then collected by side puncture, diluted 10-fold, and pelleted by centrifugation at 90,000 *g* for 1 hr. The virus pellet is resuspended in an appropriate medium or buffer for subsequent experiments.

## 3. Infection of Polarized MDCK Cells with VSV

### *a. Growth of Cells on Permeable Supports*

As described earlier, animal viruses have proven useful in studying mechanisms of vectorial transport of membrane proteins in polarized epithelial cells. It has been observed that VSV preferentially enters cells and is released by budding at the basolateral surfaces of polarized epithelial cells (e.g., MDCK, Vero C1008, and Caco-2 cells) (Rodriguez-Boulan and Sabatini, 1978; Tucker and Compans, 1993). MDCK cells retain many of the properties of differentiated kidney epithelial cells (e.g., asymmetric distribution of enzymes and vectorial transport of sodium and water from apical to basolateral surfaces). Two strains of MDCK cells vary in their transepithelial resistance (Richardson *et al.*, 1981; Balcarova-Ständer *et al.*, 1984). Strain I forms tight epithelial monolayers reaching a transepithelial resistance above 1500 ohm $cm^2$, whereas Strain II forms monolayers with lower resistance of 200–300 ohm $cm^2$.

For studies of virus infection of polarized cells, the cells are usually grown on permeable supports (filters) that facilitate access to both the apical and the basolateral surfaces. We typically use 3-cm diameter filters of mixed cellulose esters (Millicell–HA, 0.45–$\mu$m pores, Millipore, Bedford, MA). With filters of a pore size greater than 1 $\mu$m, epithelial cells can migrate through the filter (Tucker *et al.*, 1992). Thus, filters with pore size of >1 $\mu$m are recommended.

1. MDCK cells are seeded on 0.45-$\mu$m filters at a density of $1 \times 10^6$ cells per 3-cm filter. Prior to seeding, the filters should be wetted from the bottom with growth medium for at least 5 min, followed by the addition of 1 ml growth medium to the interior of each filter unit. This can be done by pouring 2 ml medium into the exterior chambers and then placing the filters into the individual chambers. It is important that the filters are completely wet prior to addition of apical medium, to prevent trapping air in the pores of the filter.
2. After seeding, the 6-well filter chamber is gently swirled to remove any air trapped beneath the filter.
3. The cells are then placed at 37°C in a humidified $CO_2$ incubator. It is important to avoid touching or puncturing the membrane during seeding.
4. Every 2–3 days, change both apical (interior) and basolateral (exterior) medium chambers, taking care not to damage the membrane during aspiration of medium. The medium from the basolateral chamber should be removed before the apical medium, followed by addition of fresh medium first to the apical chamber and then to the basolateral chamber. This prevents upward movement of medium from the basolateral chamber, which can damage the integrity of the cell monolayer.

Transepithelial resistance can be monitored using a Millipore ERS apparatus (Millipore, Bedford, MA), which is designed to measure membrane voltage and resistance of cultured epithelial cells. This apparatus contains an outer and an inner electrode. The former is somewhat longer and contains small silver pads for passing current through the membrane, whereas the latter is shorter in length and contains small Ag/AgCl voltage sensors. Follow the manufacturer's instructions when using this instrument. Resistance is measured at room temperature in medium without serum. Make sure the medium is at room temperature and the electrodes are sterile (soak in 70% EtOH). When moving the electrodes from one filter-grown monolayer to the next, rinse with PBS, if necessary, and avoid touching the membrane with the internal electrode. As a control, the resistance of the growth medium plus membrane support without cells is measured; this value is subtracted from the reading obtained with cells. Use of nonpolarized epithelial cells for comparison is highly encouraged. Cell monolayers can also be monitored for impermeability to diffusion of a macromolecule across the cell layer by using [$^3$H]inulin (Caplan *et al.*, 1986). The rate of insulin diffusion in a cell monolayer should not exceed 1%/hr for cells with intact junctional complexes.

### b. Infection and Metabolic Labeling of VSV-Infected MDCK Cells on Filters

VSV replicates rapidly in cell culture, reaching maximal viral titers depending on the multiplicity of infection and host cell type approximately 8–12 hr postinfection (Wagner, 1975). Viral protein synthesis peaks at about 4 hr postinfection using high multiplicity of infection, and can be detected as early as 1 hr postinfection. The N protein is generally the first detectable, and is also the most abundant viral protein in infected cells. The L protein is synthesized throughout a 7-hr cycle of infection, albeit at relatively low levels (Wagner *et al.,* 1970). The G and M proteins gradually increase as the infection cycle progresses.

Since VSV preferentially infects MDCK cells basolaterally, the infection of polarized MDCK cells cultivated on filters should be initiated from the basolateral surface. MDCK cells cultivated on filter-membrane supports are infected with VSV only after the cells exhibit high transepithelial resistance (typically 7–10 days postseeding) at a multiplicity of infection of 5 pfu/cell.

1. Before infection, both apical and basolateral surface are washed with PBS. This can be done removing filters and dipping them edgewise into a sterile beaker of warm PBS, followed by careful aspiration and washing of the apical side.

2a. To infect cells basolaterally, the filters are transferred to a new 6-well plate containing 1 ml virus inoculum carefully placed in the center of each well. The filter is kept at an angle and is slowly lowered onto the inoculum.

2b. Alternatively, filter-grown cells can be infected by placing a droplet of 100 $\mu$l diluted virus suspension on a sheet of parafilm overlaying a moistened piece of Whatman filter paper. The parafilm and Whatman paper are placed in a sterile petri dish, droplets of inoculum are applied, and filters are gently lowered onto the droplets at an angle to prevent air bubbles from forming. This inoculation chamber is then placed in a 37°C incubator for 60 min.

3. After the virus adsorption period, the apical and basolateral medium is replaced with 1 ml growth medium and the infection is continued at 37°C for the appropriate times.

4. To label the VSV-infected cells metabolically, both apical and basolateral surfaces are starved 30 min with DMEM without methionine and cysteine prior to labeling, followed by incubation in DMEM plus 50 $\mu$Ci [$^{35}$S]Met/Cys (Express $^{35}$S$^{35}$S-labeling mix; New England Nuclear, Boston, MA) applied to the basolateral (exterior) chamber, since methionine is more efficiently taken up from the basolateral than the apical surface (Balcarova-Ständer *et al.,* 1984).

5. To follow the time course of viral protein synthesis, the cells are labeled every 2 hr postinfection (beginning of adsorption period taken as time 0) at 37°C for 20 min.

6. After labeling, both surfaces of the membrane inserts (apically and basolateral) are washed twice with ice cold PBS and the cells are lysed by the addition of 500 $\mu$l cold lysis buffer [1.0% Triton X-100 in MNT (20 m*M* MES, 100 m*M*

NaCl, 30 m*M* Tris-HCl, pH 7.5) containing 50 m*M* iodoacetamide, 1 m*M* EDTA, 1 m*M* PMSF] to the apical (interior) side of the filter inserts.

7. The cells are scraped off the filter with a rubber policeman and are transferred to a microcentrifuge tube.

8. Cellular debris is removed by centrifugation at 10,000 rpm for 10 min in a microcentrifuge.

9. The supernatants can be immunoprecipitated with anti-VSV antiserum or can be directly analyzed by electrophoresis on 7.5% polyacrylamide gels in the presence of SDS.

## 4. Electron Microscopy of Virus-Infected Cells

The following procedure is used for embedding virus-infected cell cultures for electron microscopy.

### *a. EM-Fixation and Staining of Cell Cultures*

1. Wash cell cultures and permeable membrane filters (both sides) with PBS (300 m*M*, pH 7.2) twice for 5 min each time.
2. Fix cultures for 30 min with 1% glutaraldehyde in PBS.
3. Wash samples with PBS. Samples can be stored overnight. It is important to excise the permeable membrane filters before proceeding to the next step. Cut inserts into strips (5 × 10 mm) and place in a glass tube with PBS.
4. Wash samples with 300 m*M* cacodylate buffer, pH 7.2, twice for 30 min each time.
5. Post-fix (stain) with 1% osmium tetroxide in 300 m*M* cacodylate buffer for 60 min.
6. Wash samples with 300 m*M* cacodylate buffer twice for 30 min each time.
7. *En bloc* stain with 1% uranyl acetate in 10% methanol for 60 min.
8. Rinse with 50% methanol twice for 5 min each time to remove excess stain.
9. Dehydrate samples through a graded ethanol series (50, 75, 95, 100%) for exchanges of 10–15 min each. The 100% alcohol concentration should be stored over aluminum sodium silicate (Molecular Sieves Grade 514; Fisher Scientific, Pittsburgh, PA) to ensure dryness. If working with filters proceed to steps in Section b (see below).
10. Float cell cultures off the surface of the plastic ware with propylene oxide. (Wear gloves! Score a circle around the edge of the dish, remove the alcohol, and cover the cells with propylene oxide. Then use a gentle

swirling motion to lift the cells free from the plastic. Using an applicator stick, remove the cells and transfer to glass test tubes filled with fresh propylene oxide.)

11. Wash with propylene oxide twice 5–7 min each time. If the solution is cloudy when the cells are transferred from the plastic, wash five times. Propylene oxide should be stored over Molecular Sieves to ensure dryness.
12. Infiltrate the specimens with a 1:1 mixture of propylene oxide and epoxy resin (EMBED 812; Electron Microscopy Sciences, Fort Washington, PA) for 1–2 hr in capped vials. This step can be continued overnight.
13. Remove caps and lower the level of fluid to just above the specimen. Allow the propylene oxide to evaporate from the vials for at least 2 hr.
14. Infiltrate with 1 ml 100% epoxy resin (EMBED 812) for 2 hr or longer.
15. Prepare a conical BEEM capsule for each specimen, and insert a paper label. Place a small drop of fresh epoxy resin in each capsule; close the cover.
16. Lightly lubricate each capsule with silicon grease and press into the top of a microfuge tube (remove the cover of the microfuge tube). Spin the capsules for 5 sec to pull the epoxy resin to the tip of the capsule.
17. Using a plastic transfer pipet, collect cells and fill the capsule with cells. A brief spin at 1500 rpm will sediment the cells, making collection easier.
18. Spin the capsules in the microfuge for 1–2 min to pull the cells to the tip of the capsule; there should be a visible pellet.
19. Using pliers, pop the capsule out of the microfuge tubes and place (capsule top open) in a polymerization rack.
20. Place the rack in a dry oven at 60°C for 24–48 hr.
21. Remove the block from the capsule.

***b. Embedding Cells on Permeable Membranes***

1. Remove final alcohol (see step 9 from preceding section) and replace with 1 ml of a 1:1 mixture of 100% ethanol and Low Viscosity embedding media (Electron Microscopy Sciences, Ft. Washington, PA) for 2 hr or overnight; place caps on glass tubes.
2. Remove the 1:1 mixture and replace with 1 ml of complete Low Viscosity media for 2–4 hr.
3. Prepare three gelatin capsules (size "O") for each sample; use only the larger portion of each capsule.
4. Fill capsules with fresh Low Viscosity media and place several strips from each sample and then a label into each of the three capsules.
5. Polymerize the resin at 70°C for 24 to 48 hr. Do not cover capsules.

## Acknowledgments

We thank Drs. Judith A. Ball, Miroslav Novak, Dan R. Rawlins, and Guang-jer Wu for helpful comments. Research by the authors was supported by Grant AI 12680 from the National Institute of Allergy and Infectious Diseases and Grant CA18611 from the National Cancer Institute.

## References

Aloni, Y., Dhar, R., Laub, O., Horowitz, M., and Khoury, G. (1977). Novel mechanism for RNA maturation: The leader sequences of SV40 mRNA are not transcribed adjacent to the coding sequences. *Proc. Natl. Acad. Sci. USA* **74,** 3686–3690.

Andino, R., Rieckhof, G. E., Achacoso, P. L., and Baltimore, D. (1993). Poliovirus RNA synthesis utilizes an RNP complex formed around the 5′ end of viral RNA. *EMBO J.* **12,** 3587–3598.

Bablanian, R., Eggers, H. J., and Tamm, I. (1965). Studies on the mechanism of poliovirus-induced cell damage. I. The relation between poliovirus-induced metabolic and morphological alterations in cultured cells. *Virology* **26,** 100–113.

Balcarova-Ständer, J., Pfeiffer, S. E., Fuller, S. D., and Simons, K. (1984). Development of cell surface polarity in the epithelial Madin–Darby canine kidney (MDCK) cell line. *EMBO J.* **3,** 2687–2694.

Baltimore, D., Franklin, R. M., and Callender, J. (1963). Mengovirus-induced inhibition of host ribonucleic acid and protein synthesis. *Biochim. Biophys. Acta* **76,** 425–430.

Baltimore, D., Girard, M., and Darnell, J. E. (1966). Aspects of the synthesis of poliovirus and the formation of virus particles. *Virology* **29,** 179–189.

Berget, S. M., Moore, C., and Sharp, P. A. (1977). Spliced segments at the 5′ terminus of adenovirus 2 late mRNA. *Proc. Natl. Acad. Sci. USA* **74,** 3171–3175.

Bergmann, J. E., Kupfer, A., and Singer, S. J. (1983). Membrane insertion at the leading edge of motile fibroblasts. *Proc. Natl. Acad. Sci. USA* **80,** 1367–1371.

Bittman, R., Majuk, Z., Honig, D. S., Compans, R. W., and Lenard, J. (1976). Permeability properties of the membrane of vesicular stomatitis virions. *Biochim. Biophys. Acta* **433,** 63–74.

Both, G. W., Mattick, J. S., and Bellamy, A. R. (1983). Serotype-specific glycoprotein of simian 11 rotavirus: Coding assignment and gene sequence. *Proc. Natl. Acad. Sci. USA* **80,** 3091–3095.

Bowen, H. A., and Lyles, D. S. (1981). Structure of Sendai viral proteins in plasma membranes of virus-infected cells. *J. Virol.* **37,** 1079–1082.

Brewer, C. B., and Roth, M. G. (1991). A single amino acid change in the cytoplasmic domain alters the polarized delivery of influenza virus hemagglutinin. *J. Cell Biol.* **114,** 413–421.

Bukrinskaya, A. G., Vorkunova, N. D., Kornliayeva, G. V., Narmanbetova, R. A., and Vorkunova, G. K. (1982). Influenza virus uncoating in infected cells and effect of rimantadine. *J. Gen. Virol.* **60,** 49–59.

Bukrinsky, M. I., Sharona, N., McDonald, T. L., Pushkarskaya, T., Tarpley, W. G., and Stevenson, M. (1993). Association of integral, matrix, and reverse transcriptase antigens of human immunodeficiency virus type 1 with viral nucleic acids following infection. *Proc. Natl. Acad. Sci. USA* **90,** 6125–6129.

Caplan, M. J., Anderson, H. C., Palade, G. E., and Jamieson, J. D. (1986). Intracellular sorting and polarized cell surface delivery of $(Na^+,K^+)$ATPase, an endogenous component of MDCK cell basolateral plasma membranes. *Cell* **46,** 623–631.

Carr, C. M., and Kim, P. S. (1993). A spring-loaded mechanism for the conformational change of influenza hemagglutinin. *Cell* **73,** 823–832.

Cartwright, B., Talbot, P., and Brown, F. (1970). A spliced sequence at the 5′ terminus of adenovirus late mRNA. *J. Gen. Virol.* **29,** 332–344.

Challberg, M., and Kelly, T. J. (1979). Adenovirus DNA replication in vitro. *Proc. Natl. Acad. Sci. USA* **76,** 655–659.

Challberg, M. D., and Kelly, T. J. (1989). Animal virus DNA replication. *Annu. Rev. Biochem.* **58,** 671–717.

Chen, S.-Y., Matsuoka, Y., and Compans, R. W. (1991). Golgi complex localization of the Punta Toro virus G2 protein requires its association with the G1 protein. *Virology* **183,** 351–365.

Choi, W.-S., Pal-Ghosh, R., and Morrow, C. D. (1991). Expression of human immunodeficiency virus type 1 (HIV-1) *gag, pol,* and *env* proteins from chimeric HIV-1-poliovirus minireplicons. *J. Virol.* **65,** 2875–2883.

Choppin, P. W. (1969). Replication of influenza virus in a continuous cell line: High yield of infective virus from cells innoculated at high multiplicity. *Virology* **39,** 130–134.

Chow, L. T., Gelinas, R. E., Broker, T. R., and Roberts, R. J. (1977). An amazing sequence arrangement at the 5′ ends of adenovirus 2 messenger RNA. *Cell* **12,** 1–8.

Chow, M., Newman, J. F. E., Filman, D., Hogle, J. M., Rowlands, D. J., and Brown, F. (1987). Myristylation of picornavirus capsid protein VP4 and its structural significance. *Nature (London)* **327,** 482–486.

Ciampor, F., Baley, P. M., Nermut, M. V., Hirst, E. M. A., Sugrue, R. J., and Hay, A. J. (1992). Evidence that the amantidine-induced, M2-mediated conversion of influenza A virus hemagglutinin to the low pH conformation occurs in an acidic trans Golgi compartment. *Virology* **186,** 14–24.

Cole, C. N., Smoler, D., Wimmer, E., and Baltimore, D. (1971). Defective interfering particles of poliovirus. I. Isolation and physical properties. *J. Virol.* **7,** 478–485.

Compans, R. W. (1973). Distinct carbohydrate components of influenza virus proteins in smooth and rough cytoplasmic membranes. *Virology* **55,** 541–545.

Compans, R. W., Klenk, H.-D., Caliguiri, L. A., and Choppin, P. W. (1970). Influenza virus proteins: I. Analysis of polypeptides of the virion and identification of spike glycoproteins. *Virology* **42,** 880–889.

Copeland, C. S., Doms, R. W., Bolzau, E. M., Webster, R. G., and Helenius, A. (1986). Assembly of influenza hemagglutinin trimers and its role in intracellular transport. *J. Cell Biol.* **103,** 1179–1191.

Daniels, R. S., Downie, J. C., and Hay, A. J. (1985). Fusion mutants of the influenza virus hemagglutinin glycoprotein. *Cell* **40,** 401–439.

Darnell, J. E., and Levintow, L. (1960). Poliovirus protein: Source of amino acids and time course of synthesis. *J. Biol. Chem.* **235,** 74–77.

Department of the Army (1964). "Laboratory Procedures in Virology," TM 8-227-7. Headquarters, Department of the Army, Washington, D.C.

Doms, R. W., Lamb, R. A., Rose, J. K., and Helenius, A. (1993). Folding and assembly of viral membrane proteins. *Virology* **193,** 545–562.

Dotti, C. G., and Simons, K. (1990). Polarized sorting of viral glycoproteins to the axon and dendrites of hippocampal neurons in culture. *Cell* **62,** 63–72.

Dubois-Dalcq, M., Holmes, K. V., and Rentier, B. (1984). "Assembly of Enveloped RNA Viruses." Springer-Verlag, Wien.

Dulbecco, R., and Vogt, M. (1954). Plaque formation and isolation of pure lines with poliomyelitis viruses. *J. Exp. Med.* **99,** 167–182.

Enders, J. F., Wellers, T. H., and Robbins, F. C. (1949). Cultivation of the Lansing strain of poliomyelitis virus in cultures of various human embryonic tissues. *Science* **109,** 85–87.

Fanning, E., and Knippers, R. (1992). Structure and function of simian virus 40 large tumor antigen. *Annu. Rev. Biochem.* **64,** 55–85.

Fernandez-Tomas, G. B., and Baltimore, D. (1973). Morphogenesis of poliovirus. II. Demonstration of a new intermediate, the provirion. *J. Virol.* **12,** 1122–1130.

Fixman, E. D. (1992). Trans-acting requirements for replication of Epstein–Barr virus ori-Lyt. *J. Virol.* **66,** 5030–5039.

Freistadt, M. S., and Racaniello, V. R. (1991). Mutational analysis of the cellular receptor for poliovirus. *J. Virol.* **65,** 3873–3876.

Gallagher, P. J., Henneberry, I., Wilson, I., Sambrook, J., and Gething, M.-J. H. (1992). Glycosylation requirements for intracellular transport and function of the hemagglutinin of influenza virus. *J. Virol.* **66,** 7136–7145.

Gething, M. J., McCammon, K., and Sambrook, J. (1986). Expression of wild-type and mutant forms of influenza hemagglutinin: The role of folding in intracellular transport. *Cell* **46,** 939–950.

Gibson, R., Schlesinger, S., and Kornfeld, S. (1979). The non-glycosylated glycoprotein of vesicular stomatitis virus is temperature-sensitive and undergoes intracellular aggregation at elevated temperatures. *J. Biol. Chem.* **254,** 3600–3607.

Gibson, R., Kornfeld, S., and Schlesinger, S. (1980). A role for oligosaccharides in glycoprotein biosynthesis. *Trends Biol. Sci.* **5,** 290–293.

Griffiths, G., Pfeiffer, S., Simons, K., and Matlin, K. (1985). Exit of newly synthesized membrane proteins from the trans cisterna of the Golgi complex to the plasma membrane. *J. Cell Biol.* **101,** 949–964.

Harrison, S. C., Jumblatt, J., and Darnell, J. E. (1971). Lipid and protein organization in Sindbis virus. *J. Mol. Biol.* **60,** 523–528.

Hay, A. J., Wolstenholme, A. J., Skehel, J. J., and Smith, M. H. (1985). The molecular basis of the specific anti-influenza action of amantadine. *EMBO J.* **4,** 3021–3024.

Henderson, L. E., Krutzsch, H. C., and Oroszlan, S. (1983). Myristyl amino-terminal acylation of murine retrovirus proteins: An unusual post-translational protein modification. *Proc. Natl. Acad. Sci. USA* **80,** 339–343.

Hogle, J. M., Chow, M., and Filman, D. J. (1985). Three-dimensional structure of poliovirus at 2.9 Å resolution. *Science* **229,** 1358–1365.

Holland, J. J. (1961). Receptor affinities as major determinants of enterovirus tissue tropisms in humans. *Virology* **15,** 312–326.

Holland, J. J., McLaren, J. C., and Syverton, J. T. (1959). The mammalian cell virus relationship. IV. Infection of naturally insusceptible cells with enterovirus ribonucleic acid. *J. Exp. Med.* **110,** 65–80.

Holmes, K. V., and Choppin, P. W. (1966). On the role of the response of the cell membrane in determining virus virulence. Contrasting effects of the parainfluenza virus SV5 in two cells types. *J. Exp. Med.* **124,** 501–531.

Homma, M., and Ohuchi, M. (1973). Trypsin action on the growth of Sendai virus in tissue culture cells. III. Structural differences of Sendai viruses grown in eggs and tissue culture cells. *J. Virol.* **12,** 1457–1463.

Hubbard, S. C., and Ivatt, R. J. (1981). Synthesis of the N-linked oligosaccharides of glycoproteins: Assembly of the lipid-linked precursor oligosaccharide and its relation to protein synthesis in vivo. *Annu. Rev. Biochem.* **50,** 555–583.

Ihara, T., Smith, J., Dalrymple, J. M., and Bishop, D. H. L. (1985). Complete sequences of the glycoproteins and M RNA of Punta Toro Phlebovirus compared to those of Rift Valley fever virus. *Virology* **144,** 246–259.

Jones, L. V., Compans, R. W., Davis, A. R., Bos, T. J., and Nayak, D. P. (1985). Surface expression of the influenza neuraminidase, an amino-terminally anchored viral membrane glycoprotein, in polarized epithelial cells. *Mol. Cell. Biol.* **5,** 2181–2189.

Kalderon, D., Richardson, W. D., Markham, A. T., and Smith, A. E. (1984a). Sequence requirements for nuclear location of Simian virus 40 large T antigen. *Nature (London)* **311,** 33–38.

Kalderon, D., Roberts, B. L., Richardson, W. D., and Smith, A. E. (1984b). A short amino acid sequence able to specify nuclear location. *Cell* **39,** 4182–4189.

Kates, J., and Beeson, J. (1970). Ribonucleic acid synthesis in vaccinia virus. I. The mechanism of synthesis and release of RNA in vaccinia cores. *J. Mol. Biol.* **50,** 1–18.

Kato, N., and Eggers, H. J. (1969). Inhibition of uncoating of fowl plague virus by 1-amantadine hydrochloride. *Virology* **37,** 632–641.

Katz, F. N., and Lodish, H. F. (1979). Transmembrane biogenesis of the vesicular stomatitis virus glycoprotein. *J. Cell Biol.* **80,** 416–426.

Khoury, G., and Gruss, P. (1983). Enhancer elements. *Cell* **33,** 313–314.

Klenk, H.-D., and Choppin, P. W. (1969). Lipids of plasma membranes of monkey and hamster kidney cells and of parainfluenza virions grown in these cells. *Virology* **38,** 255–268.

Klenk, H.-D., and Choppin, P. W. (1971). Glycolipid content of vesicular stomatitis virus grown in baby hamster kidney cells. *J. Virol.* **7,** 416–417.

Klenk, H.-D., Compans, R. W., and Choppin, P. W. (1970). An electron microscopic study of the presence or absence of neuraminic acid in enveloped viruses. *Virology* **42,** 1158–1162.

Klenk, H.-D., Wollert, W., Rott, R., and Schollissek, C. (1974). Association of influenza virus proteins with cytoplasmic fractions. *Virology* **57,** 28–41.

Klenk, H.-D., Rott, R., Orlich, M., and Blödorn, J. (1975). Activation of influenza A viruses by trypsin treatment. *Virology* **68,** 426–439.

Klessig, D. F. (1977). Two adenovirus mRNAs have a common 5′ terminal leader sequence encoded at least 10 kb upstream from their main coding regions. *Cell* **12,** 9–21.

Koike, S., Ise, I., and Nomoto, A. (1991). Functional domain of the poliovirus receptor. *Proc. Natl. Acad. Sci. USA* **88,** 4104–4108.

Kolakofsky, D., Vidal, S., and Curran, J. (1991). Paramyxovirus RNA synthesis and P gene expression. *In* "The Paramyxoviruses" (D. W. Kingsbury, ed.). Plenum Press, New York. pp. 215–234.

Krug, R. M., Alonso-Caplen, F. V., Julkunen, I., and Katze, M. G. (1989). Expression and replication of the influenza virus genome. *In* "The Influenza Virus" (R. M. Krug, ed.). Plenum Press, New York. pp. 89–152.

Lamb, R. A., and Choppin, P. W. (1976). Synthesis of influenza virus proteins in infected cells: Translation of viral polypeptides, including three P polypeptides, from RNA produced by primary transcription. *Virology* **74,** 504–519.

Landsberger, F. R., Lenard, J., Paxton, J., and Compans, R. W. (1971). Spin-label electron spin resonance study of the lipid-containing membrane of influenza virus. *Proc. Natl. Acad. Sci. USA* **68,** 2579–2583.

Lanford, R. E., and Butel, J. (1984). Construction and characterization of an SV40 mutant defective in nuclear transport of T antigen. *Cell* **37,** 801–813.

Lazarowitz, S. G., and Choppin, P. W. (1975). Synthesis of influenza virus proteins in infected cells: Translation of viral polypeptides, including three P polypeptides, from RNA produced by primary transcription. *Virology* **68,** 440–445.

Leibowitz, R., and Penman, S. (1971). Regulation of protein synthesis in HeLa cells. III. Inhibition during poliovirus infection. *J. Virol.* **8,** 661–668.

Lenard, J., and Compans, R. W. (1974). The membrane structure of lipid-containing viruses. *Biochim. Biophys. Acta* **344,** 51–94.

Levintow, L. (1974). The reproduction of picornaviruses. *In* "Comprehensive Virology" (H. Fraenkel-Contrat and R. R. Wagner, eds.), Vol. 2, pp. 106–169. Plenum, New York.

Li, J. J., and Kelly, T. J. (1984). Simian virus 40 DNA replication in vitro. *Proc. Natl. Acad. Sci. USA* **81,** 6973–6977.

Lingappa, V. R., Kato, F. N., Lodish, H. F., and Blobel, G. (1978). A signal sequence for the insertion of a transmembrane glycoprotein. Similarities to signals of secretory proteins in primary structure and function. *J. Biol. Chem.* **253,** 8667–8670.

Lyles, D. S. (1979). Glycoproteins of Sendai virus are transmembrane proteins. *Proc. Natl. Acad. Sci. USA* **66,** 5621–5625.

Machamer, C., and Rose, J. K. (1987). A specific transmembrane domain of a coronavirus E1 glycoprotein is required for its retention in the Golgi region. *J. Cell Biol.* **105,** 1205–1214.

Machamer, C. E., Florkiewitz, R. Z., and Rose, J. K. (1985). A single N-linked oligosaccharide at either of the two normal sites is sufficient for transport of vesicular stomatitis virus G protein to the cell surface. *Mol. Cell. Biol.* **5,** 3074–3083.

Machamer, C. E., Mentone, S. A., Rose, J. K., and Farquhar, M. G. (1990). The E1 glycoprotein of an avian coronavirus is targeted to the *cis* Golgi complex. *Proc. Natl. Acad. Sci. USA* **87,** 6944–6948.

Machamer, C. E., Grim, M. G., Esquela, A., Chung, S. W., Rolls, M., Ryan, K., and Swift, A. M. (1993). Retention of a *cis* Golgi protein requires polar residues on one face of a predicted α-helix in the transmembrane domain. *Mol. Biol. Cell* **4,** 695–704.

Magee, A. I., Koyama, A. H., Malfer, C., Wen, D., and Schlesinger, M. J. (1984). Release of fatty acids from virus glycoproteins by hydroxylamine. *Biochim. Biophys. Acta* **798,** 156–166.
Manley, J. L., Fire, H., Cano, A., Sharp, P. A., and Gefter, M. L. (1980). DNA-dependent transcription of adenovirus genes in a soluble whole-cell extract. *Proc. Natl. Acad. Sci. USA* **77,** 3855–3859.
Marsh, M., Griffiths, G., Dean, G. E., Mellman, I., and Helenius, A. (1986). Three-dimensional structure of endosomes in BHK-21 cells. *Proc. Natl. Acad. Sci. USA* **83,** 2899–2903.
Martin, K., and Helenius, A. (1991). Nuclear transport of influenza virus ribonucleoproteins: The viral matrix protein (M1) promotes export and inhibits import. *Cell* **67,** 117–130.
Matlin, K. S., and Simons, K. (1983). Reduced temperature prevents transfer of a membrane glycoprotein to the cell surface but does not prevent terminal glycosylation. *Cell* **34,** 233–243.
Matlin, K. S., Reggio, H., Helenius, A., and Simons, K. (1982a). Pathway of vesicular stomatitis virus entry leading to infection. *J. Mol. Biol.* **156,** 609.
Matlin, K. S., Reggio, H., Helenius, A., and Simons, K. (1982b). The entry of enveloped viruses into an epithelial cell line. *Prog. Clin. Biol. Res.* **91,** 599.
Matsuoka, Y., Ihara, T., Bishop, D. H. L., and Compans, R. W. (1988). Intracellular accumulation of Punta Toro virus glycoproteins expressed from cloned cDNA. *Virology* **167,** 251–260.
Matsuoka, Y., Chen, S.-Y., and Compans, R. W. (1994). A signal for Golgi retention in the bunyavirus G1 glycoprotein. Submitted for publication.
McPherson, I., and Stoker, M. (1962). Polyoma transformation of hamster cell clones. An investigation of genetic factors affecting cell competence. *Virology* **16,** 147–151.
McSharry, J., and Benzinger, R. (1970). Concentration and purification of vesicular stomatitis virus by polyethylene glycol "precipitation." *Virology* **40,** 745–779.
Meier-Ewert, H., and Compans, R. W. (1974). Time course of synthesis and assembly of influenza virus proteins. *J. Virol.* **14,** 1083–1091.
Mendelsohn, C. L., Wimmer, E., and Racaniello, V. R. (1989). Cellular receptor for poliovirus: Molecular cloning, nucleotide sequence, and expression of a new member of the immunoglobulin superfamily. *Cell* **56,** 855–865.
Montagnier, L., and Sanders, F. K. (1963). Replicative form of encephalomyocarditis virus ribonucleic acid. *Nature (London)* **199,** 664–667.
Munro, S. (1991). Sequences within and adjacent to the transmembrane segment of $\alpha$-2,6-sialyltransferase specify Golgi retention. *EMBO J.* **10,** 3577–3588.
Murphy, F. A., Harrison, A. K., and Whitfield, S. G. (1973). Bunyaviridae: Morphologic and morphogenetic similarities of bunyamwera serologic supergroup viruses and several other arthropod-borne viruses. *Intervirology* **1,** 297–316.
Naim, H. Y., Amarneh, B., Ktistakis, N. T., and Roth, M. G. (1992). Effects of altering palmitylation sitdes on biosynthesis and function of the influenza virus hemagglutinin. *J. Virol.* **66,** 7585–7588.
Nakamura, K., and Compans, R. W. (1979). Biosynthesis of the oligosaccharides of influenza virus glycoproteins. *Virology* **93,** 31–47.
Nevins, J. R. (1983). The pathway of eukaryotic mRNA formation. *Annu. Rev. Biochem.* **52,** 441–466.
Ng, D. T. W., Hiebert, S. W., and Lamb, R. A. (1990). Different roles of individual N-linked oligosaccharide chains in folding, assembly, and transport of simian virus 5 hemagglutinin-neuraminidase. *Mol. Cell. Biol.* **10,** 1989–2001.
Nilsson, T., Jackson, M., and Peterson, P. A. (1989). Short cytoplasmic sequences serve as retention signals for transmembrane proteins in the endoplasmic reticulum. *Cell* **58,** 707–718.
Nilsson, T., Lucocq, J. M., Mackay, D., and Warren, G. (1991). The membrane spanning domain of $\beta$-1,4-galactosyltransferase specifies *trans* Golgi localization. *EMBO J.* **10,** 3567–3575.
Palese, P., Tobita, K., Ueda, M., and Compans, R. W. (1974). Characterization of temperature sensitive influenza virus mutants defective in neuraminidase. *Virology* **61,** 397–404.
Palmer, D. F., Coleman, M. T., Dowdle, W. R., and Schild, G. C. (1975). "Advanced Laboratory Techniques for Influenza Diagnosis." Centers for Disease Control, U.S. Department of Health, Education and Welfare. Atlanta, Georgia.

Pensiero, M. N., Jennings, G. B., Schmaljohn, C. S., and Hay, J. (1988). Expression of the Hantaan virus M genome segment by using a vaccinia virus recombinant. *J. Virol.* **62,** 696–702.

Percy, N., Barclay, W. S., Sullivan, M., and Almond, J. W. (1992). A poliovirus replicon containing the chloramphenicol acetyl-transferase gene can be used to study the replication and encapsidation of poliovirus RNA. *J. Virol.* **66,** 5040–5046.

Petterson, R. F., Gahmberg, N., Kuismanen, E., Kaariainen, L., Ronnholm, R., and Saraste, J. (1988). Bunyavirus membrane glycoproteins as models for Golgi-specific proteins. *Mod. Cell Biol.* **6,** 65–96.

Philipson, L., Wall, R., Glickman, G., and Darnell, J. E. (1971). Addition of polyadenylate sequences to virus-specific RNA during adenovirus replication. *Proc. Natl. Acad. Sci. USA* **68,** 2806–2809.

Pinto, L. H., Holsinger, L. J., and Lamb, R. A. (1992). Influenza virus M2 protein has ion channel activity. *Cell* **69,** 517–528.

Porter, D. C., Ansardi, D. C., Choi, W.-S., and Morrow, C. D. (1993). Encapsidation of genetically engineered poliovirus minireplicons which express human immunodeficiency virus type 1 Gag and Pol proteins upon infection. *J. Virol.* **67,** 3712–3719.

Poruchynsky, M. S., Tyndall, C., Both, G. W., Sato, F., and Bellamy, A. R. (1985). Deletions into an $NH_2$-terminal hydrophobic domain result in secretion of rotavirus VP7, a resident endoplasmic reticulum membrane protein. *J. Cell Biol.* **101,** 2199–2209.

Racaniello, V. R. (1988). Poliovirus neurovirulence. *Adv. Virus Res.* **34,** 217–246.

Ren, R., and Racaniello, V. R. (1992). Human poliovirus receptor gene expression and poliovirus tissue tropism in transgenic mice. *J. Virol.* **66,** 296–304.

Richardson, J. C., Scalera, V., and Simmons, N. L. (1981). Identification of two strains of MDCK cells which resemble separate nephron tubule segments. *Biochim. Biophys. Acta* **673,** 26–36.

Rifkin, D. B., and Compans, R. W. (1971). Identification of the spike proteins of Rous sarcoma virus. *Virology* **46,** 485–489.

Rindler, M. J., Ivanov, I. E., Plesken, H., Rodriguez-Boulan, E. J., and Sabatini, D. D. (1984). Viral glycoproteins destined for apical or basolateral membrane domains traverse the same Golgi apparatus during their intracellular transport in Madin–Darby canine kidney cells. *J. Cell. Biol.* **98,** 1304–1319.

Roberts, P. C., Garten, W., and Klenk, H.-D. (1993). Role of conserved glycosylation sites in maturation and transport of influenza A virus hemagglutinin. *J. Virol.* **67,** 3048–3060.

Rodriguez-Boulan, E., and Powell, S. K. (1992). Polarity of epithelial and neuronal cells. *Annu. Rev. Cell Biol.* **8,** 395–427.

Rodriguez-Boulan, E., and Sabatini, D. D. (1978). Asymmetric budding of viruses in epithelial monolayers: A model system for study of epithelial polarity. *Proc. Natl. Acad. Sci. USA* **75,** 5071–5075.

Rose, J. K., and Bergmann, J. E. (1982). Expression from cloned cDNA of cell-surface secreted forms of the glycoprotein of vesicular stomatitis virus in eukaryotic cells. *Cell* **30,** 753–762.

Rose, J. K., Adams, G. A., and Gallione, C. J. (1984). The presence of cysteine in the cytoplasmic domain of the vesicular stomatitis virus glycoprotein is required for palmitate addition. *Proc. Natl. Acad. Sci. USA* **81,** 2050–2054.

Roth, M. G., and Compans, R. W. (1981). Delayed appearance of pseudotypes between vesicular stomatitis virus and influenza virus during mixed infection of MDCK cells. *J. Virol.* **40,** 848–860.

Roth, M. G., Fitzpatrick, J. P., and Compans, R. W. (1979). Polarity of influenza and vesicular stomatitis virus maturation in MDCK cells: Lack of a requirement for glycosylation of viral glycoproteins. *Proc. Natl. Acad. Sci. USA* **76,** 6430–6434.

Roth, M. G., Compans, R. W., Giusti, L., Davis, A. R., Nayak, D. P., Gething, M. J., and Sambrook, J. (1983a). Influenza virus hemagglutinin expression is polarized in cells infected with recombinant SV40 viruses carrying cloned hemagglutinin DNA. *Cell* **33,** 435–442.

Roth, M. G., Srinivas, R. V., and Compans, R. W. (1983b). Basolateral maturation of retroviruses in polarized epithelial cells. *J. Virol.* **45,** 1065–1073.

Rothman, J. E., and Lodish, H. F. (1977). Synchronized transmembrane insertion and glycosylation of a nascent membrane protein. *Nature (London)* **269,** 775–780.

Rothman, J. E., Miller, R. L., and Urbani, L. J. (1984). Intercompartmental transport in the Golgi complex is a dissociative process: Facile transfer of membrane protein between two Golgi populations. *J. Cell Biol.* **99,** 260–271.

Rowlands, D. J., Shirley, M. W., Sangal, D. V., and Brown, F. (1975). A high density component in several vertebrate enteroviruses. *J. Gen. Virol.* **29,** 223–234.

Rueckert, R. (1990). Picornaviridae and their replication. *In* "Virology" (B. N. Fields and D. M. Knipe, eds.), Vol. 1, pp. 507–548. Raven Press, New York.

Scharff, M. D., Shatkin, A. J., and Levintow, L. (1963). Association of newly formed viral protein with specific polyribosomes. *Proc. Natl. Acad. Sci. USA* **50,** 686–694.

Scharff, M. D., Maizel, J. V., and Levintow, L. (1964). Physical and immunological properties of a soluble precursor of the poliovirus capsid. *Proc. Natl. Acad. Sci. USA* **51,** 329–337.

Scheid, A., and Choppin, P. W. (1974). Identification and biological activities of paramyxovirus glycoproteins. Activation of cell fusion, hemolysis and infectivity by proteolytic cleavage of an inactive precursor protein of Sendai virus. *Virology* **57,** 475–490.

Schlegel, R., Tralka, T. S., Willingham, M. C., and Pastan, I. (1983). Inhibition of VSV binding and infectivity by phosphatidylserine: Is phosphatidylserine a VSV binding site? *Cell* **32,** 639.

Schmid, S., Fuchs, R., Kielian, M., Helenius, A., and Mellman, I. (1989). Acidification of endosome subpopulations in wild-type Chinese hamster ovary cells and temperature-sensitive acidification-defective mutants. *J. Cell Biol.* **108,** 1291–1300.

Schmidt, M. F., and Schlesinger, M. J. (1979). Fatty acid binding to vesicular stomatitis virus glycoprotein: A new type of post-translational modification of the viral glycoprotein. *Cell* **17,** 813–819.

Schmidt, M. F. G., Bracha, M., and Schlesinger, M. J. (1979). Evidence for covalent attachment of fatty acids to Sindbis virus glycoproteins. *Proc. Natl. Acad. Sci. USA* **76,** 1687–1691.

Schneider, R. J., and Shenk, T. (1987). Impact of virus infection on host cell protein synthesis. *Annu. Rev. Biochem.* **56,** 317–332.

Schultz, A. M., and Rein, A. (1989). Unmyristylated Moloney murine leukemia virus Pr65gag is excluded from virus assembly and maturation events. *J. Virol.* **63,** 2370–2372.

Schulze, I. T. (1970). The structure of influenza virus 1. The polypeptides of the virion. *Virology* **47,** 181–196.

Shapiro, G. I., Gurney, T., Jr., and Krug, R. M. (1987). Influenza virus gene expression: Control mechanisms at early and late times of infection and nuclear-cytoplasmic transport of virus-specific RNAs. *J. Virol.* **61,** 764–773.

Shatkin, A. J. (1976). Capping of eucaryotic mRNAs. *Cell* **9,** 645–653.

Simpson, D. A., and Lamb, R. A. (1992). Alterations to influenza virus hemagglutinin cytoplasmic tail modulate virus infectivity. *J. Virol.* **66,** 790–803.

Skehel, J. J. (1972). Polypeptide synthesis in influenza virus-infected cells. *Virology* **49,** 23–36.

Skehel, J. J., Hay, A. J., and Armstrong, J. A. (1978). On the mechanism of inhibition of influenza virus replication by amantadine hydrochloride. *J. Gen. Virol.* **38,** 97–110.

Stirzaker, S. C., and Both, G. W. (1989). The signal peptide of the rotavirus glycoprotein VP7 is essential for its retention in the ER as an integral membrane protein. *Cell* **56,** 741–747.

Sugrue, R. J., Bahadur, G., Zambon, M. C., Hall-Smith, M., Douglas, A. R., and Hay, A. J. (1990). Specific alteration of the influenza hemagglutinin by amantidine. *EMBO J.* **9,** 3469–3476.

Summers, D. F., and Maizel, J. V. (1967). Disaggregation of HeLa cell polysomes after infection with poliovirus. *Virology* **31,** 550.

Summers, D. F., and Maizel, J. V. (1968). Evidence for large precursor proteins in poliovirus synthesis. *Proc. Natl. Acad. Sci. USA* **59,** 966–971.

Swift, A. M., and Machamer, C. E. (1991). A Golgi retention signal in a membrane-spanning domain of coronavirus E1 protein. *J. Cell Biol.* **115,** 19–30.

Thomas, S. M., Lamb, R. A., and Paterson, R. G. (1988). Two mRNAs that differ by two nontemplated nucleotides encode the amino coterminal proteins P and V of the paramyxovirus SV5. *Cell* **54,** 891–902.

Tobita, K., Sugiura, A., Enomote, C., and Furuyama, M. (1975). Plaque assay and primary isolation of influenza A viruses in an established line of canine kidney cells (MDCK) in the presence of trypsin. *Med. Microbiol. Immunol.* (*Berlin*) **162,** 9–14.

Toneguzzo, F., and Ghosh, H. P. (1978). In vitro synthesis of vesicular stomatitis virus membrane glycoprotein and insertion into membranes. *Proc. Natl. Acad. Sci. USA* **75,** 715–719.

Tooze, J., Tooze, S. A., and Fuller, S. D. (1987). Sorting of progeny coronavirus from condensed secretory proteins at the exit from the *trans*-Golgi network of AtT20 cells. *J. Cell Biol.* **105,** 1215–1226.

Tucker, S. P., and Compans, R. W. (1993). Virus infection of polarized epithelial cells. *Adv. Virus Res.* **42,** 187–247.

Tucker, S. P., Melsen, L. R., and Compans, R. W. (1992). Migration of polarized epithelial cells through permeable membrane substrates of defined pore size. *Eur. J. Cell Biol.* **58,** 280–290.

Varghese, J. N., Laver, W. G., and Colman, P. M. (1983). Structure of the influenza virus glycoprotein antigen neuraminidase at 2.9 Å resolution. *Nature* (*London*) **303,** 35–40.

Vidal, S., Curran, J., and Kolakofsky, D. (1990). Editing of the Sendai virus P/C mRNA by G insertion occurs during mRNA synthesis via a virus-encoded activity. *J. Virol.* **64,** 239–246.

Wagner, R. R. (1975). Reproduction of rhabdoviruses. *In* "Comprehensive Virology" (H. Fraenkel-Conrat and R. R. Wagner, ed.), Vol. 4, pp. 1–93. Plenum Press, New York.

Wagner, R. R., Synder, R. M., and Yamazaki, S. (1970). Proteins of vesicular stomatitis virus: Kinetics and cellular sites of synthesis. *J. Virol.* **5,** 548–558.

Wang, C., Takeuchi, K., Pinto, L. H., and Lamb, R. A. (1993). Ion channel activity of influenza A virus M2 protein: Characterization of the amantadine block. *J. Virol.* **67,** 5585–5594.

Wasmoen, T. L., Kaleach, L. T., and Collett, M. S. (1988). Rift Valley fever M segment: Cellular localization of M segment-encoded proteins. *Virology* **166,** 275–280.

Weil, P. A., Luse, D. S., Segall, J., and Roeder, R. G. (1979). Selective and accurate initiation of transcription at the Adz major late promoter in a soluble system dependent on purified RNA polymerase II and DNA. *Cell* **18,** 469–484.

Weisz, O. A., Swift, A. M., and Machamer, C. E. (1993). Oligomerization of a membrane protein correlates with its retention in the Golgi complex. *J. Cell Biol.* **122,** 1185–1196.

White, J. M. (1992). Membrane fusion. *Science* **258,** 917–924.

White, J., Kielian, M., and Helenius, A. (1983). Membrane fusion proteins of enveloped animal viruses. *Q. Rev. Biophys.* **16,** 151–195.

Wilson, I. A., Skehel, J. J., and Wiley, D. C. (1981). Structure of the hemagglutinin membrane glycoprotein of influenza virus at 3 Å resolution. *Nature* (*London*) **289,** 366–373.

Wu, C. A. (1988). Identification of Herpes simplex virus type 1 genes required for origin-dependent DNA synthesis. *J. Virol.* **62,** 435–443.

Wu, G. (1978). Adenovirus DNA-directed transcription of 5.5S RNA in vitro. *Proc. Natl. Acad. Sci. USA* **75,** 2175–2179.

Yamaguchi-Koll, U., Niegers, K. J., and Drzeniek, R. (1975). Isolation and characterization of "dense particles" from poliovirus-infected HeLa cells. *J. Gen. Virol.* **26,** 307–319.

# CHAPTER 2

# Expression of Exogenous Proteins in Mammalian Cells with the Semliki Forest Virus Vector


**Vesa M. Olkkonen,* Paul Dupree,* Kai Simons,* Peter Liljeström,[†] and Henrik Garoff[†]**

*European Molecular Biology Laboratory
69012 Heidelberg, Germany

[†] Department of Molecular Biology
Karolinska Institute
14157 Huddinge, Sweden




## I. Introduction

Expression of cloned sequences using carrier vectors is currently one of the central techniques in cellular and molecular biology. Viral vectors for eukaryotic cells have undergone enormous development in the past 10 years. For large-scale production of proteins in animal cells, the insect cell-restricted baculovirus vectors provide the most powerful tool available at this time (see Fraser, 1992). The most efficient mammalian cell vectors are based on vaccinia virus (see Moss, 1992; see also Chapter 7). Some of these require construction of recombinant vaccinia viruses; others involve transfection of plasmid DNA into cells infected with a helper virus. Episomal plasmid vectors based on the Epstein–Barr virus replicon have been used successfully in cloning eukaryotic genes using functional assays (see Margolskee, 1992). For long-term functional studies, stably transfected cell lines are widely used. The retroviral vectors currently available greatly facilitate the generation of such cell lines (see Miller, 1992; see also Chapter 5). Although the selection of expression vectors suited for higher eukaryotic cells is quite extensive, many of these systems are still hampered by restrictions in host range, insufficient levels of expression, or complicated protocols that are required for recombinant vector preparation.

However, an expression system for quick and efficient production of exogenous proteins in cultured mammalian cells has been described (Liljeström and Garoff, 1991). This vector is based on Semliki Forest virus (SFV), an insect-borne alphavirus of the family Togaviridae that is neuropathogenic in rodents (see Griffin, 1986). Under laboratory conditions, SFV infects a variety of mammalian, avian, reptilian, amphibian, and insect cells (see Griffin, 1986; Liljeström and Garoff, 1991). SFV has a capped and polyadenylated single-stranded RNA genome of positive polarity that functions directly as a message in infected cells. The genomic RNA molecule encodes its own RNA polymerase, driving highly efficient viral RNA replication and transcription. Consequently, capped transcripts of the full-length SFV cDNA are infectious when introduced into cells (Liljeström *et al.*, 1991). SFV replication takes place in the cytoplasm, which is a valuable feature considering its usefulness as a cDNA expression vector. The complications involved in expression of sequences located in the nucleus, for example, mRNA capping, splicing, and transport into the cytoplasm, are avoided.

The expression vector consists of a modified SFV cDNA lacking the region encoding the viral structural proteins (see Fig. 1). An open reading frame (ORF) under study is inserted here under the control of the viral promoter that normally drives transcription of the 26 S mRNA encoding the structural proteins (Liljeström and Garoff, 1991). Runoff transcripts of such a construct that has been linearized in the appropriate way are self-replicating and express the cloned ORF in transfected cells. If the cells are co-transfected with a helper RNA encoding the SFV structural proteins but carrying no genomic RNA packaging signal, virions are produced that contain the recombinant vector RNA. The viral stock obtained can be used for nonreproductive infection and expression of the protein under investigation in a variety of cell types.

In this chapter, we describe in detail the preparation of recombinant SFVs, discuss their usability in different cell types, and show examples of expression of exogenous proteins in baby hamster kidney (BHK) cells and in primary cultures of fetal rat hippocampal neurons.

## II. Procedure

### A. Media and Stock Solutions

All cell culture reagents were purchased from GIBCO/Life Technologies, Inc. (Eggenstein, Germany).

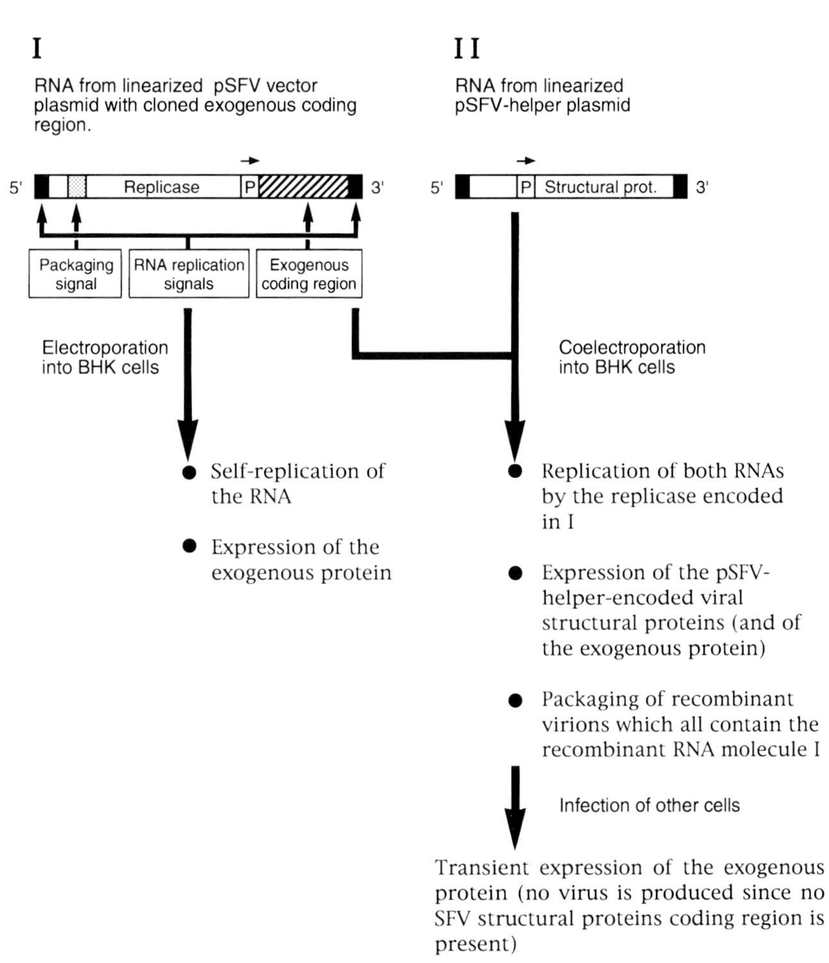


**Fig. 1** The principle of the SFV expression vector system. The recombinant vector with the inserted sequence encoding an exogenous protein and the helper plasmid are both linearized with *Spe* I; runoff transcription is performed using SP6 RNA polymerase. The *in vitro* transcripts are introduced into BHK cells by electroporation. Transfection of the recombinant vector RNA alone results in transient expression of the exogenous protein, driven by the SFV RNA polymerase encoded in the same RNA molecule. If the vector and the helper RNAs are co-introduced into the cells, the SFV structural proteins encoded by the helper will form virions that package the recombinant vector RNA and can be used for nonreproductive infection of other cells. The SFV 26S promoter driving the synthesis of the viral structural proteins (in the helper) or the exogenous protein (in the vector) is denoted by a P.

BHK medium: Glasgow's modified Eagle's medium (G-MEM), 5% fetal calf serum (FCS), 10% tryptose phosphate broth, 20 m*M* HEPES, pH 7.25, 2 m*M* L-glutamine, 0.1 U/ml penicillin, 0.1 mg/ml streptomycin. The infection medium contained 0.2% bovine serum albumin (BSA) instead of the FCS.

Neural culture medium: The plating medium (for primary culture preparation, see Goslin and Banker, 1991) consisted of Eagle's minimal essential medium (MEM) with Earle's salts, extra glucose (600 mg/liter), and 10% horse serum. The cultures were maintained in serum-free MEM containing $N_2$ supplements (Bottestein and Sato, 1979) and 1 m*M* sodium pyruvate. The infections were performed in the same medium.

10× SP6 buffer: 400 m*M* HEPES–KOH, pH 7.4, 60 m*M* MgOAc, 20 m*M* spermidine–HCl

Electroporation buffer: 0.27 m*M* KCl (0.20 g/liter), 1.47 m*M* $KH_2PO_4$ (0.20 g/liter), 137 m*M* NaCl (8.00 g/liter), 9.06 m*M* $Na_2HPO_4 \cdot 7H_2O$ (2.16 g/liter), pH 7.2 (the pH is adjusted with 1 *M* HCl)

## B. Cloning DNA into the Vector

The vector plasmid exists in three versions: pSFV1, 2, and 3 (Liljeström and Garoff, 1991; Life Technologies, Inc.). These plasmids have a polylinker with three unique restriction sites—*Bam*HI, *Sma*I, and *Xma*I—followed by three translation stop codons (in all three reading frames). The multiple cloning site in pSFB1 is situated 31 bp downstream from the transcription initiation site. When this version is used, the insert should include sequences that insure efficient translation. In pSFV2, the cloning sites are positioned immediately after the ribosome binding site (RBS) of the SFV capsid protein gene. In pSFV3, the RBS and the initiation methionine codon are provided in the vector. With respect to cloning applications, pSFV1 is the simplest version of the plasmid since it does not require adjustment of the reading frame or consideration of the distance of the insert initiation ATG from the RBS. The vector accepts a wide variety of insert sizes, but we have found that inserts in the 3-kb range (similar to the length of the endogenous region deleted from the vector) support packaging of recombinant viral stocks of the highest titers.

The cDNA inserts of the constructs used in Section II,D, *myc*–VIP21 and *myc*–rab22, were cloned bluntly into pSFV1 opened with *Sma*I. The insert sizes were 0.9 kb (*myc*–VIP21) and 1.7 kb (*myc*–rab22).

## C. Runoff Transcription

We prepare plasmid DNA for *in vitro* transcription by the alkaline lysis method (Birnboim and Doly, 1979), followed by centrifugation through 1 *M* NaCl and equilibrium centrifugation in CsCl (Sambrook *et al.*, 1989). The recombinant vector plasmid (2.5 μg per transcription) is linearized by digestion with *Spe*I. This enzyme cuts at the only restriction site that can be used, since only this digest generates a 3′ terminus of the transcript that efficiently supports the viral RNA replication process. Therefore the insert must not contain any *Spe*I restriction sites. If they exist, they should be mutagenized. It is also possible to use a partially digested template for transcription, but our experience is that high titer viral stocks are not easily obtained with this method. The cut DNA is phenol and ether extracted, precipitated with ethanol, dried, and dissolved in 5 μl diethyl pyrocarbonate (DEPC)-treated water. The transcription is performed in mixtures with the following composition:

5 μl DNA (2.5 μg)
5 μl 10× SP6 buffer
5 μl BSA (1 mg/ml; Promega, Madison, WI)
5 μl 10 m*M* m7g(5′)ppp(5′)G (Pharmacia, Piscataway, NJ)
5 μl 50 m*M* DTT
5 μl rNTP mix (10 m*M* ATP, CTP, UTP, 5 m*M* GTP; Pharmacia)
18.5 μl DEPC-treated water
1.5 μl RNAsin (50 units; Promega)
30 units SP6 polymerase (Pharmacia)

1. Mix reagents at room temperature and add to DNA. (Mixing at 0°C may cause precipitation of the DNA by the spermidine.)
2. Allow transcription to proceed at 42°C for 1 hr.
3. Remove a 1-μl sample for analysis by electrophoresis on 0.6% agarose. Snap-freeze the rest of the sample in liquid nitrogen.
4. Linearize 2.5 μg pSFV-Helper1 (Liljeström and Garoff, 1991) with *Spe*I and transcribe as just described. This reaction can be done in parallel with the runoff transcription of the vector RNA.
5. Carefully inspect the quality and the amount of RNA produced by each runoff transcription by electrophoresis on an agarose gel. The *in vitro* transcripts should run as a clean single band, the intensity of which should be at least an order of magnitude higher than that of the linearized plasmid.

When a helper construct is used, replication-competent SFV may be generated at a low frequency by recombination. A new helper construct, pSFV-Helper2, is now available (Life Technologies, Inc.). Mutations generated in this helper plasmid prevent *in vivo* cleavage of the spike protein precursor p62, which is required for the fusogenic activity of the spike and thus for SFV entry (see Lobigs and Garoff, 1990). Accordingly, any virus released during *in vivo* packaging (see Section II,C) is noninfectious until activated by chymotrypsin cleavage, which greatly reduces the theoretical risk of the vector system.

## D. *In Vivo* Packaging of Recombinant SFV Virions

The recombinant vector and the pSFV-Helper *in vitro* transcripts are introduced into BHK cells [BHK21 CCL10 (C13) cell line; CRL 8544; American

Type Culture Collection (ATCC), Rockville, MD] by electroporation, as described next. See Section II,A for a description of reagents.

1. Grow cells to late logarithmic phase and trypsinize with trypsin–EDTA in modified Puck's saline (Life Technologies, Inc.).
2. After cells detach, add complete culture medium and pump cells up and down with a 5-ml pipet until they are dispersed to a suspension of single cells.
3. Wash the cells twice with electroporation phosphate buffered saline (PBS) by centrifugation at low speed. Resuspend the cells in the same buffer to a density of $10^7$/ml.

*Note:* If the cells are not properly detached from each other and remain in clumps, only a fraction of them will receive the RNAs, which will result in a poor virus stock. On the other hand, if the trypsinization is too extensive, the cells will not withstand the electroporation well enough to support efficient production of recombinant SFV. Therefore, preparing the cells requires optimization.

4. Quickly mix 15–20 $\mu$l recombinant vector and helper RNAs with 0.8 ml cells and add to Gene Pulser cuvettes (Bio-Rad, Richmond, CA) with a 0.4-cm electrode gap.
5. Electroporate with two pulses of 25 $\mu$F, 850 V. The time constant after each pulse should be 0.4 to 0.5.
6. Collect cells and dilute 20-fold in prewarmed complete maintenance medium for BHK cells and plate on cell culture dishes.
7. Incubate cells at 37°C with 5% $CO_2$ for 36 hr. During this time the cells release virions into the medium but also express the insert to high levels.
8. Collect the culture supernatant containing viruses and remove cellular debris by centrifugation at low speed. Remove the supernatant and snap-freeze aliquots in liquid nitrogen. Store at −70°C.

*Note:* If the virus titer of the initial supernatant is too low for a specific purpose, virus can be concentrated in the following way.

9. Layer the supernatant on a 1-ml cushion of glycerol (50% glycerol in 50 m*M* Tris-HCl, pH 7.4, 100 m*M* NaCl) in an SW40 centrifuge tube.
10. Centrifuge for 90 min at 30,000 rpm at 4°C.
11. Pour off the supernatant and resuspend the virus overnight on ice by elution into a small volume of 50 m*M* Tris-HCl, pH 7.4, 100 m*M* NaCl, 0.5 m*M* EDTA.

Concentrating virus always results in losses in the total number of infectious units in the virus stock. The ratio of the vector and the helper transcripts present in the electroporation strongly affects the outcome of *in vitro* packaging. Therefore, the amounts of RNAs used often must be adjusted to obtain maximum virus titers.

We have titrated recombinant SFV virus stocks on BHK cells. Freezing the virus for storage lowers the infectious titer somewhat, so the titer of each stock should be determined after freezing but before the virus is used for experiments. We titer virus by the following procedure.

1. Grow BHK cells on 10 × 10-mm cover slips.
2. Make virus dilutions in serum-free BHK medium (containing 0.2% BSA) and spot on the cover slips, which are placed on parafilm for 1 hr at 37°C.
3. After incubation with the virus, drain the virus solution and transfer the cover slips to prewarmed complete cell maintenance medium and allow the infection to proceed for 6 hr.
4. Visualize the cells expressing the protein of interest by indirect immunofluorescence.

The virus strain used for vector construction has been routinely propagated in BHK cells, and is thus adapted for this host type. Accordingly, the titers obtained on BHK cells are maximum titers. When other cell lines are used, the stocks must be retitrated on these cells. The titers (as determined on BHK cells) obtained with this *in vivo* packaging method vary from $10^6$ to $10^{10}$ infectious units/ml.

## E. Expression of Exogenous Proteins in Cultured Cells

We have used the vector for the expression of a number of proteins in several cell lines, including BHK, VERO, HeLa, MDCK II (Louvard, 1980), mouse liver NMuLi cells (Owens *et al.*, 1974), and mouse mammary epithelial IM-2 cells (Reichmann *et al.*, 1989), and in primary cultures of fetal rat hippocampal neurons (see Goslin and Banker, 1991). The infections are usually performed by incubating the cells for 1 hr at 37°C with the viral stock diluted in serum-free medium, after which the virus solution is drained and the cells are incubated in complete medium for appropriate periods of time. The recombinant SFVs are most infectious on BHK cells (see previous section); expression of inserts is detected after 3–4 hr of infection. Comparable infection and expression efficiency is observed in cultured neurons (Olkkonen *et al.*, 1993a). In other cell types, higher amounts of virus and longer expression times are generally required. On glass-grown (not fully polarized) epithelial cells, the infection rate of the vector is very low. It is slightly higher on filter-grown epithelial cells infected from the basolateral side.

VIP21 is a small membrane protein of the *trans*-Golgi network, post-Golgi vesicles, and plasma membrane invaginations called caveolae (Kurzchalia *et al.*, 1992; Dupree *et al.*, 1993). Rab22 is a recently identified small GTPase that is localized to endosomal compartments (Olkkonen *et al.*, 1993b). We have used the SFV vector to express these proteins, epitope-tagged at their N termini with the c-*myc* epitope recognized by the monoclonal antibody 9E10 (Evan *et al.*, 1985; Munro and Pelham, 1987), in BHK cells or in rat hippocampal neurons. Figure 2A shows *myc*-tagged VIP21 in BHK cells, in which the protein appears in the Golgi region and displays typical punctate caveolar staining on the plasma membrane. The expressed Rab22 GTPase appears on the plasma membranes and on intracellular vesicular-like structures (Fig. 2B). Figure 3 shows rat hippocampal neurons expressing *myc*-tagged VIP21 (Fig. 3A,B) or *myc*-rab22 (Fig. 3C,D). VIP21 is first detected in the cell body in the Golgi region. Thereafter, the protein spreads to the plasma membranes of the cell body and to the processes, where a punctate staining similar to that observed in BHK cells appears. Rab22 displays a finer punctate staining extending out to the processes. Electron microscopy of SFV-infected BHK cells has shown that the cell morphology is well-preserved up to 5–6 hr of infection. In comparison, vaccinia virus-infected cells begin to change their morphology as soon as protein synthesis directed by the transfected cDNA is well under way.

Since the SFV RNA synthesis machinery is associated with membranes of late endocytic compartments (Froshauer *et al.*, 1988), the vector may cause some abnormalities in the endocytic pathway of infected cells. Like the virus itself, the vector shuts off host cell protein synthesis (Liljeström and Garoff, 1991). This, of course, leads to the slow demise of infected cells. Although symptoms of cytotoxicity in cultured neurons were apparent after 8 hr of infection (Olkkonen *et al.*, 1993a), other cell types may tolerate infection by the recombinant SFVs for several days. Considering all aspects of the system,

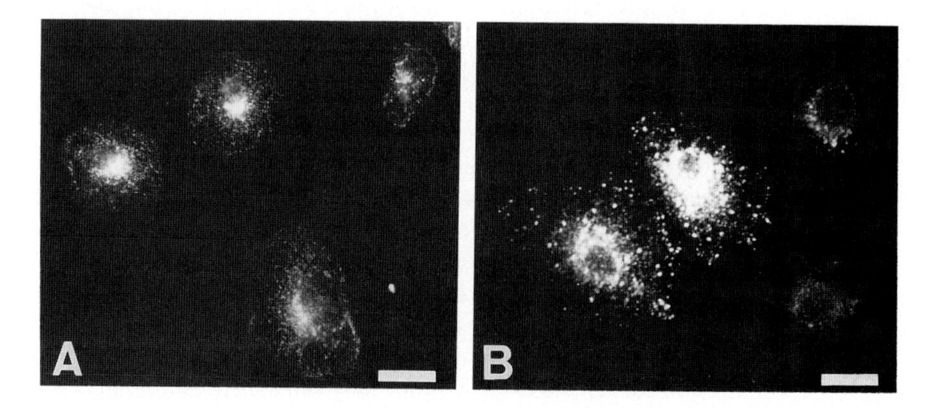


**Fig. 2** Expression of N-terminally *myc*-tagged VIP21 (A) or rab22 (B) in BHK cells. Cover slips were washed with PBS and placed cells up on parafilm after which the recombinant SFVs diluted in serum-free medium (with 0.2% BSA) were applied to the cells. After 1 hr at 37°C, the cover slips were transferred into prewarmed complete medium, and the incubation was continued for 6 hr (A) or 3 hr (B). The cells were permeabilized shortly with saponin and were fixed with paraformaldehyde; the epitope-tagged proteins were detected with the mAb 9E10. The bound antibodies were visualized using the Dianova (Hamburg, Germany) donkey anti-mouse-IgG–fluorescein isothiocyanate (FITC) conjugate. Bar, 10 $\mu$m.

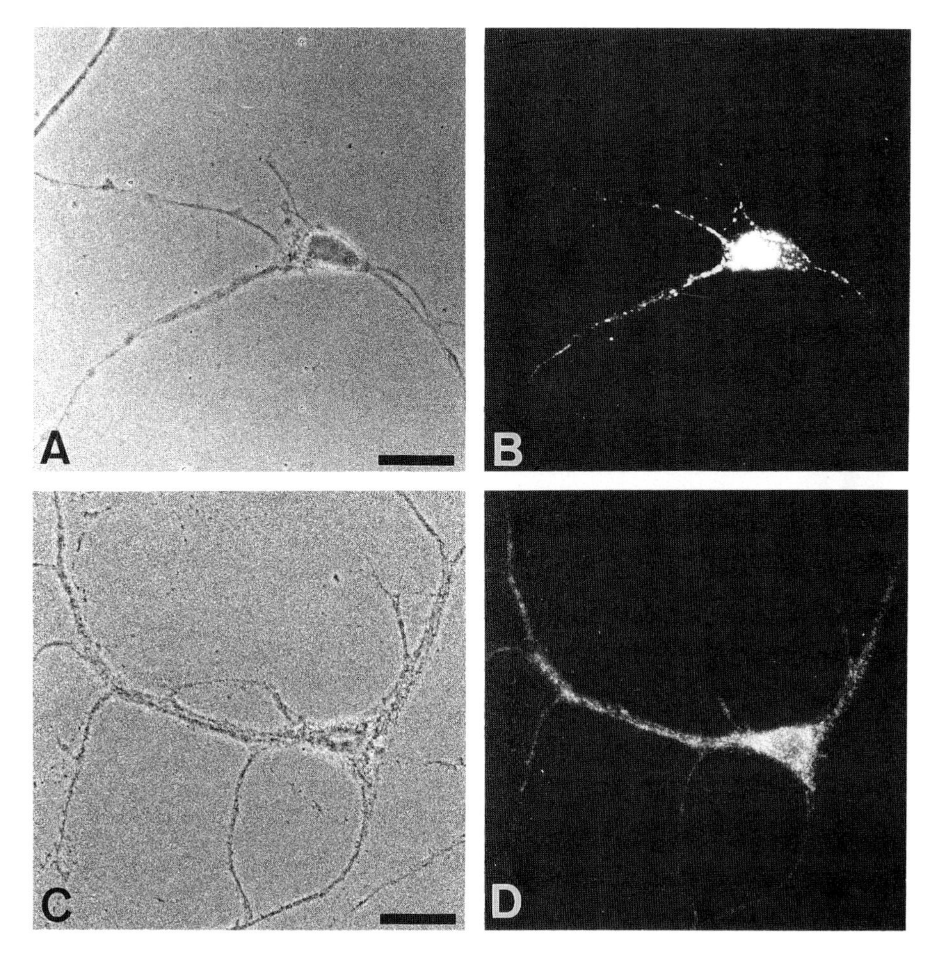


**Fig. 3** Expression of *myc*-VIP21 (A,B) or *myc*-rab22 (C,D) in fetal rat hippocampal neurons. Neuronal cultures prepared from the hippocampi of 18-day-old rat embryos (Goslin and Banker, 1991) were infected 8 days after plating. The cover slips were incubated for 1 hr at 36.5°C with the SFVs diluted in conditioned maintenance medium. The viral solutions were drained and the cover slips were transferred into prewarmed conditioned medium, in which the incubation was continued for 2 hr (*myc*-VIP21) or 5 hr (*myc*-rab22). The cells were fixed with paraformaldehyde and permeabilized with Triton X-100, and the expressed proteins were detected with 9E10. The bound antibodies were visualized using the Dianova (Hamburg, Germany) anti-mouse-IgG–rhodamine (TRITC) conjugate. Bar, 10 $\mu$m.

the SFV vector provides a convenient tool for transient expression studies. Expression of inserts is detected on the time scale of hours, and the expression levels are easily regulated by adjusting the infection time. High titer recombinant viral stocks are relatively easy and quick to prepare and can be used to infect a wide range of cell types.

## References


Birnboim, H. C., and Doly, J. (1979). A rapid alkaline extraction procedure for screening recombinant plasmid DNA. *Nucleic Acids Res.* **7,** 1513–1523.

Bottestein, J. E., and Sato, G. H. (1979). Growth of a rat neuroblastoma cell line in serum-free supplemented medium. *Proc. Natl. Acad. Sci. USA* **76,** 514–519.

Dupree, P., Parton, R. G., Kurzchalia, T. V., and Simons, K. (1993). Caveolae and sorting in the trans-Golgi-network of epithelial cells. *EMBO J.* **12,** 1597–1605.

Evan, G. I., Lewis, G. K., Ramsay, G., and Bishop, J. M. (1985). Isolation of monoclonal antibodies specific for human c-*myc* proto-oncogene product. *Mol. Cell. Biol.* **5,** 3610–3616.

Fraser, M. J. (1992). The baculovirus-infected insect cell as a eukaryotic gene expression system. *Curr. Top. Microbiol. Immunol.* **158,** 131–172.

Froshauer, S., Kartenbeck, J., and Helenius, A. (1988). Alphavirus RNA replicase is located on the cytoplasmic surface of endosomes and lysosomes. *J. Cell Biol.* **107,** 2075–2086.

Goslin, K., and Banker, G. (1991). Rat hippocampal neurons in low-density culture. *In* "Culturing Nerve Cells" (G. Banker and K. Goslin, eds.), pp. 251–281. MIT Press, Cambridge, Massachusetts.

Griffin, D. E. (1986). Alphavirus pathogenesis and immunity. *In* "The Togaviridae and Flaviviridae" (S. S. Schlesinger and M. J. Schlesinger, eds.), pp. 209–250. Plenum Press, New York.

Kurzchalia, T. V., Dupree, P., Parton, R. G., Kellner, R., Virta, H., Lehnert, M., and Simons, K. (1992). VIP21, a 21-kD membrane protein is an integral component of trans-Golgi-network-derived transport vesicles. *J. Cell Biol.* **118,** 1003–1014.

Liljeström, P., and Garoff, H. (1991). A new generation of animal cell expression vectors based on the Semliki Forest virus replicon. *BioTechnology* **9,** 1356–1361.

Liljeström, P., Lusa, S., Huylebroeck, D., and Garoff, H. (1991). In vitro mutagenesis of a full-length cDNA clone of Semliki Forest virus: The small 6,000-molecular-weight membrane protein modulates virus release. *J. Virol.* **65,** 4107–4113.

Lobigs, M., and Garoff, H. (1990). Fusion function of the Semliki Forest virus spike is activated by proteolytic cleavage of the envelope glycoprotein p62. *J. Virol.* **64,** 1233–1240.

Louvard, D. (1980). Apical membrane aminopeptidase appears at site of cell–cell contact in cultured kidney epithelial cells. *Proc. Natl. Acad. Sci. USA* **77,** 4132–4136.

Margolskee, R. F. (1992). Epstein–Barr virus based expression vectors. *Curr. Top. Microbiol. Immunol.* **158,** 67–95.

Miller, A. D. (1992). Retroviral vectors. *Curr. Top. Microbiol. Immunol.* 158, 1–24.

Moss, B. (1992). Poxvirus expression vectors. *Curr. Top. Microbiol. Immunol.* **158,** 25–38.

Munro, S., and Pelham, R. B. (1987). A C-terminal signal prevents secretion of luminal ER proteins. *Cell* **48,** 899–907.

Olkkonen, V. M., Liljeström, P., Garoff, H., Simons, K., and Dotti, C. G. (1993a). Expression of heterologous proteins in cultured rat hippocampal neurons using the Semliki Forest virus vector. *J. Neurosci. Res.* **35,** (*in press*).

Olkkonen, V. M., Dupree, P., Huber, L. A., Lütcke, A., Zerial, M., and Simons, K. (1993b). Compartmentalization of rab proteins in mammalian cells. *In* "Handbook of Experimental Pharmacology" (B. Dickey and L. Birnbaumer, eds.). Springer-Verlag, Heidelberg.

Owens, R. B., Smith, H. S., and Hackett, A. J. (1974). Epithelial cell cultures from normal glandular tissue of mice. *J. Natl. Cancer Inst.* **53,** 261–269.

Reichmann, E., Ball, R., Groner, B., and Friis, R. R. (1989). New mammary epithelial and fibroblast cell clones in coculture form structures competent to differentiate functionally. *J. Cell Biol.* **108,** 1127–1138.

Sambrook, J., Fritsch, E. F., and Maniatis, T. (1989). "Molecular Cloning. A Laboratory Manual," 2d Ed. Cold Spring Harbor Laboratory Press, Cold Spring Harbor, New York.

# CHAPTER 3

# Recombinant Sindbis Virus as an Expression System for Cell Biology

**Robert C. Piper,[*] Jan W. Slot,[†] Guangpu Li,[‡] Phillip D. Stahl,[‡] and David E. James[§]**

[*] Institute of Molecular Biology
University of Oregon
Eugene, Oregon 97403

[†] Department of Cell Biology
University of Utrecht
3508 TC Utrecht
The Netherlands

[‡] Department of Cell Biology and Physiology
Washington University
St. Louis, Missouri 63110

[§] Center for Molecular Biology and Biotechnology
University of Queensland
Brisbane, Queensland 4072
Australia

# I. Introduction

Virus-based systems for the expression of heterologous genes are becoming popular tools for cell biological studies. As with any expression system, a careful consideration of the advantages and disadvantages must be made before choosing a virus vector for study of the specific function of a protein. This chapter outlines the general features of expression vectors based on Sindbis virus, an enveloped plus-stranded RNA virus (SIN vectors). We will also provide specific examples of the use of SIN vectors for studying the function of polytopic membrane proteins, namely the facilitative glucose transporters, and peripheral membrane proteins belonging to the *ras* superfamily of GTPases. Other detailed and diverse reviews regarding the development of the SIN vectors (Huang *et al.*, 1989; Bredenbeek and Rice, 1992; Schlesinger, 1993) and the biology of Sindbis virus (Strauss and Strauss, 1986; Schlesinger and Schlesinger, 1990) have been presented elsewhere.

Our understanding of the fundamental functions of cellular organelles has been greatly enhanced by the study of viruses (see Chapter 1). Much of our knowledge about the processing and trafficking of vesicle fusion events through the secretory pathway (Pfeffer and Rothman, 1987), the biogenesis of cell polarity (Matlin and Simons, 1984), and protein folding in the endoplasmic reticulum (ER) (Hurtley and Helenius, 1989) has been derived from the study of viral envelope proteins. Thus, virally infected cells are very useful and valid culture models for a variety of basic mechanisms in the cell. Several recombinant viral expression vectors based on retroviruses, adenovirus, herpesvirus, vaccinia virus, baculovirus, and alphaviruses such as Sindbis virus and Semliki Forest virus have been developed to express heterologous genes in a variety of cell types (see Chapters 2, 4, 5, 7, 8, 9, and 10). Recombinant viral vectors offer several advantages over stable transformed cell lines or transient transfection. Viral systems can be employed not only to manufacture preparative amounts of a given protein but also to study the function of that protein within the cell. When selecting a particular viral expression system, several considerations should be made including the safety of the recombinant virus, the host cell range, the overall level of expression, the adverse effects of the virus on the cell, and the ease of techniques required to create the recombinant virus. SIN vectors offer efficient infection of a wide range of cells as well as rapid and reliable expression of heterologous proteins.

Although many of the viral systems described in this volume share some of these qualities, recombinant SIN vectors offer additional advantages. First, the genome of Sindbis virus is relatively small and can be fully manipulated by standard *in vitro* DNA/RNA technology. Second, complete recombinant infectious genomic RNA can be easily constructed *in vitro* and transfected into host cells to raise viral stocks without selection and plaque purification of recombinant virus. These viral stocks can be stored and subsequently used to infect a wide variety of cells of interest. Finally, a variety of SIN vectors as well as temperature-sensitive mutants can be used to control the level of heterologous gene expression.

## II. Sindbis Virus Biology

Sindbis virus is a member of the alphavirus genus, whose member species are transmitted to their mammalian or avian hosts via insect vectors. Sindbis virus is one of the least pathogenic of the alphaviruses, which include other viruses well known for causing encephalitic infection including Ross River virus, Semliki Forest virus, Eastern equine encephalitis virus, Western equine encephalitis virus, and Chikungunya virus. Pathogenic effects of Sindbis virus infection in humans are rare, and if there are any they are usually limited to headache or rash. Protracted disease is exceedingly rare (Griffin, 1986). Al-

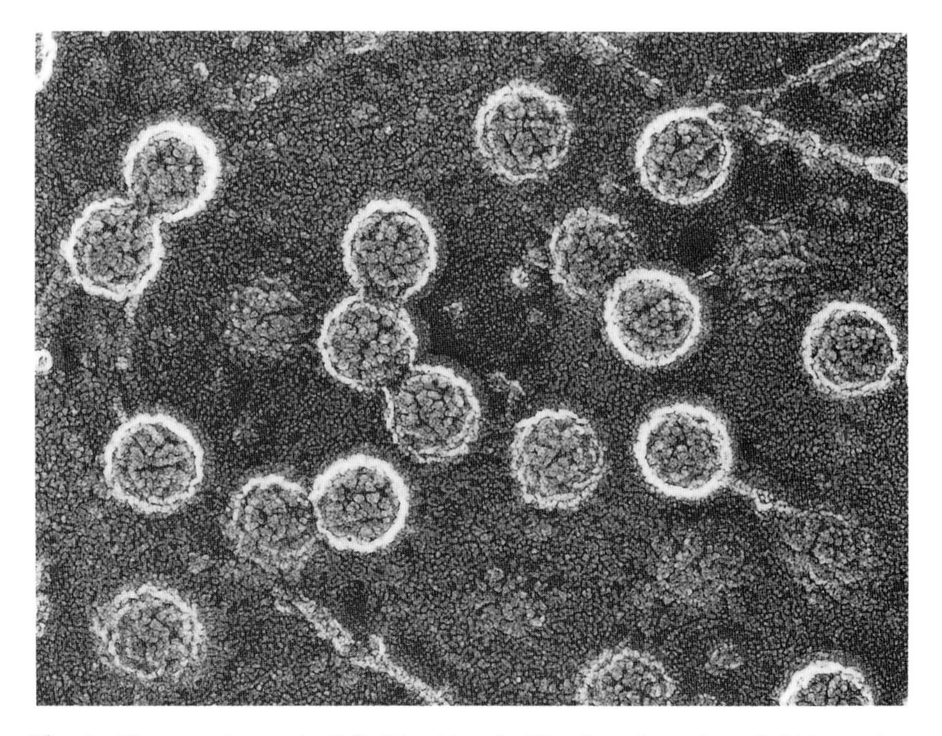

**Fig. 1** Electron micrograph of Sindbis virions budding from the surface of chicken embryo fibroblasts. Chicken embryonic fibroblasts were infected for 6 hr with Sindbis virus, freeze dried, quick frozen, and platinum replicated according to the technique of Heuser and Kirschner (1980). Sindbis virus particles can be seen acquiring their lipid bilayer envelope upon budding from the plasma membrane. (Micrograph kindly provided by Dr. J. Heuser, Washington University, St. Louis, MO).

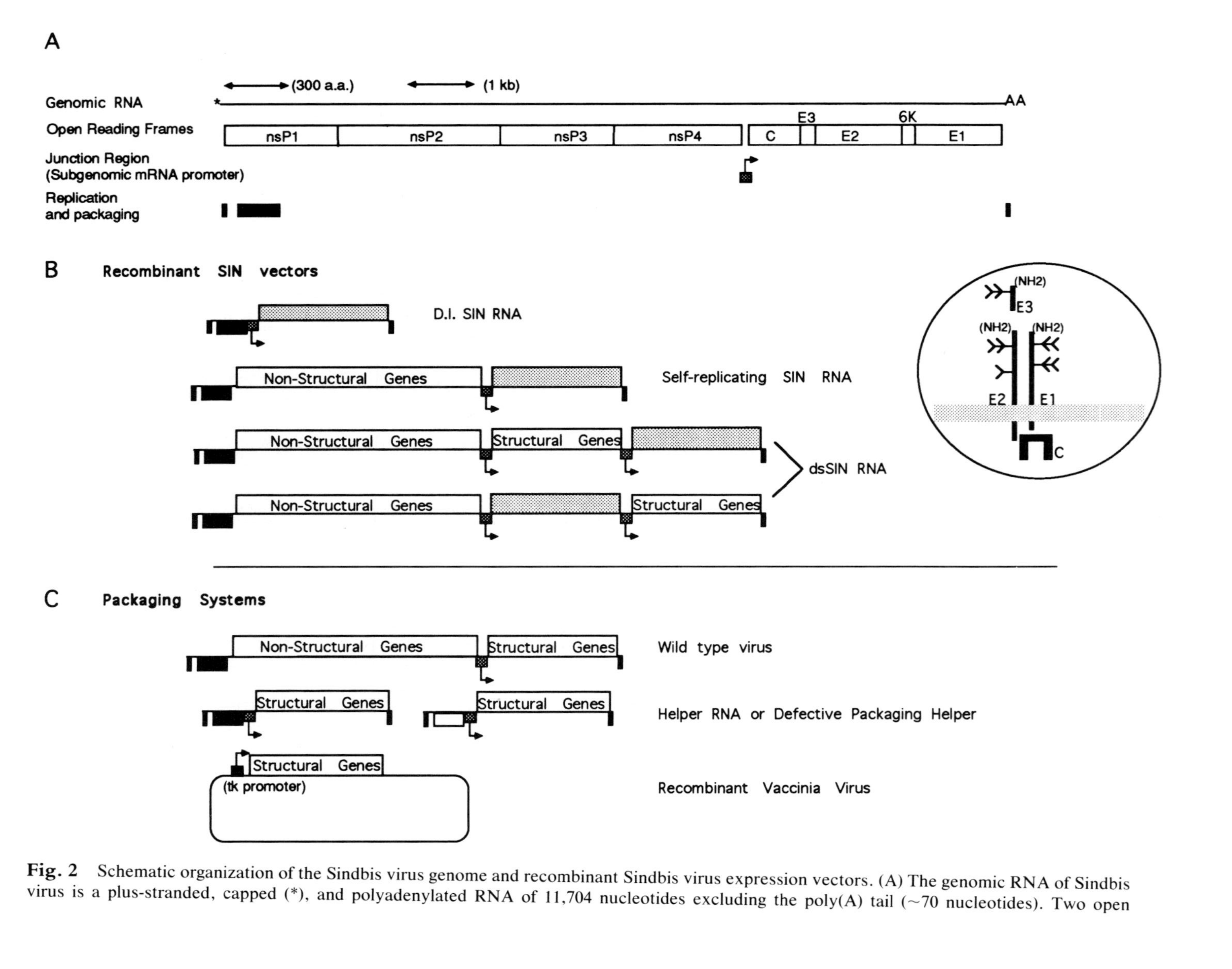


**Fig. 2** Schematic organization of the Sindbis virus genome and recombinant Sindbis virus expression vectors. (A) The genomic RNA of Sindbis virus is a plus-stranded, capped (*), and polyadenylated RNA of 11,704 nucleotides excluding the poly(A) tail (~70 nucleotides). Two open

though the SIN vectors are based on an avirulent laboratory strain of Sindbis virus, appropriate containment procedures (Biosafety Level 2) and caution must be used while working with this virus.

## A. Genomic Structure and Virus Life Cycle

As an arbovirus, Sindbis virus can infect cells of both vertebrate and invertebrate animals. However, the growth of Sindbis virus within vertebrate cells is markedly different from that in insect cell lines. Insect cells can sustain persistent infection by Sindbis, can form membrane-bound organelles that contain replicating Sindbis virus (Miller and Brown, 1993), and require concurrent host protein synthesis to promote virus growth (Brown and Condreay, 1986). Furthermore, although expression of heterologous proteins can be obtained in insect cells (Xiong *et al.*, 1989), widespread application has been mostly in vertebrate cells. Therefore, discussion of the life cycle of Sindbis virus and the use of SIN vectors will be limited to studies in vertebrate cells.

Sindbis virus particles are composed of an icosahedral nucleocapsid containing the RNA genome complexed with the capsid protein. The nucleocapsid is enclosed in a membrane envelope composed of a lipid bilayer and the two major viral structural glycoproteins E1 and E2 (Fig. 1; Harrison, 1986; Harrison *et al.*, 1992). Researchers generally believe that viral particles enter the cell by endocy-

---

**Fig. 2** (*Continued*) reading frames encode the nonstructural proteins (nsP1, nsP2, nsP3, nsP4) and the structural proteins (C, E3, E2, 6K, E1). Regions that constitute the subgenomic mRNA promoter ( ) are located in the junction region between the nonstructural and structural genes. *Cis*-acting sequences are required for RNA replication and packaging (■). Inset I shows the topology of the Sindbis virus envelope glycoproteins and relative positions of complex ( ) and core glycosylation ( ). (B) Various recombinant Sindbis virus-based RNAs for the expression of heterologous genes ( □ ). (*Top*) Vector based on the genome of defective interfering RNAs (DI RNA). This vector contains this *cis*-acting sequences necessary for replication and packaging; expression of heterologous gene is driven by subgenomic mRNA promoter. This vector requires co-expression with wild-type virus for both expression of heterologous gene and packaging into virions. (*Middle*) Self-replicating RNA that contains viral nonstructural genes encoding replicase functions. Heterologous gene expression is driven by subgenomic mRNA promoter. Packaging of these RNAs into virions requires co-expression with wild-type virus, a "helper" RNA encoding the structural genes, or structural genes expressed by recombinant vaccinia virus. (*Bottom*) dsSIN vectors contain two subgenomic mRNA promoters that drive the expression of both the structural genes and the heterologous gene. These RNAs produce infectious virus capable of propagation. Heterologous genes may be either carried at the 3′ end of the viral genome or between the nonstructural and structural genes. (C) Packaging systems for recombinant Sindbis virus RNAs that lack ability to produce structural proteins. (Top) Wild-type virus. (*Middle*) "Helper" RNA containing *cis*-acting sequences necessary for replication and packaging and encodes the viral structural proteins under the control of a subgenomic mRNA promoter. Defective helper RNA with deletion or mutation within the capsid recognition site (empty box) can be used to package recombinant Sindbis RNAs into virions capable of only one round of infection. (*Bottom*) Expression of the Sindbis virus structural proteins by recombinant vaccinia virus.

tosis and that acidification of the endosome triggers the fusion of the viral envelope with the endosomal membrane, releasing the genomic RNA into the cytoplasm (Marsh and Helenius, 1989). However, some evidence suggests that endosome acidification is not necessary for the fusion of Sindbis virus particles, which may also occur directly at the plasma membrane (Edwards and Brown, 1991). Replication of the viral genome and growth of the virus occur solely in the cytoplasm and can be completed in the presence of inhibitors of host cell mRNA production (e.g., actinomycin D or $\alpha$-amanitin; Baric *et al.*, 1983).

The Sindbis virus genome is a plus-stranded (49S) RNA molecule of 11,703 nucleotides that is polyadenylated and capped (Fig 2; Strauss *et al.*, 1984). Replication of the genome is via a minus-stranded intermediate that is synthesized soon after viral entry (Kaariainen *et al.*, 1987). The genomic RNA itself serves as the mRNA for nonstructural proteins (nsP) that are encoded in the 5′ two-thirds of the genomic RNA (Strauss and Strauss, 1986). A single open reading frame, interrupted by an opal codon, encodes four distinct nsPs that are produced by proteolytic processing of a fusion polyprotein precursor (Ding and Schlesinger, 1989; Knipfer and Brown, 1989; Shirako and Strauss, 1990). The full-length nsP1–2–3–4 fusion polyprotein is produced by read-through translation of an opal codon after the coding region of nsP3 (Strauss *et al.*, 1983; Li and Rice, 1989). Processing of the nonstructural polyprotein precursor is catalyzed by a viral protease activity present in the C-terminal half of the nsP2 protein that can function both in *cis* and in *trans*. These four nsPs, in conjunction with certain fusion protein intermediates, are produced in the early phase of infection and act as the viral RNA replication machinery (Lemm and Rice, 1993).

In addition to the full-length genomic RNA and the minus-strand RNA intermediate, a 26S subgenomic RNA is expressed at high levels in the later phase of infection (usually 2–5 hr postinfection depending on cell type). The subgenomic RNA is collinear with the 3′ one-third of the genomic RNA. Production of the 26S RNA is initiated at the subgenomic mRNA promoter located within the junction region between the nonstructural genes and the structural genes (Levis *et al.*, 1990). A single open reading frame within the subgenomic mRNA encodes all the viral structural proteins, which are synthesized as the polyprotein precursor C–pE2–6K–E1 (Schlesinger and Schlesinger, 1986). The cytoplasmic capsid protein (C) is co-translationally cleaved from the poly-protein,which allows the pE2 portion of the precursor polyprotein to insert into the ER and be cleaved from the remaining polyprotein precursor. 6K acts as a cleavable N-terminal ER signal sequence for the E1 protein (Hashimoto *et al.*, 1981). E1 and pE2 form heterodimers that associate as trimers in the ER. Both the E1 and the pE2 protein are type I membrane proteins that are glycosylated and acylated with long chain fatty acids as they progress through the secretory pathway (Fig. 2). N-Terminal cleavage of pE2 results in the mature E2 membrane glycoprotein (52 kDa) and the E3 protein (10 kDa). E3 does not remain associated with the viral spike glycoproteins or virions and is secreted (Mayne *et al.*, 1984). Final cleavage of pE2 to form E2 is thought to occur in the trans-Golgi or trans-Golgi reticulum (Green *et al.*, 1981; Hakimi and Atkinson, 1982). Viral particles are produced once an adequate supply of viral structural proteins has been generated. Sindbis virus genomic RNA rapidly associates with the cytoplasmic capsid protein to form the icosahedral nucleocapsid. The sequences required for packaging the genomic RNA are located at the 5′ end of the genome within the coding region for nsP1 (Weiss *et al.*, 1989). The resulting nucelocapsid associates with the C termini of the viral spike glycoproteins, forming an enveloped virus that buds from the plasma membrane (Fig. 1).

## B. Viral Expression Vectors

The ability to produce infectious genomic Sindbis virus RNA *in vitro* from cloned viral cDNAs (Rice *et al.*, 1987) has led to the delineation of the *cis*-acting sequences necessary for genomic RNA replication, packaging into virions, and transcription of subgenomic mRNA. These features have been combined to assemble a variety of virus-based and recombinant virus expression systems (Fig. 2). Earlier expression systems combined the *cis*-acting sequences necessary for replication and packaging with a subgenomic mRNA promoter to drive the expression of a heterologous gene (chloramphenicol acetyltransferase, CAT) (Levis *et al.*, 1987,1990; Grakoui *et al.*, 1989). Since these RNAs did not encode the viral structural proteins or nonstructural proteins, wild-type helper virus was required for replication of genomic RNAs, production of subgenomic mRNA encoding the heterologous gene, and packaging of these RNAs to produce a viral stock capable of infecting other cells. High levels of CAT expression were obtained with this system because of the production of large amounts of subgenomic mRNA encoding CAT.

Heterologous gene expression by SIN vectors was also obtained by substituting the heterologous CAT gene for the region encoding the viral structural proteins (Xiong *et al.*, 1989). Such RNAs contained the viral nonstructural (replicase) genes and were capable of self-replication once transfected into cells. These vectors produced high levels of CAT ($10^8$ enzymatically active CAT polypeptides/cell) in transfected cells. However, this system was limited by the inefficiency of the RNA transfection; the overall level of CAT within a cell population in these studies was relatively low. The packaging of the CAT-containing RNAs into viral particles required helper virus function. Helper function was supplied by complete wild-type virus (Xiong *et al.*, 1989) or by a helper RNA that contained the *cis*-acting sequences necessary for replication as well as the structural genes under the control of a subgenomic RNA promoter (Geigenmuller-Gnirke *et al.*, 1991). These helper RNAs were also packaged into the resulting virions, and repeated passage of the viral stock was used to amplify viral particles containing the CAT gene. Such packaging systems are somewhat cumbersome, since repeated passage of the viral stock was required to select for virions carrying the heterologous gene. However, extensive passaging of virus may not be necessary for this system if the initial transfection efficiency of recombinant RNAs is sufficiently high.

The development of other packaging systems for Sindbis virus RNAs that can produce recombinant Sindbis virus particles capable of only one round of infection have recently been developed. The *cis*-acting sequences required for packaging into virions have been identified (Weiss *et al.*, 1989). It has been possible to construct "helper" RNAs, similar to those just described, that contain deletions or mutations within this packaging sequence. This strategy was previously developed for alphavirus expression vectors based on Semliki Forest virus (Liljestrom and Garoff, 1991; Chapter 2) and more recently for Sindbis virus (Bredenbeek *et al.*, 1994). The resulting helper RNAs are capable of replication and production of structural genes under the control of a subgenomic mRNA promoter; however, these RNAs are not packaged into virions at detectable levels. Co-transfection of recombinant RNA containing a heterologous gene with packaging-defective helper RNA results in production of recombinant Sindbis virions capable of only one round of infection. Helper RNAs that are defective in packaging (defective helper RNAs; DH RNAs) can produce on the order of $1 \times 10^8$ infectious units per milliliter of vector that can be used in various applications when the spread of infectious virus is not desired. Vaccinia virus has also been used to produce Sindbis structural proteins for packaging vector RNAs. This approach results in the production of $10^4$–$10^7$ infectious vector particles/ml (Bredenbeek and Rice, 1992), so currently this system appears to be less efficient than the one using DH RNAs. Between these two systems, final development of an efficient packaging system for Sindbis virus "suicide" vectors seems possible.

Recombinant SIN vectors that can propagate infection without helper RNA function have also been developed. These vectors incorporate dual subgenomic mRNA promoters (dsSIN vectors), one of which initiates transcription of a subgenomic mRNA that encodes the viral structural proteins and one of which initiates transcription of the heterologous gene (Raju and Huang, 1991; C. S. Hahn *et al.*, 1992). The heterologous gene and the structural genes are expressed in tandem. Two basic vectors have been constructed: one that places the heterologous gene and subgenomic mRNA promoter 3′ to the structural genes and another that places the additional subgenomic mRNA promoter and heterologous gene between the coding regions for the nonstructural and structural proteins. The former vector architecture, which carries the heterologous gene at the 3′ end of the genome, may offer a higher level of expression (C. S. Hahn *et al.*, 1992; Schlesinger, 1993; R. C. Piper and D. E. James, unpublished data). However, the heterologous gene may be more stable during passage of the virus in the alternative "middle" position (C. S. Hahn, C. Xiong, and H. Huang, unpublished observation). A more stable vector would allow the amplification of a recombinant Sindbis virus-based cDNA expression library or would compensate for low transfection efficiency during production of initial viral stocks (H. Huang, personal communication). These dsSIN vectors are somewhat easier to use than the vectors described previously and have been employed to express a number of proteins including hemagglutinin virus antigens (C. S. Hahn *et al.*, 1992; Y. S. Hahn *et al.*, 1992), glucose transporters (Piper *et al.*, 1992,1993a,b) *rab* GTPase proteins (Li and Stahl, 1993a,b), tissue plasminogen activator (C. Xiong and H. Huang, unpublished data), and CAT.

Control over the level of heterologous gene expression can be obtained by a number of methods. Deletion and point mutations within the subgenomic mRNA promoter, as well as promoter sequences from other alphavirus junction regions, can also be incorporated into SIN vectors to obtain various expression levels. Several subgenomic mRNA variants have been engineered into dsSIN vectors and result in expression levels varying within two orders of magnitude (Raju and Huang, 1991; Hertz and Huang, 1992). Further control of the level of protein production can be obtained with mutations that are temperature sensitive for viral RNA replication. A number of temperature-sensitive mutants (RNA− mutants) that completely or partially inhibit viral RNA synthesis at high temperature (40°C) have been characterized (Hahn *et al.*, 1989a,b; Sawicki *et al.*, 1990). Incorporation of these mutations into SIN vectors allows the level of heterologous gene expression to be controlled via temperature shift and has been used to modulate expression of CAT (Xiong *et al.*, 1989).

Various Sindbus virus expression plasmids are available with different multiple cloning sites to facilitate subcloning of heterologous gene sequences and linearization of template DNA for *in vitro* RNA production. (Since many of these vectors have been developed by Rice, Huang, Schlessinger, and colleagues, specific inquiries should be addressed to them at the Department of Microbiology, Washington University, St. Louis, MO). Since the Sindbis viral genome is relatively small, the entire genome of the recombinant Sindbis virus can be encoded on standard bacterial cloning plasmids. Once the heterologous gene is subcloned into the vector, recombinant Sindbis virus genomic RNA can be generated by *in vitro* transcription. The Sindbis virus plasmid constructs must be linearized precisely at the 3′ end of the Sindbis genome to insure proper replication (Rice *et al.*, 1987). Capping of *in vitro*-synthesized RNA is also essential since the viral replicase is translated directly from tranfected genomic RNAs.

## C. Production of Recombinant Virus

1. RNA is synthesized with 0.4 μg linearized template/20 μl reaction with 0.4 m*M* rNTPs, 1 m*M* DTT, and 30 units of SP6 RNA polymerase at 37°C for 1 hr. 5′ Capping is accomplished co-transcriptionally by also including 1 m*M* 7 mGppG cap analog (New England Biolabs, Beverly, MA). This reaction yields about 2 μg RNA.

2. Production of RNA can be followed by incorporation of [$^3$H]UTP or by agarose gel electrophoresis.

3. Recombinant Sindbis virus stocks from dsSIN vectors are prepared by transfecting RNAs into rapidly growing cultures of baby hamster kidney (BHK)

cells or secondary cultures of chicken embryo fibroblast (CEF) cells (Pierce *et al.*, 1974). Lipofectin [Bethesda Research Laboratories, (BRL, Gaithersburg, MD)] may be used to transfect genomic RNA easily.

4. 200 μg RNA are mixed with 10 μl Lipofectin and incubated on ice for 10 min.

5. The RNA solution is then diluted in 250 μl phosphate-buffered saline (PBS) and added to cell monolayers (45-mm culture plate) previously washed with PBS. After a 15-min incubation, the cells are washed with PBS and incubated with 2 ml Dulbecco's modified Eagle's medium (DMEM) with 5% fetal calf serum (FCS). This procedure can be scaled up for large-scale production of virus.

6. Propagation of virus is achieved at 37°C; however, virus may be grown at 30°C if recombinant viruses are unstable or are produced at low levels. Viral infection is usually evident at 20 hr post-transfection as cells appear more spindle-like and necrotic.

7. 30–36 hr after transfection, culture supernatant can be collected and stored at −70°C for later use.

8. Viral stocks can be titered on monolayers of CEF cells (Strauss *et al.*, 1976). Titers usually obtained are $10^8$–$10^9$ plaque forming units (pfu)/ml.

The plasmid technology used to construct recombinant Sindbis virus cDNAs makes it reasonable to express many heterologous genes with SIN vectors. Since recombinant Sindbis virus RNAs are produced directly *in vitro,* recombinants need not be plaque purified. Thus, Sindbis virus may at least provide a good preliminary system to screen many different heterologous genes to identify a subset for further analysis in a more sophisticated or cumbersome expression system. No more than 2 kb of heterologous sequence should be inserted into the current dsSIN vectors. Larger inserts tend to be less stable on virus growth, and the resulting virions can suffer from significant rearrangement or deletion of the heterologous gene. Although no rigorous comparison has been made, the dsSIN vectors in which the heterologous gene lies between the nonstructural and structural genes may be more stable than recombinants containing the heterologous gene at the 3′ end of the genome. Stability of the heterologous gene can become a major issue if production of viral stock is compromised by low transfection efficiency, which requires a significant number of passages of the initial virus to achieve usable infectious titers. For this reason, high transfection efficiency of *in vitro* SIN genomic RNAs is desirable. Adequate transfection efficiency has been obtained with cationic liposomes (Lipofection) (Grakoui *et al.*, 1989); however, electroporation of RNA (Liljestrom *et al.*, 1991; see Chapter 2) can be employed for very high efficiency transfection. Replenishing viral stocks should be done directly by transfection of *in vitro* synthesized genomic RNA rather than by passaging of virus through cell cultures.

Synchronized infection of fresh cell cultures with viral stock is accomplished simply by incubating in the presence of virus in low-serum media for 10–

20 min at 37°C. Infection can then proceed in a variety of standard tissue culture media. A high multiplicity of infection (moi) is required for quantitative infection and can be estimated from titers obtained on CEF monolayers. However, the infectivity of Sindbis virus varies for any given host cell type. Therefore, the amount of virus required for quantitative infection must be determined empirically and can be checked by immunolabeling cells for viral glycoproteins or presence of virally expressed heterologous protein.

Multiple infection with two or more different recombinant Sindbis viruses can be used to express a combination of heterologous proteins within the same cell. Thus, a variety of internal controls and possible interactions between the proteins in question can be specifically addressed using double infections. Infection by Sindbis virus rapidly blocks subsequent infection (superinfection) of the same cell by another Sindbis virion; therefore, double infections can only be performed during the initial infection (Johnston *et al.,* 1974; Adams and Brown, 1985).

## D. Host Cell Range

Sindbis virus can infect a wide variety of cells including those of insect, avian, mammalian, amphibian, and reptilian origin (Clark *et al.,* 1973; Leake *et al.,* 1977; Brown and Condreay, 1986; Schlesinger and Schlesinger, 1990). In practice, however, there are limitations on the host range. For example, we have found low expression of heterologous proteins using SIN vectors in MDCK cells, 3T3-L1 adipocytes, HL-60 cells, and various insulinoma cell lines (R. C. Piper, C. M. Rice, H. Huang, and D. E. James, unpublished data). The basis for this selectivity is not completely understood and may involve several steps of the viral life cycle. One major determinant for the efficiacy of infection is the presence of Sindbis virus receptors on target host cells. Three putative Sindbis virus receptors have been described: one which promotes encephalitis and is present in mouse but not human neurons (Ubol and Griffin, 1991), a 63-kDa protein from chicken embryo fibroblasts that has been identified by producing anti-idiotypic antibodies to E2 (Wang *et al.,* 1991), and, finally, the laminin receptor, which is found on a wide variety of cells (Wang *et al.,* 1992). Thus, one way to increase the efficiency of infection is to express the laminin receptor in the host cell line of interest (Wang *et al.,* 1992). Another way to enhance infection has been to bind virus physically to the surface of macrophages coated with nonneutralizing antibodies against either E1 or E2 (Peiris and Porterfield, 1979). Virus is bound to the macrophage surface Fc receptor and is subsequently taken up by the cells, presumably via phagocytosis. This method has been used to study Sindbis virus mutants that otherwise do not possess ligands for cell surface receptors (Dubuisson and Rice, 1993).

An alternative strategy to increase infectivity is to take advantage of the high mutation rate inherent in the replication of viral RNA and to select for virus adapted to a particular cell line by repeated passage over the cell line of interest

(Symington and Schlesinger, 1975). This procedure has been used to select for a Sindbis virus strain that can grow in transformed mouse B cells (Symington and Schlesinger, 1975), as well as to select Sindbis virus strains that are more neurovirulent (Griffin and Johnson, 1977; Jackson *et al.,* 1988). Many times the molecular changes in such variants map to the E2 surface glycoprotein, indicating a response to different viral surface receptors. It has been possible to insert heterologous epitopes into the E2 protein that do not affect virus viability (London *et al.,* 1992), as well as to isolate Sindbis virus mutants that no longer possess broad vertebrate cell tropism (Dubuisson and Rice, 1993). Thus, it may soon be possible to engineer viral particles that contain defined ligands to known cell surface receptors and to tailor viruses with defined cellular specificity.

The availability of neuro-adapted strains of Sindbis virus as well as the use of part of their corresponding genome in some dsSIN vectors (C. S. Hahn *et al.,* 1992) raises the possibility that SIN vectors could be used as expression vectors *in vivo.* In intracerebrally inoculated weaned mice, neuro-adapted Sindbis virus efficiently infects all cell types of the central nervous system (CNS) including neurons, as indicated by immunostaining of viral glycoprotein in cell bodies. Suckling mice can be inoculated intraperitoneally since the virus multiplies in skeletal muscle and subsequently crosses the blood-brain barrier to infect the CNS (Griffin, 1986). Thus, recombinant Sindbis virus vectors may be able to express proteins *in vivo* in tissues such as the brain where immune surveillance is limited. Expression of foreign gene products *in vivo* with dsSIN virus vectors has been employed to a limited degree to sensitize cytotoxic T lymphocytes to heterologously expressed MHC class I-restricted antigens in mice (Y. S. Hahn *et al.,* 1992).

## E. Pathological Effects of Sindbis Virus Infection

Generally, vertebrate cell culture lines infected with Sindbis virus will eventually die because of the corruption of the host cell synthetic machinery. The pathological effects of Sindbis virus are qualitatively similar to those described for Semliki Forest virus. Death usually occurs after 10–15 hr; however, this time frame is highly dependent on cell type. Cell death is preceded by a number of other pathological effects. At the ultrastructural level, Sindbis virus infection leads to swelling of the Golgi apparatus and trans-Golgi reticulum (TGR). Late in infection, viral particles can be seen budding from the cell surface and into the lumen of the TGR. These effects are coincident with the production of viral structural proteins and may be due to the sheer mass of viral proteins moving through the secretory pathway. The secretory pathway itself is functionally intact since the viral proteins themselves are folded, assembled, acylated, glycosylated, and proteolytically processed in a temporal fashion (Schlesinger and Schlesinger, 1986). The processing of Sindbis structural proteins appears to be identical when produced either by Sindbis virus infection or via vaccinia virus recombinants (Rice *et al.,* 1985). The morphological changes associated with the secretory pathway are less evident at earlier times after infection, even though significant synthesis of structural proteins and foreign proteins is observed (Fig. 3). These morphologies can thus be avoided for some types of studies.

Another pathological effect is the inhibition of host protein synthesis, which is as much as 95% by 4–6 hr postinfection in some cells (Lachmi and Kaarianen, 1977). The mechanism for inhibition of host protein synthesis is not fully known, but probably involves competition for translation machinery by viral RNAs which can replicate to extremely high levels. Sindbis or Semliki Forest virus suicide vectors capable of only one round of infection also shut off host protein synthesis, probably because of the high level of subgenomic mRNA produced (Liljestrom and Garoff, 1991; Bredenbeek *et al.,* 1994). Thus, these vectors would also exert some pathological effects, but not those associated with virion assembly. Careful appraisal of the progress of viral pathogenesis is necessary for each host cell to estimate a window of time during which experimental parameters can be faithfully maintained. As discussed subsequently, most cellular functions are intact for the first few hours after infection, and viral protein production is high enough to complete many analyses during this initial period of infection.

The pathological effects of the virus may also be advantageous under certain experimental conditions. One advantage of the high rate of synthesis of viral envelope protein is that these proteins may be used as specific markers of the ER (via temperature block at 15°C) or the TGR (via temperature block at 19°C). Both the high rate of viral protein synthesis and the virus-induced inhibition of host protein synthesis facilitate the exclusive metabolic labeling of only virally expressed proteins (Fig. 4). Cells that do not exhibit complete inhibition of protein synthesis at the time of metabolic labeling can be pretreated for 1 hr with actinomycin D, which will deplete remaining host cell mRNA. This treatment is helpful in characterizing expressed proteins when specific antisera are not available.

The Sindbis virus fusogenic E1 protein (Omar and Koblet, 1988) has also been used as an experimental tool in some instances. Infected cell monolayers can be fused to form a syncytium simply by briefly reducing the pH of the media (pH 5–6). This approach has been used to examine organelle reorganization and movement (Xiao and Storrie, 1991). Another way to take advantage of the cell surface E1 protein is to use Sindbis virus-infected cells as microinjection vehicles. We have explored such a strategy to deliver virally expressed heterologous proteins into noninfected cells by inducing the fusion of two cell types to form heterokaryons. In this procedure, trypsinized BHK cells infected with recombinant Sindbis virus were allowed to settle on top of monolayers of uninfected 3T3-L1 adipocytes. A brief exposure to low pH was then used to fuse the BHK cells to the adipocytes, thereupon releasing the contents of the infected cells instantaneously into the acceptor cells (D. E. James, unpublished data).

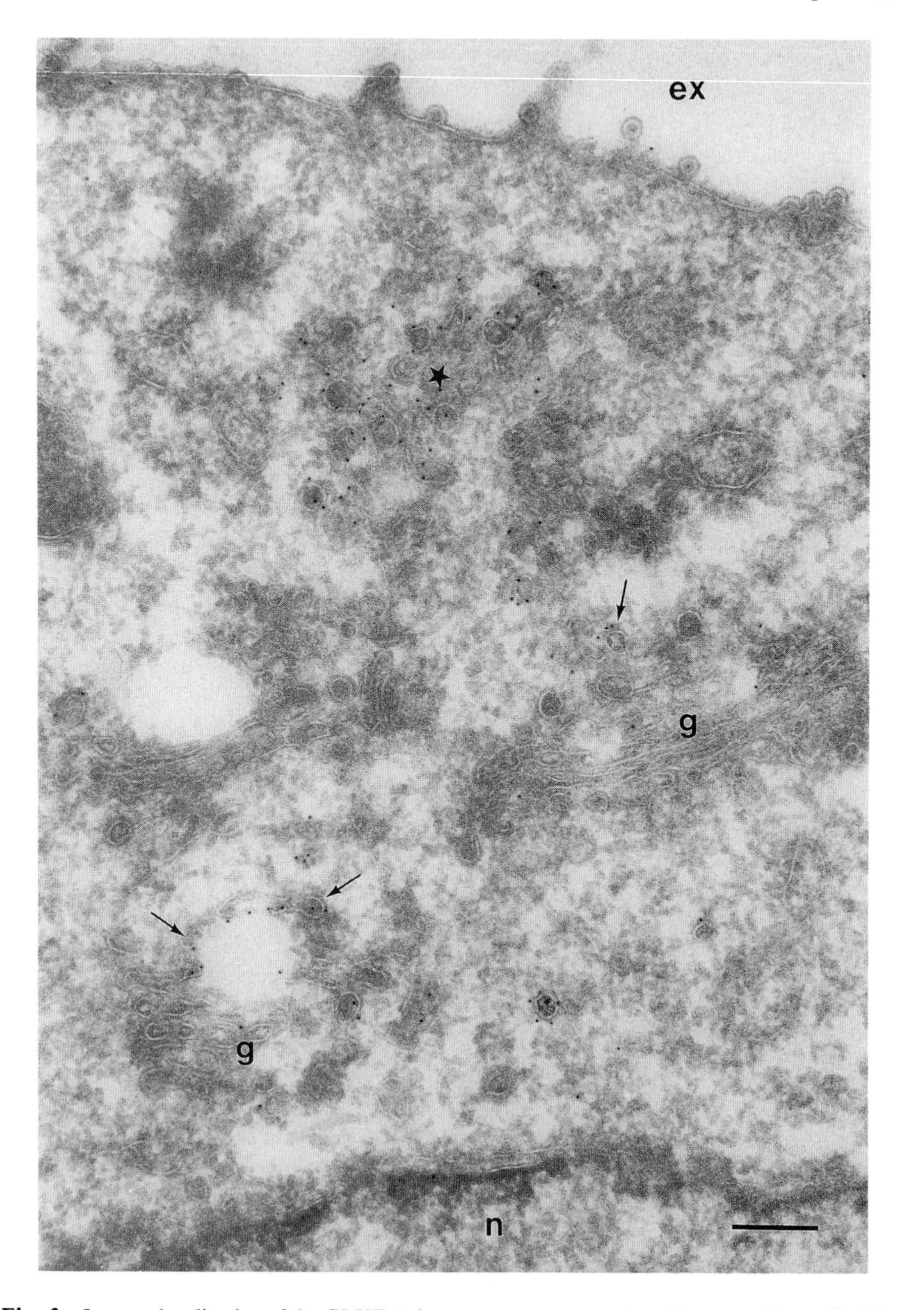


**Fig. 3** Immunolocalization of the GLUT-4 glucose transporter produced by recombinant Sindbis virus in Chinese hamster ovary (CHO) cells. CHO monolayers were infected for 6 hr with recombinant dsSIN virus containing the GLUT-4 gene. Cells were treated with cycloheximide for 1 hr, and fixed and processed for EM immunogold labeling (Slot *et al.* 1991). Gold labeling shows presence of GLUT-4 in tubulo-vesicular structures (*) and in the TGR (arrows). Virus budding is evident at the plasma membrane which is barely labeled for GLUT-4. Swelling of the Golgi and TGR is not pronounced at this stage of infection. n,Nucleus; g, Golgi complex; ex, extracellular space. Bar, 200 nm.

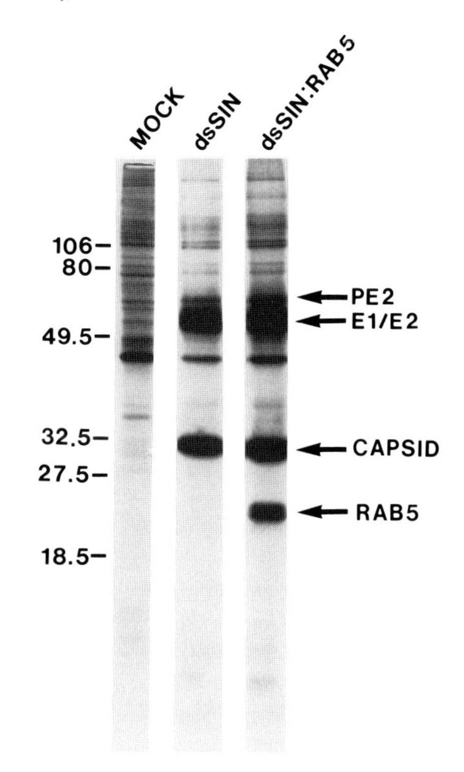


**Fig. 4** Metabolic labeling of Sindbis virus-expressed proteins. Chicken embryonic fibroblasts were mock infected, infected with wild-type virus, or infected with dsSIN virus expressing Rab5. Newly synthesized proteins were labeled with [$^{35}$S]methionine during 3 hr of continuous labeling from 3 to 6 hr postinfection. Total cell lysates were analyzed by SDS–PAGE and fluorography. Note that the majority of labeled protein is of viral origin and that the heterologous rab5 protein is expressed in proportion to the viral structural proteins.

## III. Previous Studies with Sindbis Virus Expression System

Two specific applications of recombinant Sindbis virus technology are described in this section. One is the study of glucose transporter proteins (GLUT), which facilitate the diffusion of glucose across the membrane. Another study analyzed the function and processing of an extended set of Rab protein mutants. Rab proteins are low molecular weight GTPases that bear structural homology to Ras and control membrane traffic. Sindbis virus expression of heterologous proteins in these studies was rapid and robust, and offered a tremendous benefit for structure–function studies when a large number of mutations needed to be analyzed.

### A. Functional Expression and Processing of Heterologous Proteins

High levels of protein production in a variety of eukaryotic cells is a forte of the Sindbis virus expression system and has facilitated the functional expression of glucose transporters. For these studies, a sufficient level of overexpression is required so that the specific activity of expressed transporters can be distinguished from that of endogenous glucose transporters. Using the 3′ recombinant dsSIN vectors, an 8-fold increase in glucose transport was observed when GLUT-1 was expressed in BHK cells. GLUT-1 is the endogenous isoform within these cells, and the 8-fold increase in glucose transport rate was accompanied by a concomitant increase in the total amount of GLUT-1 (Piper *et al.*, 1992). Lower levels of glucose transport activity were observed when dsSIN vectors were used that contained the heterologous gene between the structural and nonstructural proteins (2.5-fold over cells infected with wild-type virus); however, these levels were sufficient to map functionally relevant domains responsible for glucose transport inhibition caused by dibutyryl cAMP (Piper *et al.*, 1993a). SDS–PAGE analysis as well as resistance to endoglycosidase H digestion indicated that wild-type GLUTs as well as many glucose transporter chimeras composed of various complementary portions of different transporter isoforms were processed correctly (Piper *et al.*, 1992).

Recombinant Sindbis virus was very useful in studying the processing of Rab proteins. One of the problems associated with studying Rab proteins is that they are naturally expressed at only low levels; expression by Sindbis virus resulted in a 10-fold overexpression of Rab3, Rab4, Rab5, and Rab6 proteins (Li and Stahl, 1993a,b). The high rate of Rab5 synthesis facilitated its metabolic labelling with both $^{35}$S-methionine and 3H-mevalonolactone (a precursor lipid moiety that is covalently attached at the C-terminus). The robust signals achieved allowed the post-translational processing pathway of Rab5 to be mapped (Fig. 4; Li and Stahl, 1993a). Furthermore, the rapidity of analysis that Sindbis virus expression affords allowed many mutations within distinct functional domains of Rab5 to be analyzed to dissect structure–function relationships (Li and Stahl, 1993b).

## B. Effect on the Recycling Pathway

We have used Sindbis virus to express proteins such as the Rab GTPases that play a role in the recycling pathway between the plasma membrane and endosomal compartments (Li and Stahl, 1993a,b). Uptake of the fluid-phase endocytosis marker horseradish peroxidase (HRP) was measured in BHK cells infected with various recombinant Sindbis viruses for 2–10 hr (Fig. 5). Infection by wild-type Sindbis virus did not affect the rate of HRP uptake for the first 5 hr of viral infection. However, HRP uptake was inhibited by 50% by 8 hr of infection. This allowed a window of time (5 hr postinfection) to assess the effects of virally expressed Rab3, Rab4, Rab5, and Rab7. Only cells overexpressing Rab5 had elevated HRP uptake, consistent with previous experiments in which Rab5 was overexpressed through use of recombinant vaccinia virus (Bucci *et al.*, 1992). Thus, Sindbis virus infection alone did not alter the kinetics of recycling, at least at the times postinfection used in these experiments. Furthermore, overexpression of Rab proteins was found to have a specific

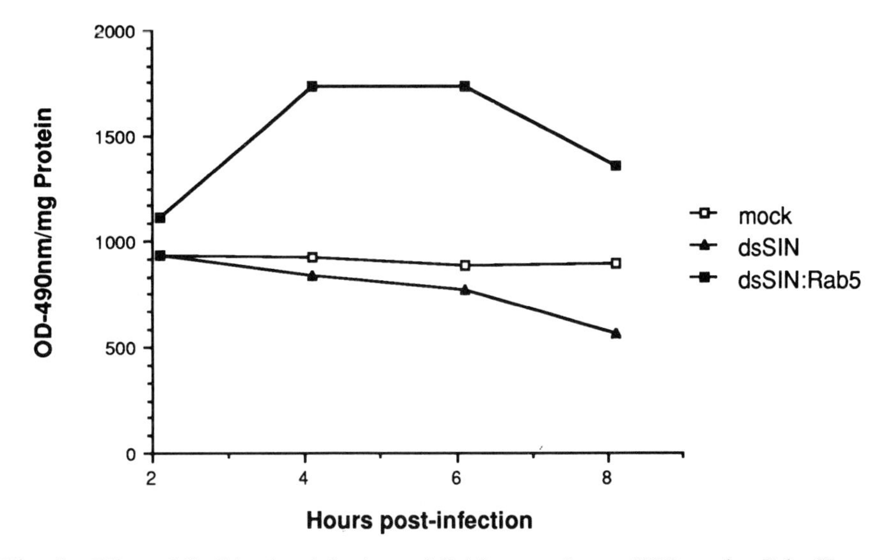


**Fig. 5** Effect of Sindbis virus infection and Rab5 expression on HRP uptake. Baby Hamster Kidney cells were infected with dsSIN virus containing no heterologous gene (dsSIN; ▲) or Rab5 (dsSIN:Rab5; ■), or mock infected (□). Uptake of HRP (1 hr) was measured at the indicated times post-infection at 37°C.

effect since only Rab5, which is localized to early endosomes and the plasma membrane, had an effect on HRP uptake via fluid-phase endocytosis. The implication from these data is that Rab proteins expressed by recombinant Sindbis virus retain the targeting specificity expected.

Another indication that the recycling pathway is intact during Sindbis virus infection comes from the colocalization of the GLUT-4 glucose transporter with clathrin-coated pits (Piper *et al.*, 1993b). Normally in 3T3-L1 adipocytes, GLUT-4 is found preferentially in areas of the plasma membrane containing clathrin lattices (Robinson *et al.*, 1992). This same colocalization was recapitulated in Chinese hamster ovary (CHO) cells expressing GLUT-4 from recombinant Sindbis viruses. We also compared the extent of colocalization in CHO cells stably transfected with GLUT-4. Both cell types showed a high level of colocalization consistent with that observed in 3T3-L1 adipocytes. Furthermore, the number and morphology of the clathrin lattices were unaltered in CHO cells at 6 hr postinfection. Some pathology of clathrin lattices was observed in Sindbis virus-infected CEF cells at 8 hr postinfection. Clathrin lattices were no longer circular but were elongated and fewer in number. Thus, care should be taken when choosing a particular cell type and infection conditions. Collectively these data show that Sindbis virus can be used to address specific questions about the recycling pathway and that, under certain conditions, Sindbis virus infection does not alter this pathway in any generalized way.

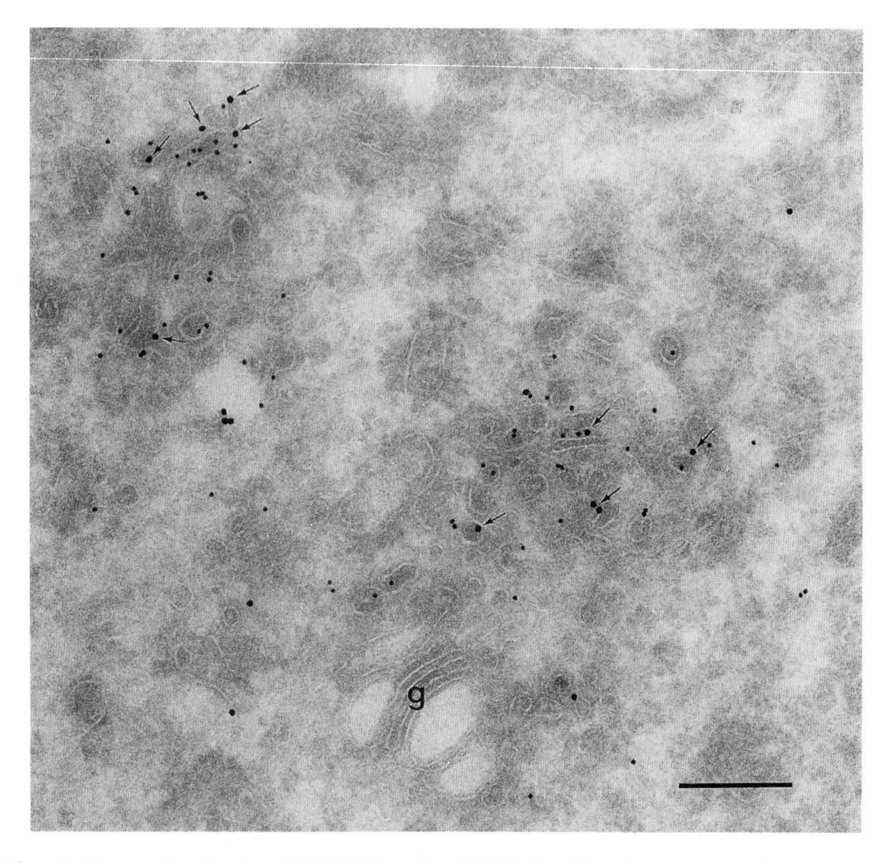


**Fig. 6** Immunolocalization of GLUT-4 and a GLUT-4/1 chimeric transporter in CHO cells. GLUT-4 was expressed in CHO cells by stable transfection using the pCMV vector as previously described (Piper *et al.* 1993). Stable transfectants expressing GLUT-4 transfectants were infected with recombinant dsSIN containing a chimeric glucose transporter composed of GLUT-4 with the C-terminal hydrophilic tail of GLUT-1 (GLUT-4/1D:GLUT-$4_{1\text{-}480}$–GLUT-$1_{466\text{-}492}$) as described in Fig. 3. Cells were double labeled using C-terminal specific antibodies to label GLUT-4 (10-nm gold) and GLUT-4/1D (15-nm gold). Both transporters are largely colocalized within tubulo-vesicular elements either in the TGR or in the cytoplasm, as indicated by the presence of a number of structures that are labeled with both 10- and 15-nm gold particles. Bar, 250 nm.

## C. Subcellular Targeting in Sindbis Virus-Infected Cells

We have also used Sindbis virus to study the differential targeting of glucose transporters in CHO cells (Piper *et al.,* 1992,1993a,b). Two mammalian glucose transporters were studied: GLUT-1, which is preferentially targeted to the plasma membrane, and GLUT-4, which is preferentially distributed intracellularly within vesicles and tubulovesicular elements near the TGR (Slot *et al.,* 1991). Sindbis virus expression of glucose transporter isoforms was able to recapitulate the differential targeting of the GLUT-1 and GLUT-4 isoforms that

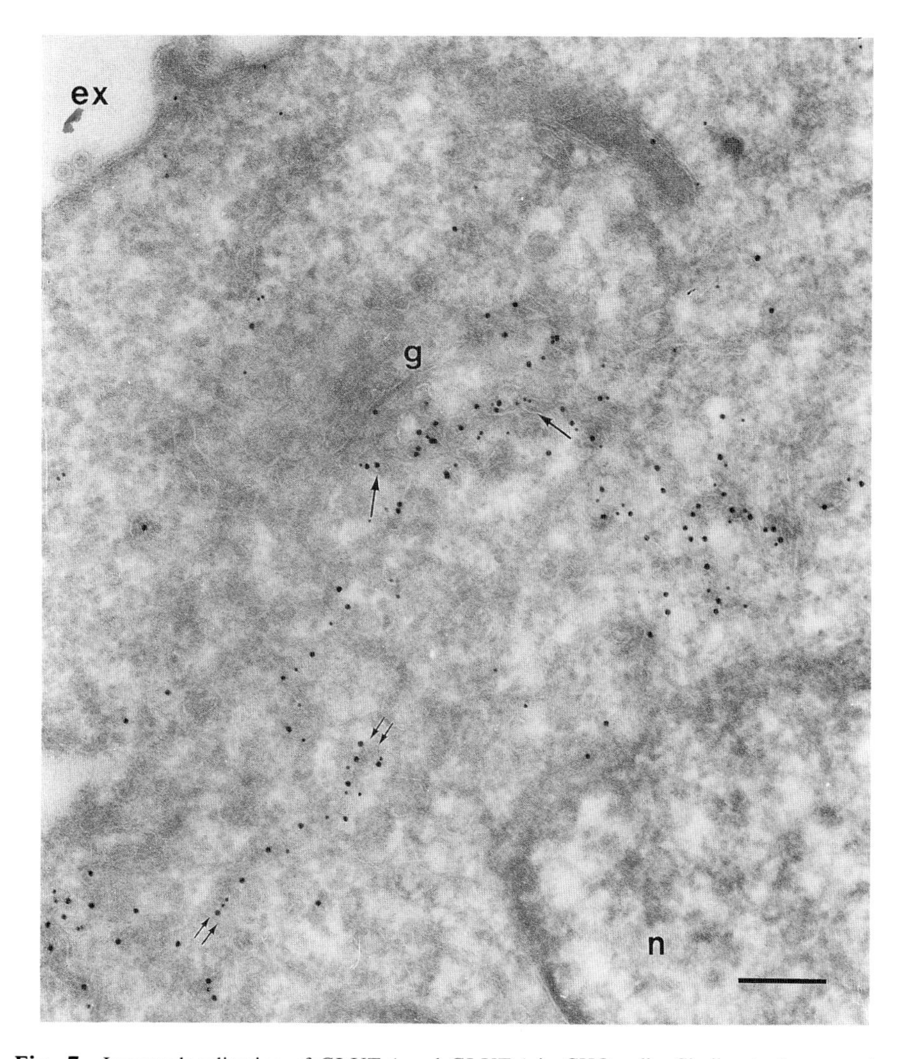


**Fig. 7** Immunolocalization of GLUT-4 and GLUT-1 in CHO cells. Similar to the procedure described in Fig. 6, GLUT-4 stable CHO transfectants were infected with dsSIN containing GLUT-1. Cells were double labeled using a GLUT-4 C-terminal specific antibody (15-nm gold) and a GLUT-1 C-terminal specific antibody (10-nm gold). Within the cell, GLUT-4 and GLUT-1 populate the same set of intracellular structures both in the TGR and in tubulo-vesicular elements clustered elsewhere in the cytoplasm. Striking colocalization of GLUT-1 and GLUT-4 is evident in both vesicles and tubules (indicated by the double arrows). Bar, 250 nm.

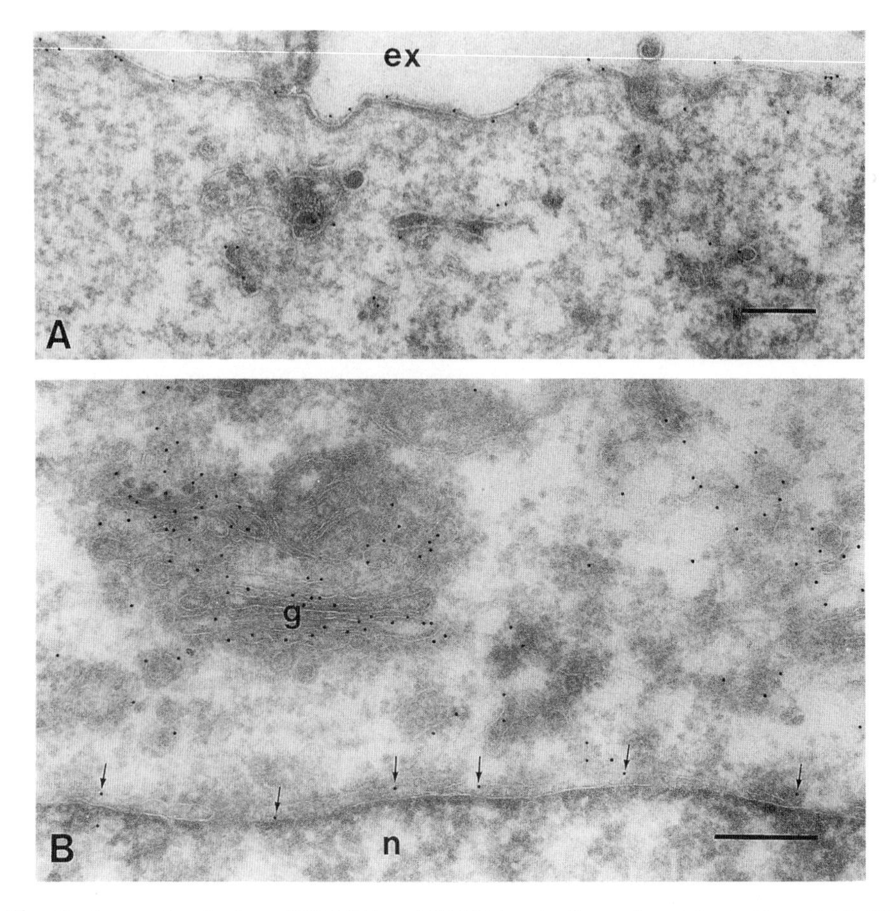


**Fig. 8** Immunolocalization of GLUT-1 and GLUT-4 and effects of cycloheximide. (A) GLUT-1 was produced by recombinant Sindbis virus in Chinese hamster ovary cells and labeled as described in Fig. 3. Predominant labeling of the cell surface is evident. (B) CHO cells were infected for 6 hr with dsSIN:GLUT-4 as in Fig. 3 except the 1 hr cycloheximide treatment was omitted. Significant labeling of Golgi complex (g) and the nuclear envelope is evident under these conditions. Bar, 250 nm.

is observed both in transfected fibroblasts and in tissues in which both are expressed *in vivo* (Piper *et al.*, 1992). Sindbis virus-expressed GLUT-4 was found intracellularly within the region of the TGR (Figs. 3, 6, and 7). In contrast, GLUT-1 was found predominantly at the cell surface (Fig. 8). The fidelity of the differential targeting of glucose transporter isoforms expressed by both stable transfection and Sindbis virus is shown in double immunoelectron microscopy labeling experiments (Figs. 6 and 7). We also found that intracellular distribution of Sindbis virus-expressed H1 subunit of the asialoglycoprotein receptor was as observed in stably transfected cells (Piper *et al.*, 1993b). Furthermore, when H1 is expressed by either stable transfection or recombinant Sindbis virus, two forms of the protein are apparent by SDS–PAGE: a lower molecular weight form that is incompletely glycosylated and resides in the ER and a higher molecular weight mature form. We found that the ratio of the two forms was also the same in both expression systems, indicating that the processing of Sindbis virus-produced H1 was correct (Piper *et al.*, 1993b).

Collectively these data show that the Sindbis virus expression system is capable of reproducing the native subcellular distribution of these proteins. One potentially complicating feature of the Sindbis virus-expressed proteins is that their biosynthetic rate was quite high. This rate was high enough to produce specific labeling of glucose transporters within the ER and Golgi apparatus by immunogold electron microscopy (Fig. 8). Thus, for studies of the targeting of endocytosed protein, we employed a 1-hr cycloheximide treatment that allowed newly synthesized proteins to exit the biosynthetic pathway (Piper *et al.*, 1992). One of the caveats of these studies is that the foreign proteins expressed were relatively young, roughly 3 hr old. The fact that the Sindbis virus expression systems can synthesize a bolus of heterologous protein may be of particular advantage for some analyses.

## Acknowledgments

We wish to thank all those who have contributed significantly to this review: Charles M. Rice, Sondra Schlesinger, John Heuser, and Henry Huang.

## References

Adams, R. H., and Brown, D. T. (1985). BHK cells expressing Sindbis virus-induced homologous interference allot the translation of non-structural genes of superinfecting virus. *J. Virol.* **54,** 351–357.

Baric, R. S., Carlin, L. J., and Johnston, R. E. (1983). Requirement for host transcription in the replication of Sindbis virus. *J. Virol.* **45,** 200–205.

Bredenbeek, P. J., and Rice, C. M. (1992). Animal RNA virus expression systems. *Sem. Virol.* **3,** 297–310.

Bredenbeek, P. J., Frolov, I., Rice, C. M., and Schlesinger, S. (1994). Sindbis virus expression vectors: Packaging of RNA replicons using defective helper RNAs. *J. Virol.* (*in press*).

Brown, D. T., and Condreay, L. D. (1986). Replication of Alphaviruses in mosquito cells. *In* "The Togaviridae and Flaviviridae" (S. Schlesinger and M. J. Schlesinger, eds.). Plenum Press, New York. pp. 5–20.

Bucci, C., Parton, R. G., Mather, I. H., Stunnenberg, H., Simons, K., Hoflack, B., and Zerial, M. (1992). The small GTPase rab5 functions as a regulatory factor in the early endocytotic pathway. *Cell* **70,** 715–728.

Clark, H. F., Cohen, M. M., and Lunger, P. D. (1973). Comparative characterization of a C-type virus-producing cell line (VSW) and a virus free cell line (VH2) from *Vipera russelli. J. Natl. Cancer Inst.* **51,** 645–657.

Ding, M. X., and Schlesinger, M. J. (1989). Evidence that Sindbis virus nsP2 is an autoprotease which processes the virus nonstructural polyprotein. *Virology* **171,** 280–284.

Dubuisson, J., and Rice, C. M. (1993). Sindbis virus attachment: Isolation and characterization of mutants with impaired binding to vertebrate cells. *J. Virol.* **67,** 3363–3374.

Edwards, J., and Brown, D. T. (1991). Sindbis virus infection of a Chinese hamster ovary cell mutant defective in the acidification of endosomes. *Virology* **182,** 28–33.

Geigenmuller-Gnirke, U., Weiss, B., Wright, R., and Schlesinger, S. (1991). Complementation between Sindbis viral RNAs produces infectious particles with a bipartite genome. *Proc. Natl. Acad. Sci. USA* **88,** 3253–3257.

Grakoui, A., Levis, R., Raju, R., Huang, H. V., and Rice, C. M. (1989). A cis-acting mutation in the Sindbis virus junction region which affects subgenomic RNA synthesis. *J. Virol.* **63,** 5216–5227.

Green, J., Griffiths, G., Louvard, D., Quinn, P., and Warren, G. (1981). Passage of viral membrane proteins through the Golgi complex. *J. Mol. Biol.* **152,** 663–698.

Griffin, D. E. (1986). Alphavirus pathogenesis and immunity. *In* "The Togaviridae and Flaviviridae" (S. Schlesinger and M. J. Schlesinger, eds.). Plenum Press, New York. pp. 209–251.

Griffin, E., and Johnson, R. T. (1977). Role of the immune response in recovery from Sindbis virus encephalitis in mice. *J. Immunol.* **118,** 1070–1075.

Hahn, C. S., Hahn, Y. S., Braciale, T. J., and Rice, C. M. (1992). Infectious Sindbis virus transient expression vectors for studying antigen processing and presentation. *Proc. Natl. Acad. Sci. USA* **89,** 2679–2683.

Hahn, Y. S., Grakoui, A., Rice, C. M., Strauss, E. G., and Strauss, J. H. (1989a). Mapping of RNA-temperature-sensitive mutants of Sindbis virus: Complementation group F mutants have lesions in nsP4. *J. Virol.* **63,** 1194–1202.

Hahn, Y. S., Strauss, E. G., and Strauss, J. H. (1989b). Mapping of RNA-temperature-sensitive mutants of Sindbis virus assignment of complementation groups A, B, and G to non-structural proteins. *J. Virol.* **63,** 3142–3150.

Hahn, Y. S., Hahn, C. S., Braciale, V. L., Braciale, T. J., and Rice, C. M. (1992). CD8 T cell recognition of an endogenously processed epitope is regulated primarily by residues within the epitope. *J. Exp. Med.* **176,** 1335–1341.

Hakimi, J., and Atkinson, P. H. (1982). Glycosylation of intracellular Sindbis virus glycoproteins. *Biochemistry* **21,** 2140–2195.

Harrison, S. C. (1986). Alphavirus structure. *In* "The Togaviridae and Flaviviridae" (S. Schlesinger and M. J. Schlesinger, eds.). Plenum Press, New York. pp. 21–35.

Harrison, S. C., Strong, R. K., Schlesinger, S., and Schlesinger, M. J. (1992). Crystallization of Sindbis virus and its nucleocapsid. *J. Mol. Biol.* **226,** 277–280.

Hashimoto, K., Erdel, S., Keranen, S., Saraste, J., and Kaariainen, L. (1981). Evidence for a separate signal sequence for the carboxy-terminal envelope glycoprotein E1 of Semliki forest virus. *J. Virol.* **38,** 34–40.

Hertz, J. M., and Huang, H. V. (1992). Utilization of heterologous alphavirus junction sequences as promoters by Sindbis virus. *J. Virol.* **66,** 857–864.

Heuser, J. E., and Kirschner, M. W. (1980). Filament organization revealed in platinum replicas of freeze-dried cytoskeletons. *J. Cell Biol.* **86,** 212–234.

Huang, H. V., Rice, C. M., Xiong, C., and Schlesinger, S. (1989). RNA viruses as genes expression vectors. *Virus Genes* **3,** 85–91.

Hurtley, S. M., and Helenius, A. (1989). Protein oligomerization in the endoplasmic reticulum. *Annu. Rev. Cell Biol.* **5,** 277–307.

Jackson, A. C., Moench, T. R., Trapp, B. D., and Griffin, D. E. (1988). Basis of neurovirulence in Sindbis virus encephalomyelitis of mice. *Lab. Invest.* **58,** 503–509.

Johnston, R. E., Wan, K., and Bose, H. (1974). Homologous interference induced by Sindbis virus. *J. Virol.* **14,** 1076–1082.

Kaariainen, L., Takkinien, K., Keranen, S., and Soderlund, H. (1987). Replication of the genome of alphaviruses. *J. Cell Sci.* **7,** 231–250.

Knipfer, M. E., and Brown, D. T. (1989). Intracellular transport and processing of Sindbis virus glycoproteins. *Virology* **170,** 117–122.

Lachmi, B., and Kaarianen, L. (1977). Control of protein synthesis in Semliki Forest virus-infected cells. *J. Virol.* **22,** 142–149.

Leake, C. J., Vama, M. G. R., and Pudney, M. (1977). Cytopathic effect and plaque formation by arboviruses in a continuous cell line (XTC-2) from the toad *Xenopus laevis*. *J. Gen. Virol.* **35,** 335–339.

Lemm, J. A., and Rice, C. M. (1993). Assembly of functional Sindbis virus RNA replication complexes: requirement for coexpression of P123 and P34. *J. Virol.* **67,** 1905–1915.

Levis, R., Huang, H., and Schlesinger, S. (1987). Engineered defective interfering RNAs of Sindbis virus express bacterial chloramphenicol acetyltransferase in avian cells. *Proc. Natl. Acad. Sci. USA* **84,** 4811–4815.

Levis, R., Schlesinger, S., and Huang, H. V. (1990). Promoter for Sindbis virus RNA-dependent subgenomic RNA transcription. *J. Virol.* **64,** 1726–1733.

Li, G., and Rice, C. M. (1989). Mutagenesis of the In-frame opal codon preceding nsP4 of Sindbis virus: Studies of translational read through and its effect on virus replication. *J. Virol.* **63,** 1326–1337.

Li, G., and Stahl, P. D. (1993a). Post-translational processing and membrane association of the two early endosome-associated rab GTP-binding proteins (rab4 and rab5). *Arch. Biochem. Biophys.* **304,** 471–478.

Li, G., and Stahl, P. D. (1993b). Structure–function relationships of the small GTPase rab5. *J. Biol. Chem.* (*in press*).

Liljestrom, P., and Garoff, H. (1991). A new generation of animal cell expression vectors based on the Semliki forest virus replicon. *BioTechnology* **9,** 1356–1362.

Liljestrom, P., Lusa, S., Huylebroeck, D., and Garoff, H. (1991). *In vitro* mutagenesis of a full-length cDNA clone of Semliki Forest virus: The small 6,000-molecular-weight membrane protein modulates virus release. *J. Virol.* **65,** 4107–4113.

London, S. D., Schmaljohn, A. L., Dalrymple, J. M., and Rice, C. M. (1992). Infectious enveloped RNA virus antigenic chimeras. *Proc. Natl. Acad. Sci. USA* **89,** 207–211.

Marsh, M., and Helenius, A. (1989). Virus entry into animal cells. *Adv. Virus Res.* **36,** 107–149.

Matlin, K. L., and Simons, K. (1984). Sorting of an apical plasma membrane glycoprotein occurs before it reaches the cell surface in cultured epithelial cells. *J. Cell Biol.* **99,** 2131–2139.

Mayne, J. T., Rice, C. M., Strauss, E. G., Hunkapiller, M. W., and Strauss, J. H. (1984). Biochemical studies of the maturation of the small Sindbis virus glycoprotein E3. *Virology* **134,** 338–357.

Miller, M. L., and Brown, D. T. (1993). The distribution of Sindbis virus proteins in mosquito cells as determined by immunofluorescence microscopy. *J. Gen. Virol.* **74,** 293–298.

Omar, A., and Koblet, H. (1988). Semliki Forest virus particles containing only the E1 envelope glycoprotein are infectious and can induce cell–cell fusion. *Virology* **166,** 17–23.

Peiris, J. S. M., and Porterfield, J. S. (1979). Antibody-mediated enhancement of Flavivirus replication in macrophage-like cell lines. *Nature* (*London*) **282,** 509–511.

Pfeffer, S. R., and Rothman, J. E. (1987). Biosynthetic protein transport and sorting by the endoplasmic reticulum and Golgi. *Annu. Rev. Biochem.* **56,** 829–852.

Pierce, J. S., Strauss, J. H. (1974). Effect of ionic strength on the binding of Sindbis virus to chick cells. *J. Virol.* **13,** 1030–1036.

Piper, R. C., Tai, C., Slot, J. W., Hahn, C. S., Rice, C. M., Huang, H., and James, D. E. (1992). The efficient intracellular sequestration of the insulin-regulatable glucose transporter (GLUT-4) is conferred by the NH2 terminus. *J. Cell Biol.* **117,** 729–743.

Piper, R. C., James, D. E., Slot, J. W., Puri, C., and Lawrence, J. C. J. (1993a). GLUT4 Phosphorylation and inhibition of glucose transport by dibutyryl cAMP. *J. Biol. Chem.* **268,** 16557–16563.

Piper, R. C., Tai, C., Kulesza, P., Pang, S., Warnock, D., Baenzinger, J., Slot, J. W., Geuze, H. J., Puri, C., and James, D. E. (1993b). The GLUT-4 N-terminus contains a phenylalanine based targeting motif that regulates intracellular sequestration. *J. Cell Biol.* **121,** 1221–1232.

Raju, R., and Huang, H. (1991). Analysis of Sindbis virus promoter recognition in vivo, using novel vectors with two subgenomic mRNA promoters. *J. Virol.* **65,** 2501–2510.

Rice, C. M., Franke, C. A., Strauss, J. H., and Hruby, D. E. (1985). Expression of Sindbis virus structural proteins via recombinant vaccinia virus: Synthesis, processing, and incorporation into mature Sindbis virions. *J. Virol.* **56,** 227–239.

Rice, C. M., Levis, R., Strauss, J. H., and Huang, H. V. (1987). Production of infectious RNA transcripts from Sindbis virus cDNA clones: mapping of lethal mutations, rescue of a temperature-sensitive marker, and in vitro mutagenesis to generate defined mutants. *J. Virol.* **61,** 3809–3819.

Robinson, L. J., Pang, S., Harris, D. S., Heuser, J., and James, D. E. (1992). Translocation of the glucose transporter (GLUT-4) to the cell surface in permeabilized 3T3-L1 adipocytes: Effects of ATP, insulin, GTP$\gamma$S and localization of GLUT4 to clathrin lattices. *J. Cell Biol.* **117,** 1181–1196.

Sawicki, D., Barkhimer, D. B., Sawicki, S. G., Rice, C. M., and Schlesinger, S. (1990). Temperature sensitive shut-off of alphavirus minus strand synthesis maps to a nonstructural protein, nsP4. *Virology* **174,** 43–52.

Schlesinger, M. J., and Schlesinger, S. (1986). Formation and assembly of Alphavirus glycoproteins. *In* "The Togaviridae and Flaviviridae" (S. Schlesinger and M. J. Schlesinger, eds.). Plenum Press, New York. pp. 121–149.

Schlesinger, S. (1993). Alphaviruses–Vectors for the expression of heterologous genes. *Trends Biotechnol.* **11,** 18–22.

Schlesinger, S., and Schlesinger, M. J. (1990). Replication of togaviridae and flaviviridae. *In* "Virology" (B. N. Fields, ed.), pp. 697–711. Raven Press, New York.

Shirako, Y., and Strauss, J. H. (1990). Cleavage between nsP1 and nsP2 initiates the processing pathway of Sindbis virus nonstructural polyprotein P123. *Virology* **177,** 54–64.

Slot, J. W., Geuze, H. J., Gigengack, S., Lienhard, G. E., and James, D. E. (1991). Immunolocalization of the insulin-regulatable glucose transporter in brown adipose tissue of the rat. *J. Cell Biol.* **113,** 123–135.

Strauss, E. G., Lenches, E. M., and Strauss, J. H. (1976). Mutants of Sindbis virus. I. Isolation and partial characterization of 89 new temperature-sensitive mutants. *Virology* **74,** 154–168.

Strauss, E. G., Rice, C. M., and Strauss, J. M. (1983). Sequence coding for the alphavirus nonstructural proteins is interrupted by an opal termination codon. *Proc. Natl. Acad. Sci. USA* **80,** 5271–5275.

Strauss, E. G., Rice, C. M., and Strauss, J. H. (1984). Complete nucleotide sequence of the genomic RNA of Sindbis virus. *Virology* **133,** 92–110.

Strauss, E. G., and Strauss, J. H. (1986). Structure and replication of the Alphavirus genome. *In* "The Togaviridae and Flaviviridae" (S. Schlesinger and M. J. Schelsinger, eds.). Plenum Press, New York. pp. 35–91.

Symington, J., and Schlesinger, M. J. (1975). Isolation of a Sindbis virus variant by passage on mouse plasmacytoma cells. *J. Viol.* **15,** 1037–1041.

Ubol, S., and Griffin, D. E. (1991). Identification of a putative alphavirus receptor on mouse neural cells. *J. Virol.* **65,** 6913–6921.

Wang, K. S., Schmaljohn, Kuhn, R. J., and Strauss, J. H. (1991). Antiidiotypic antibodies as probes for the Sindbis virus receptor. *Virology* **181,** 694–702.

Wang, K. S., Kuhn, R. J., Strauss, E. G., and Strauss, J. H. (1992). High-affinity laminin receptor is a receptor for Sindbis virus in mammalian cells. *J. Virol.* **66,** 4992–5001.

Weiss, B., Nitschko, H., Ghattas, I., Wright, R., and Schlesinger, S. (1989). Evidence for specificity in the encapsidation of Sindbis virus RNAs. *J. Virol.* **63,** 5216–5227.

Xiao, L., and Storrie, B. (1991). Behavior of a transitional tubulovesicular compartment at the *cis* side of the Golgi apparatus in vivo fusion studies of mammalian cells. *Exp. Cell Res.* **193,** 213–218.

Xiong, C., Levis, R., Shen, P., Schlesinger, S., Rice, C. M., and Huang, H. V. (1989). Sindbis virus: An efficient, broad host range vector for gene expression in animal cells. *Science* **243,** 1188–1191.

# CHAPTER 4

# Recombinant Rous Sarcoma Virus Vectors for Avian Cells

**Greg Odorizzi and Ian S. Trowbridge**
Department of Cancer Biology
The Salk Institute
San Diego, California 92186



## I. Introduction

Substantial progress has been made in the last 7 years in understanding the sorting signals for receptors that are constitutively internalized from plasma membrane clathrin-coated pits to the endocytic pathways (reviewed by Trowbridge *et al.,* 1993). Major factors contributing to these advances have been the extensive use of recombinant DNA techniques to generate mutant receptors and the development of quantitative assays to analyze their function. Internalization signals of recycling receptors have defined a family of structural motifs specified by short linear arrays of amino acids that differ in specific sequence, but share a common three-dimensional conformation and chemistry (Collawn *et al.,* 1990; Trowbridge *et al.,* 1993). Tyrosine-containing internalization signals now serve as paradigms for intracellular sorting signals, including endosomal signals that are recognized by the sorting machinery of trans-Golgi clathrin-coated pits and basolateral sorting signals that determine membrane protein trafficking patterns in polarized epithelial cells.

Our studies of receptor-mediated endocytosis have focused on the human transferrin receptor (TR) as a prototype of constitutively recycling receptors (Collawn *et al.,* 1990,1991,1993; Jing *et al.,* 1990). The TR is a favorable model for several reasons: it is abundantly expressed on most cultured cells, quantitative assays are available to determine receptor internalization and recycling precisely, and a variety of effective mouse monoclonal antibodies and monospecific rabbit antisera against both the external and the cytoplasmic domains of human TR have been developed for biochemical and morphological studies (Trowbridge and Omary, 1981; White *et al.,* 1990,1992; D. Domingo and I. S. Trowbridge, unpublished results). For reasons described later, we wished to analyze the function of mutant human TRs in avian cells. To achieve this goal, we have made extensive use of a Rous sarcoma virus (RSV) vector system developed by Hughes and colleagues (NCI–Frederick Cancer Facility, Frederick, MD), generously provided to us prior to publication (Hughes *et al.,* 1987,1990). Here we describe the properties of the RSV vectors and their use in expressing recombinant proteins in chick embryo fibroblasts (CEFs) and other cultured avian cells. This expression system has significant experimental advantages and has been invaluable for the structure–function analysis of the human TR. Various related RSV vectors have now been generated and can also be used to express recombinant proteins in cultured cells and in chickens (Hughes *et al.,* 1990; Petropoulos and Hughes, 1991; Petropoulos *et al.,* 1992; Boerkoel *et al.,* 1993).

## II. Replication-Competent Rous Sarcoma Virus Vectors

RSV is the only known replication-competent retrovirus that carries a host-derived oncogene, *src*. Since the virus does not need the host-derived *src* oncogene for replication, *src* can be removed and replaced by other genes. This feature of RSV enables it to be used as a replication-competent helper-independent retroviral vector.

An understanding of the development and potential uses of RSV vectors requires some knowledge of the discovery and life cycle of the virus. RSV is a member of the family of avian leukosis and sarcoma viruses (ALSV) of chickens, one of four groups of avian C-type retroviruses. RSV was originally isolated by Peyton Rous in 1911 as the filterable infectious agent causing fibrosarcomas in chickens, but the name is now used for independently isolated strains of virus that all carry closely related versions of *src* and transform cells by the same genetic mechanism. Many of the different strains are named after the investigators using them [e.g., Bryan high-titer (BH-RSV), Schmidt-Ruppin

(SR-RSV) strains] or the locations where they were used [e.g., Prague (PR-RSV) and Bratislava (B77-RSV]. The two RSV strains used for the development of retroviral vectors are SR-RSV, which is a replication-competent strain from which all the retroviral vectors are derived, and BH-RSV, a replication-defective strain. Although BH-RSV is replication defective, in the presence of helper virus it gives higher virus titers than other strains. The ability of BH-RSV to replicate more efficiently than other strains of RSV in the presence of helper virus appears to map within the *pol* region of the virus; substitution of this region into RSV retroviral vectors can significantly increase expression of recombinant proteins in CEFs.

Retroviruses recognize and infect target cells that display virus-specific cell surface receptors. The interaction between viral envelope glycoproteins encoded by the *env* gene and host cell surface receptors leads to fusion of virus and cell membranes, so viral nucleoprotein core particles are introduced into the host cell cytoplasm. Retroviruses have evolved to use proteins with very different structures and functions as cellular receptors: human immunodeficiency virus binds CD4, a member of the Ig superfamily that is involved in T-cell signal transduction (Maddon *et al.*, 1986), whereas the ecotropic murine leukemia virus receptor is a cationic amino acid transporter (Albritton *et al.*, 1989; Kim *et al.*, 1991; Wang *et al.*, 1991), and the gibbon ape leukemia virus receptor is structurally similar to a phosphate transporter (O'Hara *et al.*, 1990; Johann *et al.*, 1992). As described subsequently, the receptor for one subgroup of RSV has been shown to be related to the low density lipoprotein receptor (Bates *et al.*, 1993).

In chicken ALSV, the five major subgroups are designated A through E. Viruses within each subgroup have the same host range, encode immunologically related envelope glycoproteins, and demonstrate cross-interference (Vogt, 1965; Dorner *et al.*, 1985; Weiss *et al.*, 1985; Dorner and Coffin, 1986). Chickens have been classified by their susceptibility to different ALSV subgroups (see Hughes *et al.*, 1990; Young *et al.*, 1993); C/O chickens, for example, can be infected by all five virus subgroups, whereas C/A chickens are resistant to ALSV-A. Because C/O CEFs can be infected with all subgroups of RSV, they are the preferred recipient cells for expression of recombinant proteins encoded by RSV vectors. The gene encoding the likely receptor for the ALSV-A subgroup has been isolated (Young *et al.*, 1993), and the structure of this molecule has been deduced for the quail homolog (Bates *et al.*, 1993). Two alternatively spliced processed genes were isolated, the larger of which encodes a protein consisting of an N-terminal extracellular domain of 83 residues, a transmembrane region of 23 residues, and a cytoplasmic domain of 32 residues. The smaller gene encodes a protein with an identical extracellular domain and 15 additional, mainly hydrophobic, residues. These residues may represent a transmembrane region or may be cleaved; this form of the receptor is tethered to the plasma membrane by a non-membrane-spanning anchor such as a glycosylphosphatidylinositol tail. The N-terminal half of the extracellular domain of the receptor is related to the ~40-residue ligand-binding repeat of the low density lipoprotein receptor (Bates *et al.,* 1993). Receptors for the other ALSV subgroups have yet to be characterized.

Although *in vivo* studies have shown that RSV is capable of infecting a broad range of cell types (Dolberg and Bissell, 1984; Howlett *et al.,* 1987; Stoker *et al.,* 1990; Morgan *et al.,* 1992), its cellular tropism is not well characterized. Only subgroup D virus can normally infect mammalian cells, and the efficiency of infectivity is much lower than for CEFs (Hughes *et al.,* 1990). However, expression of the subgroup A RSV receptor in monkey Cos cells and BALB/c 3T3 cells confers susceptibility to infection by subgroup A RSV (Young *et al.,* 1993). This observation provides the prospect of extending the use of RCAS(A) vectors to other avian cell types and mammalian cells not currently susceptible to infection. However, evidence suggests that susceptibility to ALSV infection depends on viral penetration through the cell membrane, not merely binding to the cell surface (Piraino, 1967). Thus, unless this step is also mediated by the receptor or an endogenous protein, expression of the subgroup A receptor may not be sufficient to confer susceptibility on all cell types.

### A. Development and Structure of RCAS Vectors

The prototype retroviral vector for RCAS (RC, replication-competent; A, avian leukemia virus long terminal repeat; S, with splice acceptor) was derived from the SR-A strain of RSV essentially by removing the *src* gene and introducing a *Cla* I cloning site (Hughes and Kosik, 1984). The *src* gene of wild-type RSV is flanked by direct repeats, which in the SR-A strain of virus are 110 bp in length (Czernilofsky *et al.,* 1980,1983). To suppress recombination events between these direct repeats that could eliminate the cloned cDNA to be expressed, the upstream repeat was removed (Hughes and Kosik, 1984). This vector was derived from circularly permutated retroviral DNA containing tandem long terminal repeats (LTRs). The *src* splice acceptor site was also reintroduced upstream of the *Cla* I cloning site so the mRNA from the foreign gene could be expressed as a spliced product. A series of recombinant DNA manipulations was performed to create a proviral form of the retroviral vector in which the *Escherichia coli* replicon is positioned between the LTRs (Hughes *et al.,* 1987,1990). This vector produces infectious virus when transfected directly into CEFs without the need for digestion to remove the *E. coli* replicon and religation to cirularize the retroviral DNA. The structure of the RCAS(A) vector and the mRNA products produced are shown in Fig. 1. To facilitate cloning of cDNAs into the RCAS vector, two adaptor plasmids—*Cla*12 and *Cla*12*Nco*—were also constructed that contain a pUC12 polylinker region flanked by *Cla* I sites (Hughes *et al.,* 1987,1990). *Cla*12*Nco* contains a eukaryotic leader and an ATG initiation codon as part of an *Nco*I recognition site to facilitate the expression of cDNAs lacking their own initiation site. To produce retroviral vectors of a different subtype, a DNA fragment containing the *env* gene was removed from RCAS(A) vector and replaced with the corresponding DNA fragment encoding

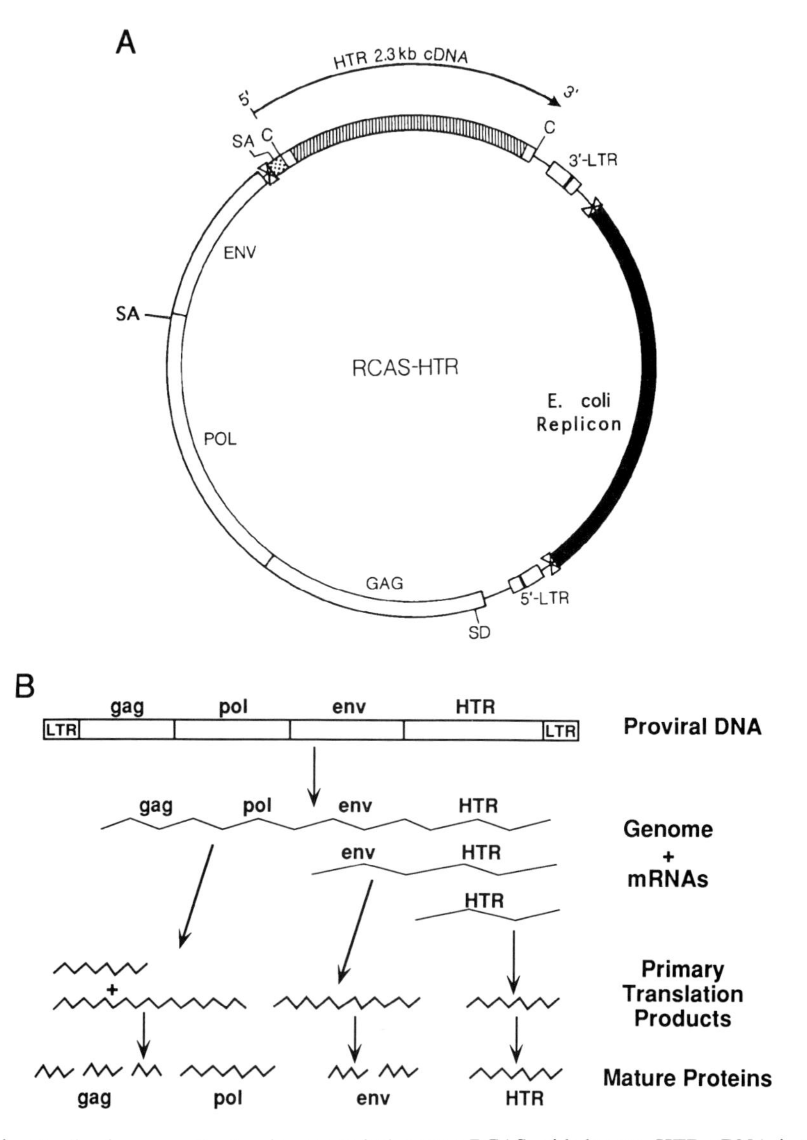


**Fig. 1** Replication-competent avian retroviral vector RCAS with human HTR cDNA insert. (A) Circularly permuted proviral vector (approximately 11.5 kb) containing the 2.3-kb cDNA insert encoding the human transferrin receptor (HTR). SD, splice donor site; SA, splice acceptor site; LTR, long terminal repeat sequences; C, *Cla* I cloning site; GAG, POL, ENV, viral genomic sequences encoding the gag, pol, and env products. (For further details, see Hughes *et al.*, 1987,1990.) (B) A simplified schematic representation of the production of viral proteins and recombinant HTR from the linearized proviral DNA which contains the HTR cDNA insert.

the *env* gene from either the B or the D subtype RSV to generate the RCAS(B) and RCAS(D) retrovirus vectors, respectively. Finally, to increase viral replication and expression of the recombinant protein encoded by the foreign DNA insert, the *pol* gene from the BH strain of RSV was introduced into the RCAS vector to produce the derivative RCAS BP, formerly known as BH-RCAS (Sudol *et al.*, 1986; Nemeth *et al.*, 1989; Petropoulos and Hughes, 1991; S. H. Hughes, unpublished results). The RCAS BP vector is the expression vector that we now routinely use. It has given consistently high-level expression of wild-type and mutant human TRs in CEFs. As described in Section III,C,3, we have used the RCAS(A) and RCAS(D) vectors to express two mammalian plasma membrane proteins concomitantly in CEFs. Expression of *ras* by RCAS(D) in CEFs already infected by RCAS(A) with a chloramphenicol acetyltransferase (CAT) insert has also been reported (Hughes *et al.*, 1987).

## B. Advantages and Limitations

The major advantage of the RCAS vectors is the fact that they are replication-competent retroviral vectors containing all the viral genes necessary to produce infectious virus (Hughes *et al.*, 1987,1990). Consequently, transfection of a few cells in a CEF culture is sufficient to produce enough infectious virus that within ~2 wk virtually all cells in the culture become infected and express the recombinant protein encoded by the foreign gene carried by the RCAS vector. As a result, it is unnecessary to use drug selection protocols to eliminate untransfected cells, so sufficient cells expressing the recombinant protein can be rapidly obtained (usually within 2–3 wk of transfection) for most experiments. If analysis of many mutant proteins is necessary, this can be rapidly performed using the RCAS expression system. Another advantage of the RCAS vectors is that high protein expression levels are reproducibly obtained from independent transfections. Since the virus is replication competent, the final protein expression level of infected cells is independent of the initial transfection efficiency. In an extensive series of experiments over a 2-yr period with the wild-type human TR and several mutant constructions, the expression levels of the receptor ranged from 3 to 8 $\times$ $10^5$ molecules per cell (i.e., ~100–200 ng protein per $10^6$ cells) as assessed by surface binding of human Tf at 4°C (Jing *et al.*, 1990). The reproducible high-level expression of recombinant human TR has greatly facilitated functional and biochemical analyses, as well as morphological studies at the light and electron microscope levels (Jing *et al.*, 1990; Miller *et al.*, 1991). Within a culture, however, the expression of recombinant proteins by individual cells can be variable. We have not investigated the reason for this heterogeneous expression, but expression is primarily influenced by the site of integration of the RCAS vector as well as by copy number. It has been possible to exploit the variation in the expression of internalization-defective human TR mutants to select overexpressing cells by growing transfected CEFs in tissue culture medium containing human transferrin as the only source of iron (Jing *et al.*, 1990). Under these conditions, only cells expressing large numbers of internalization-defective TRs ($1-2 \times 10^6$ receptors per cell) can obtain sufficient iron to survive and grow (Jing *et al.*, 1990). A small population of cells, usually ~1–2%, that expresses very high levels of human TR ($6-8 \times 10^6$ molecules per cell) has been consistently observed. These cells have allowed the relationship between receptor expression and clathrin lattice formation to be studied (Miller *et al.*, 1991). However, the cells expressing such high levels of human TR may not be viable long term, since attempts to isolate them by fluorescence-activated cell sorting and to culture them have been unsuccessful. We have most experience expressing mutant human TR and chimeric proteins consisting of the transmembrane region and external domain of the human TR fused to the cytoplasmic domains of other membrane proteins, including the major histocompatibility complex (MHC) class II invariant chain, the T-cell receptor $\gamma$ chain, and the $\beta$-amyloid precursor protein, all of which are expressed well. However, we have also used the RCAS BP vector to obtain high level expression of murine CD44 and the human $\beta$-amyloid precursor protein in CEFs. As described in detail in Section III,C,3, RCAS BP(A) and RCAS BP(D) vectors have been used to superinfect CEFs already infected with the other subtype of RCAS vector, allowing simultaneous expression of two recombinant proteins in the same cell.

Use of the RCAS vector for the efficient expression of recombinant proteins is constrained by the fact that there is an upper limit on the size of the cDNA insert that can be introduced into the vector. This limitation is because production of infectious virus depends on packaging of the viral RNA, which in turn depends on the size of the foreign DNA inserted into the vector. The coding region of the human TR spans ~2.4 kb; it and the 770-amino-acid isoform of the human $\beta$-amyloid precursor protein (also encoded by an ~2.4-kb DNA fragment) are the largest proteins that we have successfully expressed using RCAS BP. We failed to detect expression of E-cadherin encoded by a 2.7-kb cDNA insert introduced into RCAS BP. This failure may reflect the size of the E-cadherin cDNA but other possibilities cannot be ruled out. To our knowledge, the limit on the size of the cDNA insert that can be introduced into RCAS without affecting the ability to produce infectious virus has not been clearly defined. As described in Section II,C, however, it may be possible to overcome this limitation by using defective derivatives of RCAS in combination with a helper cell line.

Another constraint is that a unique *Cla* I site is the only cloning site in RCAS for the introduction of foreign DNA. Retroviral vectors in general are relatively large and contain sites for most of the commonly used restriction enzymes; RCAS is no exception. The development of adaptor plasmids that permit the conversion of essentially any DNA segment into a fragment with *Cla* I ends reduces the problems associated with this limitation (Hughes *et al.*, 1987,1990).

Since the recognition sequence for *Cla* I is 6 bp, there are rarely more than one or two *Cla* I sites in a 1- to 2-kb cDNA, so minimal modification of the foreign cDNA insert is usually required.

Finally, the RCAS vector has been used almost exclusively to express proteins in primary cultures of CEF, which have a finite life-span. The use of other cell types is limited by the fact that, with the exception of a variety of hematopoietic tumor cell lines, there are few permanent chick cell lines derived from other tissues. One chicken hepatocellular carcinoma cell line, LMH, has been established by Kawaguchi *et al.* (1987) but repeated attempts to express the human TR in these cells using RCAS BP(A) have been unsuccessful (C. G. Odorizzi and I. S. Trowbridge, unpublished results). RCAS BP can be used to express proteins in other avian cells such as the methylcholanthrene-induced Japanese quail fibrosarcomas cell line QT6 (Moscovici *et al.*, 1977). However, after transfection, high level expression of the human TR was only obtained by repeated selection of positive cells by fluorescence-activated cell sorting (S. Jing and I. S. Trowbridge, unpublished results). We have not extensively investigated the feasibility of expressing recombinant proteins in mammalian cells using the RCAS(D) vector, but attempts to infect Madin–Darby canine kidney cells were unsuccessful (C. G. Odorizzi and I. S. Trowbridge, unpublished results). The *ras* oncogene has been expressed in NIH 3T3 cells using RCAS(D), but viral titer in these cells was $\sim 10^{-3}$–$10^{-4}$ of the titer in CEFs (S. Hughes, personal communication). As described earlier in Section II, however, the availability of the receptor genes for subgroup A RSV may quickly extend the range of cells that can be infected by RCAS(A).

## C. Related Retroviral Vectors and Their Uses

The RCAS vectors are suitable for unregulated high level expression of recombinant proteins in cultured CEFs. Other related vectors have been developed by Hughes and colleagues for other purposes, including tissue-specific expression in transgenic chickens (reviewed by Hughes *et al.*, 1990). RCOS vectors have been constructed from RCAS by replacing the ALV LTRs with LTRs derived from the endogenous Rous-associated virus type O (RAV-O). This retroviral vector has little if any oncogenic potential and, therefore, is suitable for insertion into the germ line of chickens and for long-term infections *in vivo* (Greenhouse *et al.*, 1988). This retroviral vector is replication competent, but its replication efficiency is only 5–10% that of the RCAS vector and it gives low level expression of proteins (Greenhouse *et al.*, 1988). RCAN and RCON vectors are related to RCAS and RCOS, respectively. They lack the splice acceptor site that allows expression of the foreign cDNA to be driven by the LTR promoter and are suitable for tissue-specific expression of recombinant proteins from an internal promoter. (Hughes *et al.*, 1990). For example, muscle-specific expression has been obtained in chickens using the RCAN vector containing the $\beta$-actin promoter (Petropoulos *et al.*, 1992). In addition, replication-defective derivatives of RCAN and RCON lacking the *env* gene—known as BOS, BAS, and BBAN—have been constructed (Hughes *et al.*, 1990; Boerkoel *et al.*, 1993). In principle, these vectors used in conjunction with an appropriate packaging cell line should allow expression of larger foreign cDNAs than the RCAS vectors.

## III. Expression of Human Transferrin Receptor in Chick Embryo Fibroblasts

The utility of the RCAS vectors and the methods employed to obtain high level expression of recombinant proteins can be illustrated by expression of the human TR in CEFs. Because TR is a cell surface receptor, the level of expression in individual cells can be easily monitored by immunofluorescence analysis using monoclonal antibodies specific for the human TR (see Section III,C,1). Virtually all CEFs usually express high levels of human TR within 2 wk of transfection of a culture with RCAS BP containing a 2.4-kb human TR cDNA insert. Infection with a high titer viral supernatant will give comparable expression within 7 days.

### A. Rationale

The rationale for expressing the wild-type and mutant human TRs in CEFs is based on the demonstration by Shimo-Oka *et al.* (1986) that the growth of avian cells cannot be supported by most mammalian transferrins because they do not bind to ovotransferrin receptors. Human TRs can, therefore, be subjected to functional analysis in CEFs without a direct contribution to the assays of endogenous chick TRs. Although it was possible that chick ovotransferrin receptors could form heterodimers with human TRs, which normally are disulfide-bonded homodimers, there is strong evidence that this does not occur to any significant extent (Jing *et al.*, 1990). The plasma membrane clathrin-coated pits of CEFs have also been well characterized and, because of their large size, are ideal for morphological studies using conventional thin sections, whole mounts, and replica membrane preparations (Miller *et al.*, 1991).

### B. Methods

#### 1. Culture of Chicken Embryo Fibroblasts

Unpassaged CEFs prepared from C/O chicken embryos are obtained from SPAFAS (Norwich, CT).

1. Plate the cells at $12 \times 10^7$ cells/ml in 10-cm tissue culture dishes and maintain in a 37°C $CO_2$ incubator in Dulbecco's modified Eagle's medium (DMEM) supplemented with 1% (v/v) chicken serum, 1% (v/v) defined calf

serum (DCS) (Hyclone Laboratories, Logan, UT), 2% (v/v) tryptose phosphate broth (Difco Laboratories, Detroit, MI), 100 units/ml penicillin, and 100 μg/ml streptomycin.

2. Cells are typically passaged with trypsin–EDTA every 3–4 days. For long-term storage, freeze CEFs in DMEM containing 16% (v/v) DCS, 10% (v/v) dimethyl sulfoxide (DMSO; Sigma, St. Louis, MO) in liquid nitrogen. Frozen cells with high viability on thawing can be obtained by simply wrapping vials in cotton and placing in a −70°C freezer overnight before transfer to liquid nitrogen.

3. Frozen cells are rapidly thawed in a 37°C water bath and the freezing medium diluted slowly with 10 volumes of growth medium.

4. Primary CEFs are not immortal; usually their growth slows after 6–8 wk. For this reason, freeze stocks of CEFs after they reach confluence during the first passage (3–5 days) and perform transfections and infections as soon as possible after thawing the first passage cells.

5. Make frozen stocks of cells expressing the recombinant protein of interest during early passage (2–3 wk after transfection).

## 2. Transfection of CEF with RCAS Vectors

Since cellular DNA synthesis is required for the stable integration of the retroviral vector into the host genome (Weiss *et al.*, 1984), transfection efficiency correlates with the percentage of CEFs undergoing cell cycling.

1. Split confluent 10-cm tissue culture dish of early passage CEFs (usually less than 10 days old) 1 : 5 and replate 12–24 hr before use so cells are ~40% confluent when transfected.

2. Perform transfection with recombinant RCAS BP retroviral constructs by the method of Kawai and Nishizawa (1984). Remove growth medium from the dish of CEFs, and add 30 μg recombinant retroviral DNA in 3 ml fresh complete growth medium containing 25 μg/ml polybrene (Aldrich Chemicals, Milwaukee, WI) to the center of the dish.

3. Place the dish in the 37°C $CO_2$ incubator for 6 hr, with gentle rocking of the dish each hour to insure widespread exposure of the cells to the DNA.

4. After the 6-hr incubation, remove the medium and shock the cells at room temperature for 5 min with 6 ml ice-cold complete growth medium containing 30% (v/v) DMSO.

5. Remove the shock medium and gently rinse the cells with 5 ml prewarmed (37°C) complete growth medium. Grow in 10 ml medium in the 37°C $CO_2$ incubator for an additional 2–3 days to allow integration of the transferred DNA.

6. At the first passage after transfection, split the cells 1 : 3 into 10 ml fresh growth medium. To accelerate the infection of the CEF culture by recombinant

virus produced by transfected cells, 5 ml media removed during the first passage is added to the freshly split cells. Typically, after 10–14 days of culture, virtually all cells express the foreign gene product.

### 3. Infection of CEF with RCAS Vectors

Spent culture supernatant containing infectious virus from CEFs expressing recombinant RCAS BP constructs can be used to infect or superinfect CEFs that have not been previously infected by a retrovirus of the same subgroup (Weiss *et al.*, 1984). This method is useful to obtain more rapid expression of a recombinant protein (typically within 7 days) in CEFs than is seen with transfection. Superinfection can be used to express two different recombinant proteins in the same cells, using retroviral vectors that have different subgroup specificities (see Section III,C,3).

1. Centrifuge 10 ml of spent culture supernatant from a confluent dish of CEFs, transfected with a RCAS BP construct and known to express high levels of the recombinant gene product, at 3000 *g* for 10 min and pass through a 0.45-$\mu$m filter (e.g., Uniflo; Schleicher & Schuell, Keene, NH) to remove cell debris.
2. Split a confluent dish of early passage (usually less than 10 days old), uninfected CEFs 1:4 into a 10-cm dish, and add the precleared retrovirus-containing supernatant to the cells along with 8 ml fresh growth medium.
3. After 3 days in culture, remove the medium and split the infected CEFs 1:3.
4. After an additional 3–4 days in culture, again passage the infected CEFs. Analyze for retroviral infection by the reverse transcriptase assay (see Section III,B,4) or for expression of the foreign cDNA by functional or biochemical analysis. It is important to use fresh virus stocks to infect cells, since repeated passage of virus (as opposed to infected cells) can lead to the accumulation of variants in which all or part of the foreign gene is deleted.

### 4. Reverse Transcriptase Assay

The *pol* gene carried by replication-competent retroviruses encodes reverse transcriptase (RT), an RNA-directed DNA polymerase, the activity of which is present in mature virions (Weiss *et al.*, 1984). CEFs that have been infected with retroviral supernatant or transfected with recombinant RCAS BP constructs can be assayed for viral infection by measuring the activity of RT in the supernatant. Similarly, the infectivity of retroviral supernatant to be used for infection can be estimated by this method (see Fig. 2). The RT assay exploits the capacity of the enzyme to synthesize DNA from an RNA template.

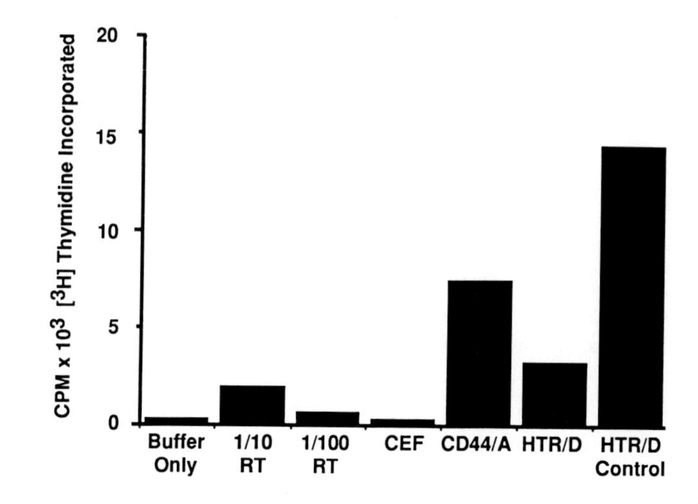


**Fig. 2** Reverse transcriptase assay of retroviral supernatants from chick embryo fibroblasts (CEF) 7 days postinfection. Uninfected CEF were incubated with supernatants from CEF transfected with recombinant retroviral constructs. After 7 days of culture, the supernatants from the infected CEF were assayed for reverse transcriptase activity. Supernatants from confluent dishes of CEF were cleared of cell debris and centrifuged at 35,000 rpm for 1 hr at 4°C in a Beckman SW40 Ti rotor. The concentrated virions were resuspended in 100 $\mu$l reaction buffer with 1 $\mu$Ci [$^3$H]TTP (10 $\mu M$) and incubated at 37°C for 1 hr. RT, Serial dilutions of 100 $\mu$l of reverse transcriptase (200 u/ml; BRL, Gaithersburg, MD); CEF, supernatant from uninfected CEF; CD44/A, supernatant from CEF infected with CD44/RCAS BP (subgroup A); HTR/D, supernatant from CEF infected with HTR/RCAS BP (subgroup D); HTR/D control, supernatant from CEF transfected with HTR/RCAS BP (subgroup D).

1. Centrifuge the supernatant from a confluent dish of CEFs at 3000 *g* and pass through a 0.45-$\mu$m filter to remove cell debris.

2. Concentrate the virions by centrifuging the precleared supernatant in a Beckman SW40 Ti rotor at 35,000 rpm for 1 hr at 4°C.

3. Resuspend the pelleted virions in 100 $\mu$l reaction buffer: 50 m$M$ Tris-Cl, 20 m$M$ dithiothreitol, 12 m$M$ $MgCl_2$, 60 m$M$ NaCl, 0.1% Nonidet P-40, 10 $\mu$g/ml poly(rA)-$dT)_{12-18}$ template–primer (Pharmacia, Piscataway, NJ), and 10 $\mu M$ [$^3$H]TTP (1 mCi/ml, 1 m$M$; Amersham, Arlington Heights, IL).

4. Incubate the reactions at 37°C for 1 hr. Use commercially available reverse transcriptase [Bethesda Research Laboratories (BRL), Gaithersburg, MD] as a positive control by incubating serial dilutions of the enzyme in reaction buffer free of virions.

5. Following the incubation, spot 10 $\mu$l reaction on a Whatman GF/C filter (2.4-cm diameter). Wash the filter with ice-cold 10% trichloroacetic acid and dry with a 70% ethanol wash followed by an acetone wash. Air-dry the filter and place in 5 ml Ecolume liquid scintillation cocktail (ICN Biomedicals, Irvine, CA); determine the counts per minute using a scintillation counter.

## C. Results

### 1. Expression of Human Transferrin Receptor

The utility of the RCAS retroviral expression system can be illustrated by our studies of the human TR expressed in CEFs. As shown in Fig. 3, 2 wk after transfection with RCAS containing a human TR cDNA, virtually all cells in a CEF culture are infected and express large amounts of the human TR on

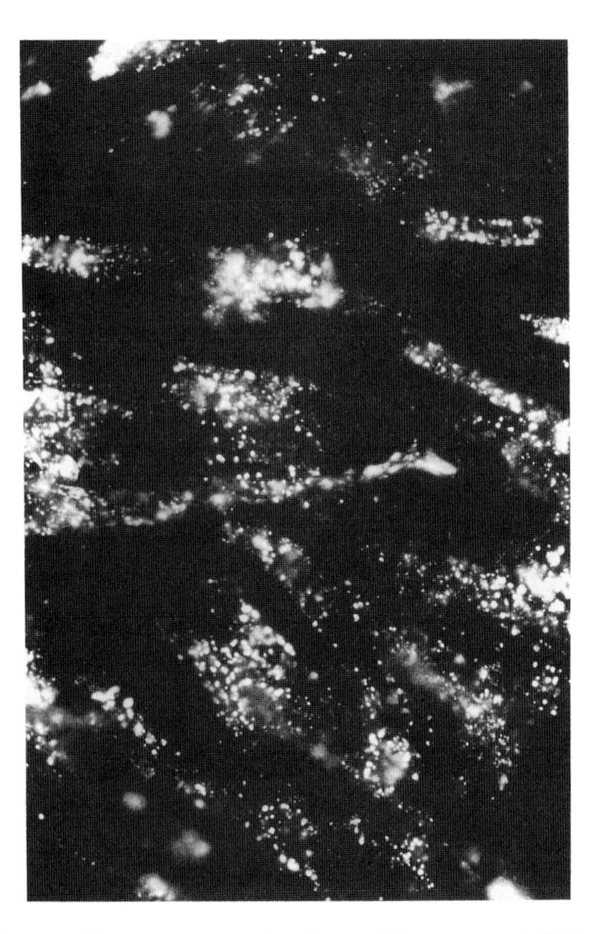

**Fig. 3** Indirect immunofluorescence analysis of human TR expression in CEF. CEF are plated at 70% confluence in tissue culture dishes containing No. 1 glass cover slips (sterilized by flaming with 70% ethanol) and cultured overnight. Cells are then stained for the human TR by incubating the cover slips with 100 $\mu$l culture supernatant containing B3/25 anti-human TR monoclonal antibody for 30 min at room temperature in a 30-mm culture dish. The cover slips are washed 3× by rinsing with 2 ml phosphate-buffered saline (PBS) containing 0.2% (w/v) bovine serum albumin and then incubated with FITC–rabbit anti-mouse Ig (Chappel, Durham, NC) for 30 min at room temperature. The cover slips are washed 3× again, mounted in 10% (v/v) PBS–90% glycerol containing 0.1% *p*-phenylenediamine (Johnson *et al.*, 1981), and sealed with nail polish.

their cell surface (see also Fig. 6). Similar results have been obtained for many mutant TR molecules, human β-amyloid precursor protein, and CD44; as described earlier (Section II,B), the major limitation on expression is the size of the foreign cDNA that can be inserted into the RCAS vector and still yield infectious virus. All the molecules we have expressed using RCAS BP are produced in amounts on the order of $1–5 \times 10^5$ molecules per cell.

## 2. Biochemical and Functional Studies of Human Transferrin Receptor

Using the RCAS expression system it has been possible to perform a variety of biochemical and functional studies of the human TR, including defining

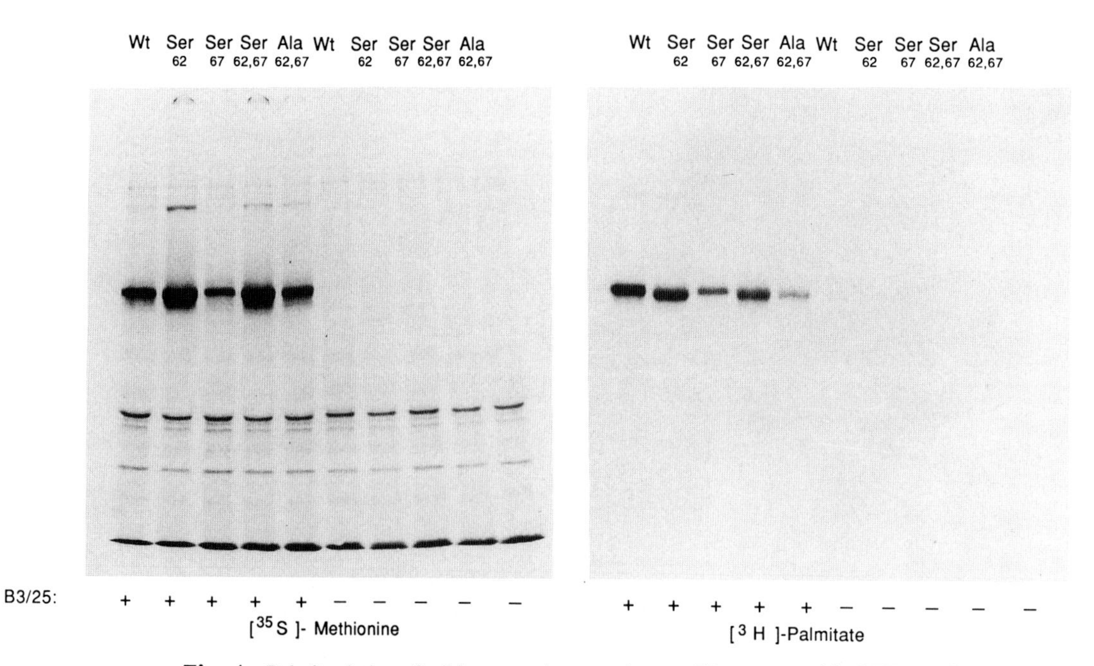


**Fig. 4** Palmitoylation of wild-type and mutant human TRs expressed in CEF. Replicate cultures of CEF expressing wild-type or mutant human TRs were metabolically labeled with [$^{35}$S]methionine and [$^3$H]palmitate. For metabolic labeling, 10-cm tissue culture dishes of cells at ~80% confluency were labeled for 4 hr with either 100 μCi/ml L-[$^{35}$S]methionine (translabel >1000 Ci/mmol; ICN, Irvine, CA) or 9,10-[$^3$H]palmitic acid (54 Ci/mmol; Amersham, Arlington Heights, IL) in a total volume of 3.0 ml of either methionine-free DMEM for methionine labeling or DMEM with 1 m*M* pyruvate and 1× nonessential amino acids for palmitate labeling, supplemented with 2% dialyzed fetal calf serum. Cell lysates were prepared and human TRs immunoprecipitated from equal cell equivalents using B3/25 monoclonal antibody. Immunopercipitates were then analyzed by SDS–polyacrylamide gel electrophoresis on 7.5% acrylamide gels. Fluorography was for 16 hr for [$^{35}$S]methionine-labeled TRs and 3 days for [$^3$H]palmitate-labeled TRs. Data from Jing and Trowbridge (1990). Reproduced with permission from the *Journal of Biological Chemistry*. The small amount of radioactivity (19% of wildtype) incorporated into the Ala 62,67 mutant TRs from [$^3$H]palmitate was resistant to hydroxylamate treatment and, therefore, is the result of metabolic conversion of palmitate to amino acids (Jing and Trowbridge, 1990).

the internalization signal within the cytoplasmic domain of the receptor. We illustrate here the sort of data that can be obtained from studies of the functional significance of the lipid moiety of the human TR (Jing and Trowbridge, 1990). The human TR was shown some years ago to be palmitolated; by expression of mutant TR molecules in monkey Cos cells, Cys 62 was shown to be the major acylation site (Omary and Trowbridge, 1981; Jing and Trowbridge, 1987). However, it was not possible to investigate whether acylation of the human TR was important for internalization or recycling of the receptor because of the functional endogenous receptors expressed on Cos cells. Consequently, mutant TRs in which the two cysteine residues proximal to the cytoplasmic face of the plasma membrane (Cys 62 and Cys 67) were modified either independently or together to serine or alanine were expressed in CEFs using the RCAS vector. Metabolic labeling studies with [$^{35}$S]methionine or [$^{3}$H]palmitate were then performed that demonstrated that both cysteine residues were acylated in CEFs and that a serine residue at position 67 but not position 62 could also serve as a lipid-attachment site (Fig. 4). Sufficient amounts of the wild-type and mutant TRs were synthesized by the infected cells that autoradiographs of the palmitylated receptors could be obtained by exposures of only a few days (Fig. 4).

By altering both Cys 62 and Cys 67 to alanine, mutant TRs were produced that lacked a lipid moiety. Functional analysis of these mutant receptors was

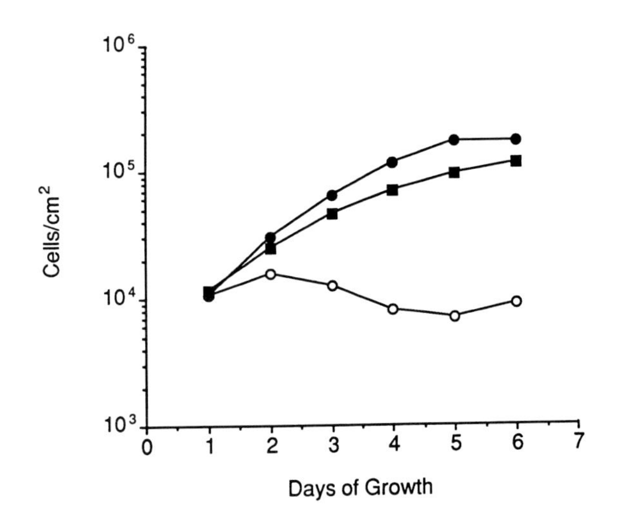


**Fig. 5** Growth of CEF (○) expressing wild-type (●) and Ala$^{62,67}$ (■) mutant receptors in tissue culture medium supplemented with human transferrin. Cells were plated at a cell density of $1.5 \times 10^4$ cells/cm$^2$ in 6-well Costar cluster dishes (Cambridge, MA) containing 3 ml DMEM supplemented with 3% (v/v) horse serum, 2% tryptose phosphate broth, and 50 $\mu$g/ml diferric human transferrin (Miles Scientific, Naperville, IL). Cells were removed each day from triplicate dishes with 0.05% trypsin in versene buffer and counted in a Coulter counter. Data from Jing and Trowbridge (1990). Reproduced with permission from the *Journal of Biological Chemistry*.

possible in CEFs because, as discussed earlier, ovotransferrin receptors do not bind human transferrin. The Ala[62,67] mutant receptors were shown to be internalized as efficiently as wild-type receptors (Jing and Trowbridge, 1990), and were able to supply sufficient iron from human transferrin for CEFs displaying the mutant receptors to grow as rapidly as those expressing wild-type human TRs (Fig. 5).

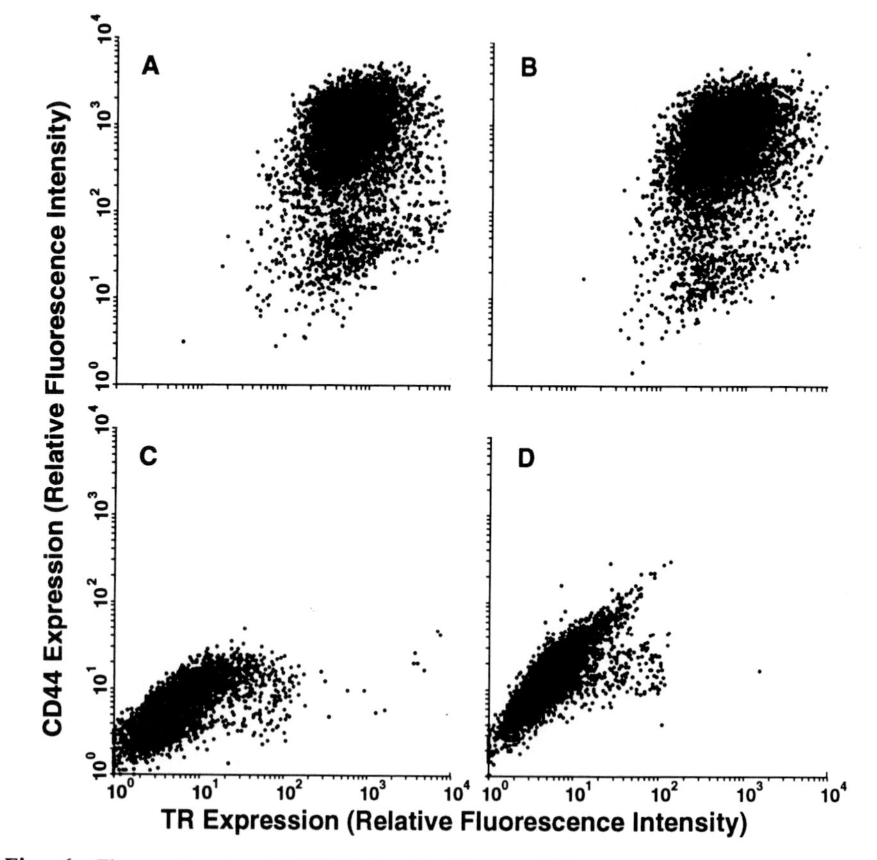


**Fig. 6** Flow cytometry of CEF infected with two different recombinant proteins using RCAS BP retroviral vectors that have different subgroup specificities. Retroviral supernatant from CEF transfected with murine CD44/RCAS BP (subgroup A) was incubated with CEF transfected with human transferrin receptor (TR)/RCAS BP (subgroup D) and vice versa. The number of cells in a single population of CEF expressing both recombinant proteins was determined by fluorescence activated cell sorting using a mouse monoclonal antibody against murine CD44 (IM7) directly conjugated with fluorescein isothiocyanate and rabbit polyclonal antisera against human TR plus goat anti-rabbit conjugated to R-phycoerythrin (Molecular Probes, Eugene, OR). (A) CEF transfected with CD44/RCAS BP(A), superinfected with TR/RCAS BP(D) and then stained for CD44 and TR. (B) CEF transfected with TR/RCAS BP(D), superinfected with CD44/RCAS BP(A) and then stained for CD44 and TR. (C) Negative control; CEF transfected with CD44/RCAS BP(A) stained for TR. (D) Negative control: CEF transfected with TR/RCAS BP(D) stained for CD44.

### 3. Co-expression of a Second Recombinant Protein by Superinfection with an RCAS Vector of Different Subgroup

The host range of RCAS vectors has been expanded by the replacement of the *env* gene encoding subgroup A specificity by the corresponding regions from subgroup B and subgroup D viruses (Hughes *et al.*, 1987). The availability of vectors with different subgroup specificities makes it possible to coexpress different recombinant proteins within the same host cell. We have used this system to coexpress the human TR and murine CD44 in CEFs. The strategy involves individual transfection of CEFs with either CD44/RCAS BP(A) or HTR/RCAS BP(D) recombinant vector constructs. Supernatant from each transfected line of CEFS was assayed for RT activity 2 wk after transfection; the retrovirus-containing supernatant from CD44/RCAS BP(A) transfectants was used to infect HTR/RCAS BP(D) transfectants, and vice versa. Each population of CEFs was analyzed by fluorescence-activated cell sorting 1 wk after superinfection to detect coexpression of CD44 and HTR on the cell surface of individual cells. As shown in Fig. 6, high expression of both molecules was obtained on all cells, regardless of whether the A or D subtype RCAS vector was used for superinfection of CEFs infected with RCAS of the other subtype.

## Acknowledgments

This work was supported by Grant CA34787 from the National Cancer Institute. We especially thank Dr. Stephen Hughes (Frederick Cancer Research and Development Centre, Frederick, MD) for providing the RCAS vectors used in this work, often prior to publication. We also thank Joe Trotter for help with flow cytometry.

## References

Albritton, L. M., Tseng, L., Scadden, D., and Cunningham, J. M. (1989). A putative murine ecotropic retrovirus receptor gene encodes a multiple membrane-spanning protein and confers susceptibility to virus infection. *Cell* **57,** 659–666.

Bates, P., Young, J. A. T., and Varmus, H. E. (1993). A receptor for subgroup A Rous sarcoma virus is related to the low density lipoprotein receptor. *Cell* **74,** 1043–1051.

Boerkoel, C. F., Federspiel, M. J., Salter, D. W., Payne, W., Crittenden, L. B., Kunk, H.-J., and Hughes, S. H. (1993). A new defective retroviral vector system based on the Bryan strain of Rous sarcoma virus. *Virology* **195,** 669–679.

Collawn, J. F., Stangel, M., Kuhn, L. A., Esekogwu, V., Jing, S., Trowbridge, I. S., and Tainer, J. A. (1990). Transferrin receptor internalization sequence YXRF implicates a tight turn as the structural recognition motif for endocytosis. *Cell* **63,** 1061–1072.

Collawn, J. F., Kuhn, L. A., Liu, L.-F. S., Tainer, J. A., and Trowbridge, I. S. (1991). Transplanted LDL and mannose-6-phosphate receptor internalization signals promote high-efficiency endocytosis of the transferrin receptor. *EMBO J.* **10,** 3247–3253.

Collawn, J. F., Lai, A., Domingo, D., Fitch, M., Hatton, S., and Trowbridge, I. S. (1993). YTRF is the conserved internalization signal of the transferrin receptor, and a second YTRF signal at position 31–34 enhances endocytosis. *J. Biol. Chem.* **268,** 1–7.

Czernilofsky, A. P., Levinson, A. D., Varmus, H. E., Bishop, J. M., Tischer, E., and Goodman, H. M. (1980). Nucleotide sequence of an avian sarcoma virus oncogene (*src*) and proposed amino acid sequence for gene product. *Nature (London)* **287,** 198–203.

Czernilofsky, A. P., Levinson, A. D., Varmus, H. E., Bishop, J. M., Tischer, E., and Goodman, H. M. (1983). Corrections to the nucleotide sequence of the *src* gene of Rous sarcoma virus. *Nature* (*London*) **301,** 736–738.

Dolberg, D. S., and Bissell, M. J. (1984). Inability of Rous sarcoma virus to cause sarcomas in the avian embryo. *Nature* (*London*) **309,** 552–556.

Dorner, A. J., and Coffin, J. M. (1986). Determinants for receptor interaction and cell killing on the avian retrovirus glycoprotein gp85. *Cell* **45,** 365–374.

Dorner, A. J., Stoye, J. P., and Coffin, J. M. (1985). Molecular basis of host range variation in avian retroviruses. *J. Virol.* **53,** 32–39.

Greenhouse, J. J., Petropoulos, C. J., Crittenden, L. B., and Hughes, S. H. (1988). Helper-independent retrovirus vectors with Rous-associated virus type O long terminal repeats. *J. Virol.* **62,** 4809–4812.

Howlett, A. R., Cullen, B., Hertle, M., and Bissell, M. J. (1987). Tissue tropism and temporal expression of Rous sarcoma virus in embryonic avian limb *in ovo*. *Oncogene Res.* **1,** 255–263.

Hughes, S., and Kosik, E. (1984). Mutagenesis of the region between *env* and *src* of the SR-A strain of Rous sarcoma virus for the purpose of constructing helperindependent vectors. *Virology* **136,** 89–99.

Hughes, S. H., Greenhouse, J. J., Petropoulos, C. J., and Sutrave, P. (1987). Adaptor plasmids simplify the insertion of foreign DNA into helper-independent retroviral vectors. *J. Virol.* **61,** 3004–3012.

Hughes, S. H., Petropoulos, C. J., Federspiel, M. J., Sutrave, P., Forry-Schaudies, S., and Bradac, J. A. (1990). Vectors and genes for improvement of animal strains. *J. Reprod. Fert. Suppl.* **41,** 39–49.

Jing, S., and Trowbridge, I. (1987). Identification of the intramolecular disulfide bonds of the human transferrin receptor and its lipid attachment site. *EMBO J.* **6,** 327–331.

Jing, S., and Trowbridge, I. S. (1990). Nonacylated human transferrin receptors are rapidly internalized and mediate iron uptake. *J. Biol. Chem.* **265,** 11555–11559.

Jing, S., Spencer, T., Miller, K., Hopkins, C., and Trowbridge, I. S. (1990). Role of the human transferrin receptor cytoplasmic domain in endocytosis; Localization of a specific signal sequence for internalization. *J. Cell Biol.* **110,** 283–294.

Johann, S. V., Gibbons, J. J., and O'Hara, B. (1992). GLVR1, a receptor for gibbon ape leukemia virus, is homologous to a phosphate permease of *Neurospora crassa* and is expressed at high levels in the brain and thymus. *J. Virol.* **66,** 1635–1640.

Johnson, G. D., and de C. Nogueira Araujo, G. M. (1981). A simple method of reducing the fading of immunofluorescence during microscopy. *J. Immunol. Methods* **43,** 349–350.

Kawaguchi, T., Nomura, K., Hirayama, Y., and Kitagawa, T. (1987). Establishment and characterization of a chicken hepatocellular carcinoma cell line, LMH[1]. *Cancer Res.* **47,** 4460–4464.

Kawai, S., and Nishizawa, M. (1984). New procedure for DNA transfection with polycation and dimethyl sulfoxide. *Mol. Cell. Biol.* **4,** 1172–1174.

Kim, J. W., Closs, E. I., Albritton, L. M., and Cunningham, J. M. (1991). Transport of cationic amino acids by the mouse ecotropic retrovirus receptor. *Nature* (*London*) **352,** 725–728.

Maddon, P. J., Dalgleish, A. G., McDougal, J. S., Clapham, P. R., Weiss, R. A., and Axel, R. (1986). The T4 gene encodes the AIDS virus receptor and is expressed in the immune system and the brain. *Cell* **47,** 333–348.

Miller, K., Shipman, M., Trowbridge, I. S., and Hopkins, C. R. (1991). Transferrin receptors promote the formation of clathrin lattices. *Cell* **65,** 621–632.

Morgan, B. A., Izpisa-Belmonte, J.-C., Duboule, D., and Tabin, C. J. (1992). Targeted misexpression of *Hox-4.6* in the avian limb bud causes apparent homeotic transformations. *Nature* **358,** 236–239.

Moscovici, C., Giovannela Moscovici, M., Jiminez, H., Lai, M. M. C., Haymann, M. J., and Vogt, P. K. (1977). Continuous tissue culture cell lines derived from chemically induced tumors of Japanese quail. *Cell* **11,** 95–103.

Nemeth, S. P., Fox, L. G., DeMarco, M., and Brugge, J. S. (1989). Deletions within the amino terminal half of the c-*src* gene product that alter the functional activity of the protein. *Mol. Cell. Biol.* **9,** 1109–1119.

O'Hara, B., Johann, S. V., Klinger, H. P., Blair, D. G., Rubinson, H., Dunn, K. J., Sass, P., Vitek, S. M., and Robins, T. (1990). Characterization of a human gene conferring sensitivity to infection by gibbon ape leukemia virus. *Cell Growth Diff.* **1,** 119–127.

Petropoulos, C. J., and Hughes, S. H. (1991). Replication-competent retrovirus vectors for the transfer and expression of gene cassettes in avian cells. *J. Virol.* **65,** 3728–3737.

Petropoulos, C. J., Payne, W., Salter, D. W., and Hughes, S. H. (1992). Appropriate in vivo expression of a muscle-specific promoter by using avian retroviral vectors for gene transfer. *J. Virol.* **66,** 3391–3397.

Piraino, F. (1967). The mechanism of genetic resistance of chick embryo cells to infection by Rous sarcoma virus-Bryan strain (BS-RSV). *Virology* **32,** 700–707.

Shimo-Oka, T., Hagiwara, Y., and Ozawa, E. 91986). Class specificity of transferrin as a muscle trophic factor. *J. Cell. Physiol.* **126,** 341–351.

Stoker, A. W., Hatier, C., and Bissell, M. J. (1990). The embryonic environment strongly attenuates v-*src* oncogenesis in mesenchymal and epithelial tissues, but not in endothelia. *J. Cell Biol.* **111,** 217–228.

Sudol, M. T., Lerner, T. L., and Hanafusa, H. (1986). Polymerase-defective mutant of the Bryant high-titer $\pi$ strain of the Rous sarcoma virus. *Nucleic Acids Res.* **14,** 2391–2405.

Trowbridge, I. S., and Omary, M. B. (1981). Human cell surface glycoprotein related to cell proliferation is the receptor for transferrin. *Proc. Natl. Acad. Sci. USA* **78,** 3039–3043.

Trowbridge, I. S., Collawn, J. F., and Hopkins, C. R. (1993). Signal-dependent membrane protein trafficking in the endocytic pathway. *Annu. Rev. Cell Biol.* **9,** 129–161.

Vogt, P. K. (1965). A heterogeneity of Rous sarcoma virus revealed by selectively resistant chick embryo cells. *Virology* **25,** 237–247.

Wang, H., Kavanaugh, M. P., North, R. A., and Kabat, D. (1991). Cell surface receptor for ecotropic murine retroviruses is a basic amino acid transporter. *Nature* (*London*) **352,** 729–731.

Weiss, R., Teich, N., Varmus, H., and Coffin, J. (1984). *RNA Tumor Viruses.* Vol. 1. Cold Spring Harbor.

Weiss, R., Teich, N., Varmus, H., and Coffin, J. (1985). *RNA Tumor Viruses.* Vol. 2. Cold Spring Harbor.

White, S., Taetle, R., Seligman, P. A., Rutherford, M., and Trowbridge, I. S. (1990). Combinations of anti-transferrin receptor monoclonal antibodies inhibit human tumor cell growth *in vitro* and *in vivo:* Evidence for synergistic antiproliferative effects. *Cancer Res.* **50,** 6295–6301.

White, S., Miller, K., Hopkins, C., and Trowbridge, I. S. (1992). Monoclonal antibodies against defined epitopes of the human transferrin receptor cytoplasmic tail. *Biochim. Biophys. Acta* **1136,** 28–34.

Young, J. A. T., Bates, P., and Varmus, H. E. (1993). Isolation of a chicken gene that confers susceptibility to infection by subgroup A avian leukosis and sarcoma viruses. *J. Virol.* **67,** 1811–1816.

# CHAPTER 5

# Generation of High-Titer Pseudotyped Retroviral Vectors with Very Broad Host Range

**Jiing-Kuan Yee,* Theodore Friedmann,† and Jane C. Burns†**

* Department of Pediatrics
City of Hope
Duarte, California 91010

† Department of Pediatrics
Center for Molecular Genetics
University of California, San Diego
La Jolla, California 92093



## I. Introduction

Retroviral vectors derived from Moloney murine leukemia virus (MoMLV) mediate efficient gene transfer into mammalian cells and are approved for use in human gene therapy trials (reviewed by Miller, 1992; Mulligan, 1993). The successful use of retroviral vectors for gene transfer is related to the availability of packaging cell lines that constitutively express retrovirus-encoded gag, pol, and env proteins (Mann *et al.*, 1983; Miller and Buttimore, 1986). Replication-defective retroviral vectors free of wild-type virus are produced when the genome of a retroviral vector is introduced into such packaging cells. Current approaches to human clinical trials or to the evaluation of preclinical models of gene therapy have focused primarily on introducing foreign genes into appropriate primary cells *in vitro*, followed by grafting the genetically modified cells into a physiologically relevant site in humans or animals. Direct gene transfer into target tissues or organs *in vivo* is limited by several disadvantages of retroviral vectors, which include the inability of the retroviral vectors to infect nondividing cells, the low virus titers generated from packaging cells, and the poor efficiency of infection in certain human cells (Adams *et al.*, 1992). The low virus titers can be partially attributed to the instability of the virus-encoded envelope protein, and all attempts to concentrate retroviruses by physical methods such as ultracentrifugation and ultrafiltration have resulted in a severe loss of the infectious virus.

To overcome some of these problems, we generated pseudotyped retroviral vectors with an altered host range (Emi *et al.*, 1991). Researchers have shown that mixed infection of a cell by vesicular stomatitis virus (VSV) and retroviruses results in the production of progeny virions bearing the genome of one virus encapsidated by the envelope proteins of the other (Weiss *et al.*, 1977; Witte and Baltimore, 1977). This phenomenon, termed "pseudotype formation," occurs in cells co-infected with VSV and viruses such as retrovirus, herpes simplex virus, paramyxovirus, or togavirus (reviewed by Weiss, 1980). Whereas the retroviral envelope protein must bind to a specific cell surface protein receptor to mediate successful infection (Albritton *et al.*, 1989), the G glycoprotein of VSV interacts with a phospholipid component of the plasma membrane (Mastromarino *et al.*, 1977). Since virus entry seems not to be dependent on the presence of specific protein receptors, VSV has an extremely broad host range and can infect almost all species of vertebrate and insect cells. In an attempt to broaden the host range and to increase the stability and the titer of the infectious retroviral vector, we created retroviral pseudotypes and demonstrated that the VSV G protein can be incorporated efficienctly into the virions of a retroviral vector in the absence of other VSV-encoded proteins (Emi *et al.*, 1991). The pseudotyped retroviral vectors not only have an altered host range characteristic of VSV, but also infect some human cells more efficiently than retroviral vectors bearing the amphotropic envelope protein. Most importantly, the VSV G-pseudotyped retroviral vectors can be concentrated at least 2000-fold by ultracentrifugation without significant loss in infectivity (Burns *et al.*, 1993). Using the VSV G-pseudotyped retroviral vectors, we have shown that foreign genes can be efficiently delivered into mammalian as well as nonmammalian cells such as those of *Xenopus*, insects, and fish (Burns *et al.*, 1993). Preparations of the high-titer VSV G-pseudotyped retroviral vectors should also allow investigators to test the possibility of direct *in vivo* gene delivery into target tissues in animals or humans. This chapter will discuss the procedures to generate such pseudotyped retroviral vectors and to use these vectors to infect nonmammalian cells.

## II. Principle of Method

All attempts to obtain stable cell lines expressing the VSV G protein have been unsuccessful because expression of the fusogenic VSV G protein on the cell surface causes syncytium formation and subsequent cell death in the several mammalian cell lines tested. It is therefore necessary to produce virus by transient expression of the gene encoding VSV-G in transfected cells. To generate high-titer stocks of retroviral vectors, the retroviral vector of interest is first introduced by infection into the 293GP cell line derived from the human Ad-5-transformed embryonal kidney cell line 293, stably expressing the gag and pol proteins of MoMLV (Fig. 1; Burns *et al.*, 1993). Retroviral vectors generated from such cells are noninfectious since the virions contain none of the envelope protein required for cell entry. To generate infectious pseudotyped retroviral vectors, plasmid pHCMV-G (Fig. 1) containing the VSV-G gene under the control of the strong immediate early promoter of human cytomegalovirus (HCMV) is transfected into 293GP cells harboring the genome of the retroviral vector of interest (Fig. 1). The VSV G-pseudotyped retroviral vector produced in this transient transfection procedure is collected over a period of 3–4 days before death of the producer cells due to syncytium formation. Since approximately 90% of the 293GP cells can be transiently transfected with pHCMV-G as determined by the expression of VSV G protein on the cell surface, we have been able to obtain retroviral vectors with a titer of greater than $10^6$ colony forming units (cfu)/ml with this approach. The pseudotyped retroviral vectors can be concentrated further by ultracentrifugation without significant loss of infectious virus titer. With two cycles of ultracentrifugation, we recently concentrated a pseudotyped retroviral vector to a titer greater than $10^9$ cfu/ml (Burns *et al.*, 1993).

To simplify the description of generating high-titer pseudotyped retroviral vectors, we will use a retroviral construct, plasmid pLZRNL (Fig. 2; Xu *et al.*, 1989), as an example throughout this chapter. Plasmid pLZRNL contains the gene encoding *Escherichia coli* β-galactosidase (LacZ) controlled by the 5′ long terminal repeat (LTR) of Moloney murine sarcoma virus followed by the gene encoding neomycin phosphotransferase (Neo) controlled by the promoter of Rous sarcoma virus (RSV). Cells infected with LZRNL will generate two virus-specific transcripts: the transcript initiated from the 5′LTR encodes β-galactosidase whereas the transcript initiated from the internal RSV promoter encodes Neo; both transcripts are polydenylated at the 3′LTR.

## III. Methods

### A. Generation of Pseudotyped Retroviral Vectors

#### 1. Cell Lines, Reagents, and Solutions

PA317 (Miller and Buttimore, 1986), 293GP (Burns *et al.*, 1993), and 208F (Quade, 1979) cells are grown in Dulbecco's modified Eagle's medium (DMEM) with high glucose supplemented with 10% fetal calf serum (FCS). PA317 cells may be obtained from the American Type Culture Collection (ATCC, Rockville, MD). 293GP cells may be obtained from Dr. Douglas D. Jolly (Viagen Inc.,

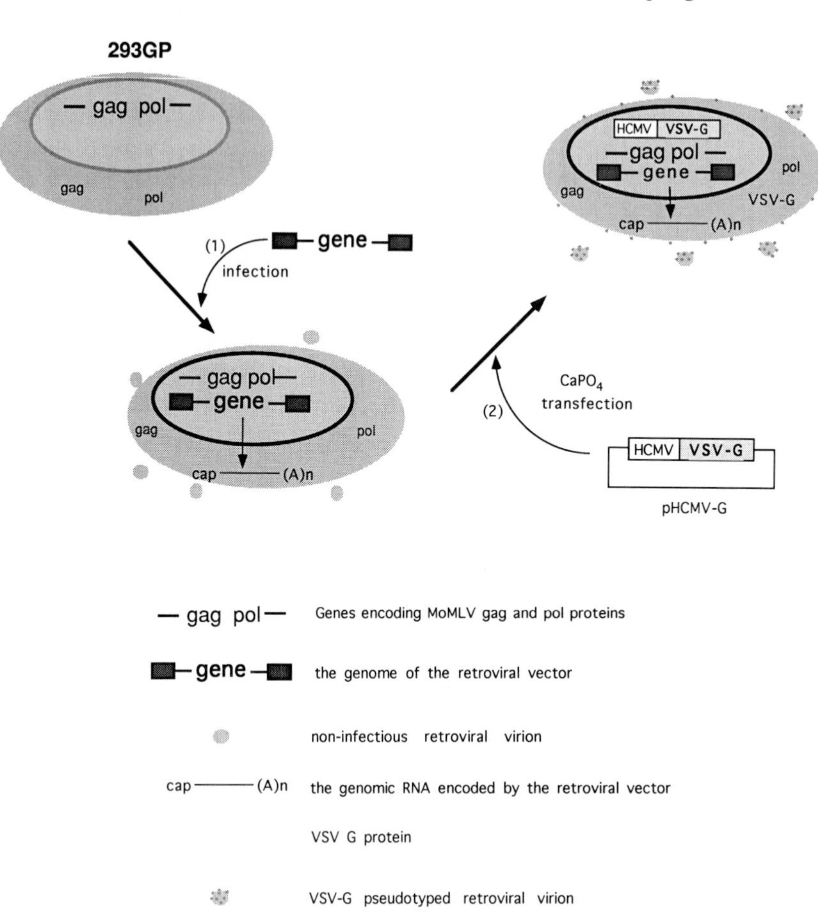


**Fig. 1** Schematic diagram outlining the generation of the pseudotyped retroviral vector from 293GP cells. (1) The retroviral vector (■) derived from the amphotropic packaging line PA317 cells is used to infect 293GP cells and the 293GP clone harboring the integrated retroviral vector is picked for subsequent production of the pseudotyped virus as described in the text. The closed boxes in the retroviral genome represent the LTRs. Virions containing the retroviral vector of interest can be generated from these clones because of the presence of the gag and pol proteins of MoMLV in 293GP cells as indicated by "gag" and "pol" in the cytoplasm of 293GP cells. However, these virions are noninfectious ( ○ ) because of the absence of the envelope protein on the cell surface. (2) To produce the pseudotyped virus, expression plasmid pHCMV-G for the VSV G protein (·) is transfected by calcium phosphate co-precipitation into the 293GP clone harboring the retroviral vector of interest. Infectious pseudotyped virus ( ⊛ ) is collected 24–96 hr after transfection as described in the text.

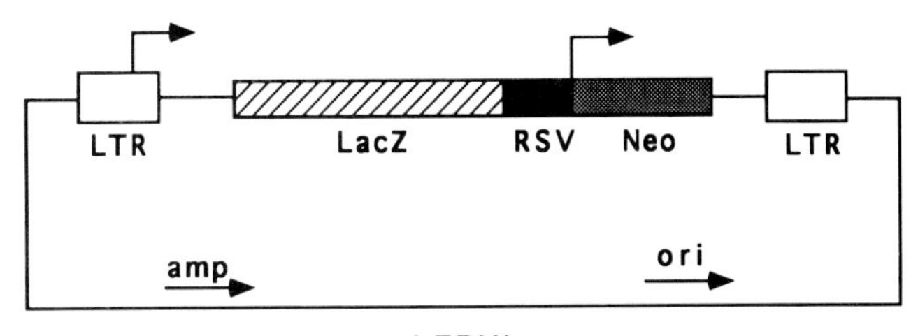

**Fig. 2** The structure of plasmid pLZRNL as described previously (Xu *et al.*, 1989). Open boxes represent the LTRs derived from murine leukemia virus. The hatched box represents the gene encoding *E. coli* β-galactosidase. The closed box represents the promoter derived from Rous sarcoma virus and the shaded box represents the gene derived from transposon Tn5 encoding neomycin phosphotransferase. The arrows above the 5 LTR and the RSV promoter indicate the approximate positions of transcription initiation sites in the retroviral vector.

San Diego, CA). 208F cells may be obtained from the authors, or NIH 3T3 Cells may be used for determination of virus titers.

Polybrene (H-9268; Sigma, St. Louis, MO) stock solution (4 mg/ml), prepared by dissolving 0.4 g hexadimethrine bromide in 100 ml sterile water, and stored at 4°C

G418 stock solution (40 mg active component/ml), prepared by dissolving 4 g active Geneticin (GIBCO/BRL, Grand Island, NY) in 100 ml 0.1 *M* HEPES, pH 7.9, sterilized by filtering through a 0.2-μm filter, and stored at 4°C

Phosphate-buffered saline (PBS) (137 m*M* NaCl/2.6 m*M* KCl/8.1 m*M* $Na_2HPO_4$/1.5 m*M* $KH_2PO_4$), prepared by dissolving 8 g NaCl, 0.2 g KCl, 1.15 g $Na_2HPO_4$ and 0.2 g $KH_2PO_4$ in 1 liter of water, autoclaved, and stored at room temperature

Cell lysis solution (250 m*M* Tris-HCL, pH 7.8), prepared by mixing one part 1 *M* Tris-HCl, pH 7.8, with three parts sterile water

β-Gal buffer (50 m*M* Tris-HCl, pH 7.5/100 m*M* NaCl/10 m*M* $MgCl_2$), prepared by mixing 5 ml 1 *M* Tris-HCl, pH 7.8, 2 ml 5 *M* NaCl, 1 ml 1 *M* $MgCl_2$, and 92 ml water

*O*-Nitrophenyl-β-galactopyranoside (ONPG), obtained from Sigma

10 m*M* EDTA solution, prepared by diluting 1 *M* EDTA stock 1 : 100 with water, autoclaved, and stored at room temperature

## 2. Procedures

1. Day 1. Approximately $5 \times 10^5$ PA317 cells are seeded in a 100-mm tissue culture plate.

2. Day 2. The PA317 cells are transfected using standard calcium phosphate co-precipitation (Graham and Van der Eb, 1973) with 20 μg pLZRNL DNA (Fig. 1).

3. Day 3. Approximately 1 × $10^5$ 293GP cells are seeded in a 100-mm tissue culture plate 24 hr before the harvest of the amphotropic virus from the transfected PA317 cells.

4. Day 4. The culture medium from pLZRNL-transfected PA317 cells is filtered through a 0.45-μm filter 48 hr after transfection, and is used to infect 293GP cells in the presence of 8 μg/ml polybrene.

5. Day 5. The culture medium of the infected 293GP cells is replaced with fresh medium containing 400 μg/ml G418 16 hr after infection, and the medium is changed every 3 days until G418-resistant colonies appear 2 wk later. Since 293GP cells attach to tissue culture plates rather loosely, care must be taken not to disturb the G418-resistant colonies during the change of culture medium.

6. The G418-resistant 293GP colonies are picked with an automatic pipettor and are transferred directly into 24-well plates without the aid of trypsin solution.

7. To choose the 293GP/LZRNL clone that will produce the highest titer of pseudotyped LZRNL virus, the clones must be screened for the expression of the *LacZ* gene. The clones in the 24-well plates are transferred into 100-mm tissue culture plates. Protein extracts from confluent plates are prepared by first washing the cell monolayer once with 10 ml PBS; the cells are scraped off in 2 ml 250 m*M* Tris-HCl, pH 7.8, with a rubber policeman. The cells are pelleted by low-speed centrifugation at room temperature and are resuspended in 100 μl 250 m*M* Tris-HCl, pH 7.8. The cells are subjected to four rapid freeze/thaw cycles and the cell debris is removed by low-speed centrifugation at room temperature.

8. The activity of β-galactosidase in the protein extracts is determined by mixing 5 μl cell extract with 500 μl β-gal buffer containing 0.75 mg/ml ONPG. The mixture is incubated at 37°C until yellow color appears; the reaction is stopped by addition of 500 μl 10 m*M* EDTA solution. The optical density of the reaction is determined at a wavelength of 420 nm.

9. The 293GP/LZRNL clone generating the highest amount of β-galactosidase activity is then expanded and used for subsequent production of the pseudotyped LZRNL virus.

10. To generate pseudotyped LZRNL virus, 1 × $10^6$ 293GP/LZRNL cells seeded 24 hr earlier in a 100-mm tissue culture plate are transfected with 20μg pHCMV-G using the method of calcium phosphate co-precipitation. The calcium–DNA precipitate is replaced with fresh culture medium without G418 6–8 hr after transfection. Overnight transfection will result in the detachment of most of the 293GP/LZRNL cells from the plate and therefore should be avoided.

11. The pseudotyped virus generated from the transfected 293GP/LZRNL cells can be collected at least once a day between 24 and 96 hr after transfection. We have observed that the highest virus titer is generated approximately 48–72 hr after initial pHCMV-G transfection. Although syncytium formation in the

majority of the transfected cells becomes visible 48 hr after transfection, the cells continue to generate the pseudotyped virus for at least another 48 hr as long as they remain attached to the tissue culture plate.

12. The collected culture medium is pooled, filtered through a 0.45-$\mu$m filter, and stored at $-70°C$.

13. To determine virus titer, $5 \times 10^5$ rat 208F fibroblasts (or NIH 3T3 cells) seeded in a 100-mm culture plate 24 hr before infection are infected with serial dilutions of the LZRNL virus in the presence of 8$\mu$g/ml polybrene. The medium is replaced with fresh medium containing 400 $\mu$g/ml G418 16 hr after infection, and selection is continued for 14 days until G418-resistant colonies become visible.

## 3. Comments

1. Since 293GP cells do not always attach to the tissue culture plates quickly and firmly, it may be necessary to seed the cells 48 hr before transfection with pHCMV-G to allow sufficient time for cells to attach.

2. Plasmid pHCMV-G ranging from 10 to 40 $\mu$g has been transfected into 293GP/LZRNL cells under the conditions just described above, and the titers of the virus generated varied less than 2-fold. Since transfection of 10$\mu$g pHCMV-G consistently produced the lowest titer of pseudotyped virus whereas transfection of 40 $\mu$g pHCMV-G only slightly increases the virus titer, we chose to transfect 20 $\mu$g pHCMV-G routinely into 293GP/LZRNL cells for generation of pseudotyped virus.

3. An alternative method for identifying the 293GP clone that will generate the highest titer of the retroviral vector of interest is to transfect each individual 293GP clone with pHCMV-G. The titer of pseudotyped virus conferring G418 resistance generated 48 hr after transfection is then determined on 208F cells, as just described. This method is especially useful when there is no convenient system for assaying the expression of the gene of interest in 293GP cells.

4. We have determined the infectivity of the pseudotyped virus in the presence of various polycations—including polybrene, protamine sulfate, and poly-L-lysine—that are important for viral attachment and cell entry. Infection as measured by the number of G418-resistant colonies for either the pseudotyped virus or the conventional retroviral vectors containing the amphotropic envelope protein is most efficient in the presence of 8 $\mu$g/ml polybrene (Burns *et al.*, 1993). Substitution of protamine sulfate or poly-L-lysine results in a several-fold decrease in virus titer. Protamine sulfate, however, is a drug that has been approved for human use and for human gene therapy model studies. It is also less toxic than polybrene to some primary cells such as human peripheral blood mononuclear cells and primary human leukemic T cells. Omission of polycations during infection results in a 100- to 1000-fold decrease in infection efficiency (Burns *et al.*, 1993).

5. We have compared the stability of the pseudotyped virus with that of the conventional virus containing the amphotropic envelope protein under different environmental conditions. Our results indicate that, despite the enhanced stability of the pseudotyped virus during ultracentrifugation, the difference in the envelope proteins does not significantly affect the stability of the two viruses at different temperatures or following multiple freeze/thaw cycles (Burns *et al.*, 1993).

## B. Concentration of Pseudotyped Retroviral Vectors

### 1. Solutions

Hank's balanced salt solution (0.1 ×), prepared by mixing one part 1 × Hank's balanced salt solution (1.3 m*M* $CaCl_2$/5 m*M* KCl/0.3 m*M* $KH_2PO_4$/0.5 m*M* $MgCl_2 \cdot 6H_2O$/0.4 m*M* $MgSO_4 \cdot 7H_2O$/138 m*M* NaCl/4 m*M* $NaHCO_3$/0.3 m*M* $NaH_2PO_4 \cdot H_2O$) with 9 parts PBS, autoclaved, and stored at 4°C

TNE (50 m*M* Tris-HCl, pH 7.8/130 m*M* NaCl/1 m*M* EDTA), prepared by mixing 50 ml 1 *M* Tris-HCl, pH 7.8, with 950 ml water containing 7.6 g NaCl and 0.4 g EDTA, autoclaved, and stored at 4°C

### 2. Procedures

1. The frozen culture medium harvested from pHCMV-G-transfected 293GP/LZRNL cells is thawed at 37°C in a water bath and transferred to Beckman (Palo Alto, CA) ultraclear centrifuge tubes (14 × 89 mm) for the SW41 rotor. The tubes are previously sterilized overnight in a laminar flow hood by UV light.

2. The virus is sedimented in an SW41 rotor at 50,000 *g* (25,000 rpm) at 4°C for 90 min. The culture medium is removed in a laminar flow hood and the tubes are drained well.

3. The virus pellet is resuspended to 0.5–1% of the original volume of culture medium in either TNE or 0.1 × Hank's balanced salt solution overnight without swirling at 4°C. The virus pellet can be dispersed with gentle pipetting after overnight incubation without significant loss of infectious virus.

4. If the virus stock is to be used for direct injection into embryos or animals, any visible cell debris or aggregated virions that cannot be resuspended under the conditions just described should be removed by low-speed centrifugation in a microfuge for 5 min at 4°C.

5. To concentrate the virus further, a second cycle of ultracentrifugation can be performed. The resuspended virus from the first cycle of concentration is pooled and pelleted by ultracentrifugation a second time, as just described.

6. The virus titers of pre- and postconcentration fluids are determined by infection of 208F cells and selection of G418-resistant colonies, as described already.

## 3. Comments

1. Depending on the volume of TNE or 0.1× Hank's balanced salt solution used for resuspension of the pelleted virus, the virus titers can be increased 100- to 300-fold routinely after one cycle of ultracentrifugation. The efficiency of infectious virus recovery, however, varies between 30 and 100%. The virus titer can be increased approximately 2000-fold after two cycles of ultracentrifugation (Burns *et al.,* 1993).

2. Concentration of the pseudotyped retroviral vectors by ultracentrifugation is based on the observation that infectious VSV can be purified routinely by this method (Wunner, 1985). Numerous other methods are described for the purification of VSV, including precipitation of the virus by zinc acetate, ammonium sulfate, and polyethylene glycol or banding the virus in a sucrose density gradient (Wunner, 1985). We have not performed extensive studies on these alternative methods for concentration of the pseudotyped virus.

## C. Infection of Nonmammalian Cells with Pseudotyped Retroviral Vectors

One application of the VSV-G pseudotyped retroviral vectors is the stable transfer of foreign genes into cells of nonmammalian origin. Retroviral vectors containing the amphotropic envelope protein of MoMLV have a limited host range determined by the expression of the receptor protein on the surface of the cell to be infected (Albritton *et al.,* 1989). The broadened host range of the VSV-G pseudotyped vectors is conferred by the VSV-G glycoprotein, which interacts with phospholipid components of the plasma membrane and permits virus entry into a wide variety of cells. We have shown that the VSV-G pseudotyped vectors can stably deliver genes into cell lines derived from fish (Burns *et al.,* 1993), mosquito, Lepidoptera, and *Xenopus*.

Pseudotyped retroviral vectors for infection of nonmammalian species are produced in the same manner as those for mammalian cells. Consideration must be given, however, to the selection of suitable promoters for the species to be infected. In the construction of vectors for use in mammalian cells, genes may be expressed from the MoMLV 5′LTR, which is a strong promoter in many mammalian cells. We have demonstrated, however, that this promoter is not functional in a variety of nonmammalian cell lines including fish (Burns *et al.,* 1993) and *Xenopus* cell lines. Genes of interest must therefore be expressed from suitable internal promoters chosen to be compatible with the species to be infected. Mono- or dicistronic vectors can be created with gene expression from the internal promoter (Adam *et al.,* 1991; Levine *et al.,* 1991). Similarly, genes encoding selectable markers required only for selection of the mammalian packaging cells can be put under the control of the 5′LTR whereas the genes of interest can be expressed from an internal promoter appropriate for the cell line to be infected.

An additional consideration in working with nonmammalian cells is the concentration of G418 required to kill uninfected cells. We have found many species to be highly resistant to the toxic effects of G418. Killing curves to establish the optimal concentration of G418 needed to kill uninfected cells must be developed for each cell line prior to initiating experiments that require selection in G418. Additional parameters that must be determined empirically for each cell line are the time required to select G418-resistant colonies and the efficiency of infection. For example, G418-resistant clones were detected in the zebrafish cell line ZF4 only after 4–6 wk of selection in media containing 800 $\mu$g/ml G418. The titer of the pseudotyped virus on ZF4 cells as 100-fold less than the titer on the same viral stock on mammalian cells (Burns *et al.*, 1993).

To illustrate the method of infecting nonmammalian cell lines with the VSV-G pseudotyped retroviral vector, we will describe infection of ZF4 cells with the pseudotyped LZRNL virus.

## 1. Cell Lines and Solutions

ZF4 cells derived from embryonic zebrafish (Driever and Rangini, 1993) were obtained from Dr. Wolfgang Driever. Cells are maintained in 50 : 50 (v/v) DME/F12 medium supplemented with 10% FCS, 2 m*M* L-gultamine, 100 $\mu$g/ml streptomycin, and 100 units/ml penicillin. ZF4 cells are grown at room temperature in 25-cm$^2$ flasks in which the ambient air has been replaced with 5% $CO_2$ and 95% air.

Trypsin solution, prepared by mixing 1 ml 2.5% trypsin in saline with 25 ml PBS

## 2. Procedures

1. Day 1. Approximately $5 \times 10^5$ ZF4 cells are seeded in a 25-cm$^2$ tissue culture flask. Cells can be suspended by treating monolayers with 1 ml trypsin solution. Use of solutions containing EDTA should be avoided since cells are very sensitive to this compound.

2. Day 2. One day after the cells are seeded, they are infected with an aliquot of the pseudotyped retroviral vector stock in the presence of 8 $\mu$g/ml polybrene. The volume of virus added will depend on the titer of the stock and the number of colonies desired.

3. Day 3. The culture medium is aspirated and replaced with fresh medium containing 800 $\mu$g/ml G418 16 hr after infection. The medium is replaced every 5–7 days with fresh G418-containing medium until colonies appear 4–6 wk later.

4. Selection of individual colonies is difficult with this particular cell line since cells will not grow when plated at low densities. If individual clones are desired, colonies can be picked and grown in a 24-well plate in the presence of noninfected ZF4 cells. After 1 wk, the cultures are placed in G418 selection and the resistant cells are allowed to grow.

### 3. Comments

1. The titer of virus used to infect the cells can vary depending on the goal of the experiment. In ZF4 cells, the efficiency of infection by the pseudotyped retroviral vectors is reduced by a factor of 100–1000 compared with mammalian cells (Burns *et al.*, 1993). Sufficient virus must therefore be applied to overcome this inefficiency to obtain a confluent plate of infected cells. Whether the inefficient infection is due to factors relating to viral attachment or uncoating or to factors affecting reverse transcription and integration of the viral DNA into host genome has not been established.

2. If a low titer of virus is used to infect the monolayer, the time for visualization of individual colonies will be increased since the growth rate of the cells is directly proportional to their density.

3. To facilitate picking of individual colonies, cells can be grown in 100-mm tissue culture plates in a 5% $CO_2$ incubator at ambient temperature. Care must be taken, however, to insure that the internal temperature in the incubator does not exceed 27°C since the cells will die above this temperature. The $CO_2$ circulation fans in many incubators generate sufficient heat to preclude this approach as a viable option for zebrafish cell culture.

## D. Detection of Retroviral Genome in Infected Cells

Retroviral genomes, especially those containing promoters not known to function well in a particular cellular background, may be detected by polymerase chain reaction (PCR). As an example, the detection of LZRNL virus using a PCR assay for the *neo* gene in ZF4 cells is presented. Note that the following assay documents the presence of retroviral DNA in the infected cells but does not provide information about integration of the viral DNA into the host genome.

### 1. Solutions and Reagents

Cell lysis buffer, prepared as 100 m*M* NaCl, 10 m*M* Tris-HCl, pH 7.4, 25 m*M* EDTA, 0.5% SDS, and 200 $\mu$g/ml proteinase K

10× *Taq* buffer, prepared by combining 500 $\mu$l 1 *M* KCl, 100 $\mu$l 1 *M* Tris, pH 8.4, 15 $\mu$l 1 *M* $MgCl_2$, and 385 $\mu$l $H_2O$ to yield 1 ml solution (500 m*M* KCl, 100 m*M* Tris, 15 m*M* $MgCl_2$)

PCR primers for the *neo* gene, prepared by diluting the primers with water to a final concentration of 6.25 $\mu M$

Upstream primer sequence for the *neo* gene: 5′-GCA TTG CAT CAG CCA TGA TG-3′

Downstream primer sequence for the *neo* gene: 5′-GAT GGA TTG CAC GCA GGT TC-3′

10× dNTP stock solution, prepared by diluting 100 m*M* stocks of dCTP, dTTP, dGTP, and dATP with $H_2O$ to a final concentration of 2 m*M*

2. Procedures

1. Approximately $1 \times 10^4$–$1 \times 10^5$ cells are lysed in 500 $\mu$l cell lysis buffer at 37°C overnight.
2. The DNA is extracted once with an equal volume of phenol:chloroform: isoamyl alcohol (25:24:1) followed by extraction once with an equal volume of chloroform:isoamyl alcohol (24:1). The DNA is precipitated with an equal volume of isopropanol at −20°C for 1 hr.
3. The precipitated DNA is pelleted in a microfuge tube at maximum speed for 10 min.
4. The DNA pellet is resuspended in 25–100 $\mu$l sterile water.
5. For PCR amplification, 1 $\mu$l DNA is placed in a 500-$\mu$l eppendorf tube suitable for use in a thermal cycling device.
6. The DNA sample is overlayed with 20 $\mu$l light mineral oil, heated to 100°C for 1 min, and chilled on ice.
7. The 25-$\mu$l PCR reaction is prepared by adding, to the chilled DNA, 24 $\mu$l chilled reaction mix containing 1 $\mu$l each primer stock (the final concentration for each primer is 0.5 $\mu M$), 2.5 $\mu$l 10 × dNTP stock, 0.5 $\mu$l *Taq* DNA polymerase (4 units/$\mu$l), and 19 $\mu$l sterile water. The solutions are mixed by spinning the tube in a microfuge for a few seconds.
8. The tube is placed in a thermal cycling device and DNA strands are melted at 94°C for 3 min, followed by 40 cycles of 94°C for 1 min, 60°C for 40 sec, and 72°C for 40 sec. The reaction is terminated with a final extension step at 72°C for 5 min.
9. Load 10 $\mu$l PCR product onto a 2% agarose gel. The predicted 349-bp product may be visualized from amplification of the infected cells. If further sensitivity is desired, the DNA can be transferred to a nylon membrane and hybridized with a $^{32}$P-labeled *neo* probe.

3. Comments

Sensitivity of this PCR assay in a background of 1 $\mu$g genomic DNA is 100 copies by ethidium bromide-stained gel and 1 copy by Southern blot hybridization. It may be necessary, therefore, to perform Southern blot hybridization if the number of PCR templates (the viral DNA in infected cells) is low.

## IV. Summary

Encapsidation of the VSV G protein into the virions of MoMLV-derived retroviral vectors in the absence of other VSV-encoded proteins is shown to be an efficient process, although the exact mechanism for this process is currently unclear. Unlike the conventional retroviral vectors bearing the amphotropic envelope protein, the pseudotyped virus has the ability to withstand the shearing forces encountered during ultracentrifugation. This property of the pseudotyped virus enables the generation of high-titer retroviral vector stocks and has potential application for *in vivo* gene therapy studies. We have found as many as four copies of a pseudotyped vector to integrate into the genome of a single cell when a high multiplicity of infection was used to infect the cells. Multiple integration events were not observed with amphotropic retroviral vectors, probably because of their low virus titers. In addition, when retroviral vectors are pseudotyped with the VSV G protein, they acquire the host range of VSV and are able to infect nonmammalian cells derived from fish, *Xenopus,* mosquito, and Lepidoptera. Since techniques for efficient gene transfer in some of these nonmammalian systems are not currently available, retrovirus-mediated gene transfer described here should be useful for transgenic and other genetic studies in lower vertebrate species. The inability to establish a stable cell line expressing the VSV G protein, however, limits large-scale production of the pseudotyped retroviral vectors. Generation of stable packaging cell lines for the pseudotyped retroviral vectors is a major challenge for the future.

## Acknowledgments

We thank Patricia La Porte for her technical assistance and Barbara L. Sullivan for tissue culture assistance. This work is supported in part by NIH Grants HL-#01855 and HD-#20034 and NCI Grant CA-58317.

## References

Adam, M. A., Ramesh, N., Miller, A. D., and Osborne, W. R. A. (1991). Internal initiation of translation in retroviral vectors carrying picornavirus 5′ nontranslated regions. *J. Virol.* **65,** 4985–4990.

Adams, R. M., Soriano, H. E., Wang, M., Darlington, G., Steffen, D., and Ledley, F. D. (1992). Transduction of primary human hepatocytes with amphotropic and xenotropic retroviral vectors. *Proc. Natl. Acad. Sci. USA* **89,** 8981–8985.

Albritton, L. M., Tseng, L., Scadden, D., and Cunningham, J. M. (1989). A putative murine ecotropic retrovirus receptor gene encodes a multiple membrane-spanning protein and confers susceptibility to virus infection. *Cell* **57,** 659–666.

Burns, J. C., Friedmann, T., Driever, W., Burrascano, M., and Yee, J.-K. (1993). VSV-G pseudotyped retroviral vectors: Concentration to very high titer and efficient gene transfer into mammalian and nonmammalian cells. *Proc. Natl. Acad. Sci. USA* **90,** 8033–8037.

Driever, W., and Rangini, Z. (1993). Characterization of a cell line derived from zebrafish (*Brachydanio rerio*) embryos. *In Vitro Cell Dev. Biol.* (*in press*).

Emi, N., Friedmann, T., and Yee, J. K. (1991). Pseudotype formation of murine leukemia virus with the G protein of vesicular stomatitis virus. *J. Virol.* **65,** 1202–1207.

Graham, F. L., and van der Eb, A. J. (1973). A new technique for the assay of infectivity of human adenovirus 5 DNA. *Virology* **52,** 456–467.

Levine, F., Yee, J. K., and Friedmann, T. (1991). Efficient gene expression in mammalian cells from a dicistronic transcriptional unit in an improved retroviral vector. *Gene* **108,** 167–174.

Mann, R., Mulligan, R. C., and Baltimore, D. (1983). Construction of a retrovirus packaging mutant and its use to produce helper-free defective retrovirus. *Cell* **33,** 153–159.

Mastromarino, P., Conti, C., Goldoni, P., Hauttecoeur, B., and Orsi, N. (1987). Characterization of membrane components of the erythrocyte involved in vesicular stomatitis virus attachment and fusion at acidic pH. *J. Gen. Virol.* **68,** 2359–2369.

Miller, A. D. (1992). Human gene therapy comes of age. *Nature (London)* **357,** 455–460.

Miller, A. D., and Buttimore, C. (1986). Redesign of retrovirus packaging lines to avoid recombination leading to helper virus production. *Mol. Cell. Biol.* **6,** 2895–2902.

Mulligan, R. C. (1993). The basic science of gene therapy. *Science* **260,** 926–932.

Quade, K. (1979). Transformation of mammalian cells by avian myelocytomatosis virus and avian erythroblastosis virus. *Virology* **98,** 461–465.

Weiss, R. A. (1980). Rhabdovirus pseudotypes. In "Rhabdoviruses" (D. H. L. ed.), pp. 52–65. CRC Press, Boca Raton, Florida.

Weiss, R. A., Boettiger, D., and Murphy, H. M. (1977). Pseudotypes of avian sarcoma viruses with the envelope properties of vesicular stomatitis virus. *Virology* **76,** 808–825.

Witte, O. N., and Baltimore, D. (1977). Mechanism of formation of pseudotypes between vesicular stomatitis virus and murine leukemia virus. *Cell* **11,** 505–511.

Wunner, W. H. (1985). Growth, purification and titration of rhabdoviruses. *In* "Virology: A Practical Approach" (B. W. J. Murphy, ed.), pp. 79–93. IRL Press, Oxford.

Xu, L., Yee, J.-K., Wolff, J. A., and Friedmann, T. (1989). Factors affecting long-term stability of Moloney murine leukemia virus-based vectors. *Virology* **171,** 331–341.

# CHAPTER 6

# SV40 Virus Expression Vectors

**Hussein Y. Naim and Michael G. Roth**

Department of Biochemistry
University of Texas Southwestern Medical Center
Dallas, Texas 75235



## I. Introduction

### A. Brief History of SV40 Virus Vectors

Simian virus 40 (SV40) is a small DNA virus that was the first animal virus for which the complete genome was sequenced. The small size of the viral DNA (5243 bp in the standard 776 strain) facilitated its genetic analysis; this virus was one of the first to have a complete genetic and physical map. This mapping in turn led to the use of SV40 in the construction of the first mammalian expression vectors. Two strategies were adopted—plasmid vectors employing the SV40 early transcriptional control elements and viral vectors (Gething and Sambrook, 1981; Sveda and Lai, 1981; Hartman *et al.*, 1982) in which either early or late viral proteins were replaced with exogenous proteins. The Plasmid vectors had the advantage that the size of the exogenous DNA that could be expressed was not limited and they could be introduced into a variety of cells. They had the disadvantage that protein expression from the SV40 early promoter was quite poor in many cell types and the efficiency of DNA transfection tended to be low. Viral vectors were limited to the narrow host range of the virus (Rhesus and African Green monkey cells) and could package no more DNA than 100% of the wild-type genome size, but could achieve uniform infection of an entire cell population. The viral vectors could also replicate to approximately 1000 copies per cell and produced large amounts of protein from the late promoter (Gething and Sambrook, 1981; Sambrook *et al.*, 1986).

Because of the relative ease of plasmid construction, use of plasmid-based SV40 vectors expanded rapidly after the introduction of COS monkey cell lines (Gluzman, 1981) that supplied the SV40 early gene product, large T antigen, in *trans* and allowed plasmid vectors based on SV40 to replicate to levels nearly as high as those in virus-infected cells. However, the cost of this high-level expression was the imposition of a host range restricted to a single transformed cell line in which the levels of production of T antigen varied considerably from cell to cell and also during passage of the COS cell population. Unfortunately, these latter characteristics of COS cells were not widely publicized and some laboratories new to the use of COS cells transfected with SV40-based expression plasmids have been plagued with the difficulty of reproducing results when both transfection efficiencies and vector replication varied independently. Thus, transfected COS cells have been most useful when the assay, such as immunofluorescence, measured expression in single cells. COS cells are more difficult to use for biochemical techniques that measure protein expression of the entire cell population and that require the comparison of the results of separate transfections. For the latter type of experiment, SV40 viral vectors are superior.

### B. Early-Replacement Vectors and Late-Replacement Vectors

The use of SV40 viral vectors has remained less extensive than that of the plasmid-based vectors, primarily because of the constraints on DNA packaging and host range. No more than 2500 bp of exogenous DNA can be inserted into the late region of SV40 and only a few cell lines are good hosts. However, there are many advantages to the use of SV40 viral vectors; for certain experiments, they may be the best choice of expression system. The outstanding advantage of this virus is the slow development of cytopathic effects which is quite unusual for a lytic virus capable of extremely high levels of protein expression. Without inhibiting host protein synthesis, SV40 vectors can express an exogenous protein at levels that equal or exceed the production of viral proteins by enveloped viruses such as influenza virus or vesicular stomatitis virus (Chapter 1), which divert cellular synthesis totally to their own products. Since SV40 replicates in the nucleus and does not encode viral glycoproteins, SV40 vectors are particularly good for studies of the biosynthesis of membrane proteins. For the first 48 hr of infection, both exocytosis and endocytosis are quite normal in CV-1 cells infected with SV40 vectors. The uniform infections possible with high-titer stocks of SV40 vectors lead to predictable levels of protein expression that rise as a function of the time postinfection. In addition, although classified as a tumor virus, the use of laboratory-adapted SV40 is safe. SV40 was first detected in 1960 when the primary Rhesus monkey kidney cells that were used to grow poliovirus for vaccine production were screened for possible contaminating viruses. A large group people who had been inoculated during the 1950s with poliovirus vaccine contaminated with SV40 was followed in a long-term study and was found to have no significant increase in tumors. Thus, use of SV40 in the laboratory does not entail safety precautions beyond those that are routine for standard practice with viruses such as vesicular stomatitis virus or influenza virus (Chapter 1).

In this chapter, characteristics of SV40 infection and the use of various types of SV40-derived virus vectors will be briefly discussed. Advantages and disadvantages of early- or late-replacement vectors will be presented. The types of experiments in which recombinant SV40 vectors have proven useful will described, followed by protocols for preparation, titration, and storage of high-titer virus stocks. Finally, examples of the assays to which this vector system is well suited will be presented.

## II. Uses of SV40 Virus Expression Vectors

### A. Characteristics of SV40 Virus Infection

The SV40 virion is small and nonenveloped, with an icosahedral protein capsid that is relatively heat resistant. SV40 stocks can be stored for many years at $-20°C$ and retain most of their titer after storage for months at 4°C, providing care is taken to keep the medium at neutral pH. The virus is stable in high salt and can be purified and concentrated on CsCl gradients. The organization of the SV40 genome is relatively simple but compact, as one would expect of a very small virus. In the region containing the origin of replication, the circular double-stranded DNA genome of 5243 nucleotides contains two overlapping promoters in opposite orientations, so transcription occurs on both DNA strands. One of these promoters controls early gene expression, which results in two products by differential splicing—the large T antigen and small t antigen. T antigen is a multifunctional protein necessary for viral DNA replication (Tooze, 1981) that also stimulates entry of the infected cell into S phase and prolongs $G_2$ and M phase (Sladek and Jacobberger, 1992) so cellular replication machinery can be used by the virus. Small t probably aids in this process, but is dispensable for virus growth in cell culture. T antigen binds to three sites at the origin of replication; this represses early gene transcription and stimulates late gene expression. The late genes encode three capsid proteins—VP1, VP2, and VP3—and a small nonstructural protein called the agnoprotein (Jay *et al.*, 1981). By convention (Fiers *et al.*, 1978), nucleotides are counted from the unique *Bgl* I cleavage site at the origin of replication on the strand encoding the late genes. Both early and late messages terminate close together at nucleotides 2586 (early) and 2674 (late), very near a unique *Bam* HI site at nucleotide 2533 (Fig. 1). Sequences necessary for early promoter function extend to nucleotide 250. The late promoter does not require sequences beyond 5171.

In cultured African Green or Rhesus monkey kidney cells, SV40 undergoes a conventional replication cycle in which the expression of the viral early genes is followed by replication of the viral DNA, synthesis of the late viral proteins, and finally assembly of progeny virions in the nucleus (for review, see Tooze, 1981). Because the virus requires host polymerases for replication, the onset of SV40 replication can be quite asynchronous, depending on the multiplicity of infection (and, thus, the levels of T antigen produced) and the stage of the host cell replication cycle during which the virus entered the cell. In a typical culture of rapidly growing CV-1 African Green monkey kidney cells, viral

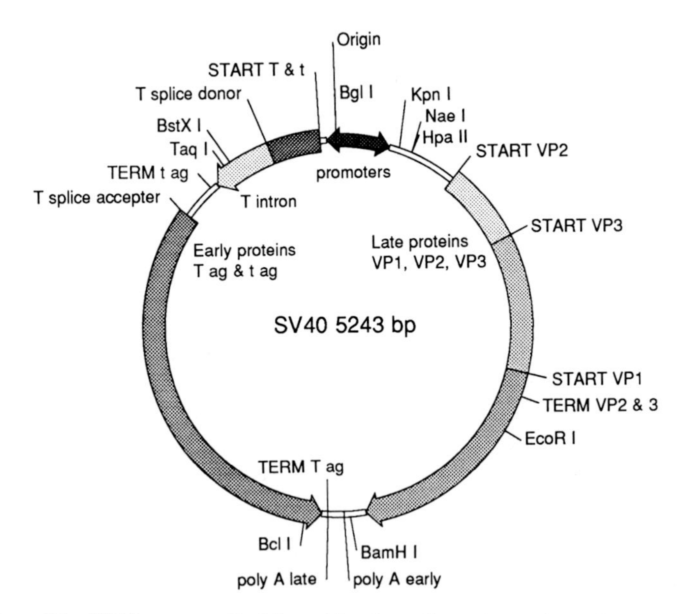


**Fig. 1** Map of the SV40 genome. Both large T and small t antigens share the same translational start site, but differ in carboxy-terminal sequences because of differential splicing. The unique *BstX* I and *Taq* I sites in the T intron, and the *Bam*H I and *Kpn* I sites in the late genes, have been used to clone the viral DNA into bacterial plasmids. The structural proteins VP2 and VP3 start at different sites in the same reading frame. VP1 is made from a different message.

replication occurs between 6 and 18 hr after infection, with some assembly of virions occurring after 24 hr. During this period, cellular metabolism is relatively unchanged and protein synthesis, secretion, endocytosis, and post-translational modification of glycoproteins is not detectably affected between 24 and 36 hr postinfection. There is very little effect on these processes for the first 2 days, even when infections are made with strong virus stocks. After 2 days of infection with a strong virus stock, one can detect isolated cells containing large vacuoles; on the third day many cells will start to round up and vacuolated cells will be common. Because SV40 requires active cellular replication, contact-inhibited quiescent uninfected cells in a monolayer are only slowly infected. Thus, under conditions of a low multiplicity of infection, secondary rounds of infection follow a much longer cycle than described for a single round of infection. In fact, on CV-1 cell monolayers SV40 plaques typically become large enough to count only after 10 days.

CV-1 fibroblasts are the best hosts for virus production, but the commonly available BSC-1, VERO, and MA104 cell lines also support infection. The latter two lines are of epithelial origin and sublines retaining some of the features of polarized epithelial cells have been isolated that support SV40 infection. Although SV40 assembles in the nucleus and can form paracrystalline hexagonal arrays there late in infection, SV40 can also exit the infected cell by an unknown process long before cytopathic effects become visible (Clayson *et al.,* 1989). The degree to which SV40 is recovered in the medium or is retained in lysed cells differs for the different host cell lines. More than 90% of the virus can be recovered from the medium of infected VERO cells, whereas about half the virus remains inside lysed CV-1 cells.

## B. Advantages and Disadvantages of SV40 Vectors for Experiments in Cell Biology

The ease with which high titer SV40 virus stocks can be made and stored for long periods allows one to infect an entire population of cells reproducibly at a multiplicity high enough that the time course of protein expression is quite predictable. During the period when protein expression is rising from undetectable levels to upward of $10^7$ molecules/cell, the cellular environment is not detectably altered. This allows both biochemical and morphological study of vastly different amounts of protein in the same cellular background. The levels of expression possible with SV40 late-expression vectors are well suited for techniques such as the detection of immunocytochemical reagents by electron microscopy. For techniques that require relatively high concentrations of proteins at the cell surface, for example, studies of membrane fusion (Clague *et al.,* 1991; Schoch and Blumenthal, 1993) or fluorescence photobleaching recovery for measurements of protein mobility (Fire *et al.,* 1991), SV40 late-replacement vectors are the best expression system. In addition, vector construction is easy in comparison with most other viral vectors.

Balancing these advantages are some important limitations on the use of SV40 beyond the mentioned restrictions of narrow host range and the limited size of exogenous DNA that can be inserted. Since SV40 stimulates cells to enter S phase, it may cause differentiated cells such as the polarized MA104.11 or VERO 6000 epithelial lines to begin to dedifferentiate during the period when protein expression is rising (Roth *et al.*, 1986). In addition, there is a limit to how synchronous an infection with SV40 can be. With other lytic viruses, the infectious cycle can be synchronized to differences on the order of 1 hr, whereas we have observed that cells infected with SV40 vectors commonly appear to differ by 6–12 hr in the onset of protein expression. Since the amount of protein expression can be quite high, cells can differ dramatically in the levels of protein expressed during early periods of expression from the late promoter. The practical consequence of this asynchronicity is that experiments that compare the expression of different proteins from different virus stocks require that all the virus stocks be essentially of the same titer. Biochemical experiments that average events in the entire population (such as standard pulse–chase protocols) should be performed relatively early in the infection to minimize the contribution of cells that are most advanced in the infectious cycle.

## III. Design of SV40 Expression Vectors

### A. Expression by Early-Replacement Vectors

SV40 vectors have been made by replacing either the early or the late genes of the virus with an exogenous gene. In the case of early-replacement vectors, the late genes encoding the capsid proteins are present and the virus capsid is merely being used as an effective plasmid delivery system. Virus replication requires that T antigen be supplied, so COS cells are used to produce virus stocks. Typically, this method requires multiple passages to create a virus stock that, as measured by the onset of cytopathic effects, is of lower titer than can be achieved with late-replacement vectors (Settleman and DiMaio, 1988; O'Banion *et al.*, 1993). Once this virus stock is prepared, the experiment may be performed in any cell line that supports expression from the SV40 early promoter, since the narrow host range of the virus is a function of an undefined block in replication and not in virus entry (Tooze, 1981). However, if one also wishes to increase protein expression through virus replication, COS cells must be used. Protein expression under these conditions is 5- to 10-fold less than expression from the late SV40 promoter in COS cells because of the weaker early promoter and the fact that T antigen inhibits some of the early transcription. Although not capable of high levels of protein expression, early-replacement vectors are quite useful for biochemical analysis of events such as RNA splicing (Valcarcel *et al.*, 1993), cellular transformation (Settleman *et al.*, 1989), and other effects of a transgene on cellular metabolism (O'Banion *et al.*, 1993) without complications caused by viral replication.

For replacement of the early genes, usually a *Hin*d III site at 5171 just upstream from the initiation codon for the T and t antigens is used as the 5′ cloning site and a unique *Bcl* I site at 2770 is used as the 3′ side (Figs. 1 and 2). This allows replacement of 2401 nucleotides. SV40 can package as little as 70% of the wild-type genome size (approximately 3800 bp) so the minimum size of insert used to replace the *Hin*d III–*Bcl* I early fragment is 960 bp. Since production of the initial virus vectors and of DNA for transfections requires growth in bacteria, a bacterial plasmid is inserted into one of the unique restriction sites in the late region of the virus. The choice of this site is determined by the restriction sites present in the exogenous gene to be inserted; however, the *Bam* HI, *Eco* RI, and *Kpn* I sites have been commonly used.

## B. Expression by Late-Replacement Vectors

In late-replacement vectors the late genes are replaced and early gene function is maintained, so the viral capsid proteins must be supplied *in trans*. Attempts to make a cell line to supply SV40 capsid proteins have not been successful, possibly because capsomeres collect in the nucleus and are toxic. Thus, the capsid proteins are usually provided by a helper virus deficient in T antigen function. A primary requirement of the helper virus is that it not recombine with the vector to produce wild-type SV40, which would then grow with single hit kinetics and overtake the virus stock. Several helper viruses have been used, including one with a temperature-sensitive T antigen. However, the most useful helper has proven to be the Dl1055 virus identified by Pipas and coworkers (Pipas *et al.*, 1979) as having undetectable frequency of reversion to wild-type. First used by Gething and Sambrook (1981), this helper virus has been used for 15 years without detectable recombination of helper and vector.

SV40 produces a 19S and a 16S RNA from the late promoter. Each has heterogeneous start sites, almost all of which are located after nucleotide 120. Since early promoter function requires the region to nucleotide 250, a convenient 5′ restriction site for subcloning is the *Kpn* I/*Ban* I site at nucleotide 294 (Figs. 1 and 2). Another good position for subcloning is at the *Nae* I site at 345 or the *Hpa* II site at 346. These sites are downstream from the start site for a small late protein called the agnoprotein, so care must be taken at this position to avoid interference between inititation of translation of the agnoprotein and the inserted protein (Perez *et al.*, 1987). A particularly highly expressing series of vectors was constructed by Gething and Sambrook, who inserted the influenza hemagglutinin (HA) gene at this site in a way that joined the agnoprotein reading frame to that of the full-length HA, which simply added a few amino acids to the N terminus of the HA signal sequence. However, Perez and colleagues needed to remove the agnoprotein initiation codon to get high expression of the Rous sacroma virus glycoprotein at this site. A virus stock expressing high levels of protein has also

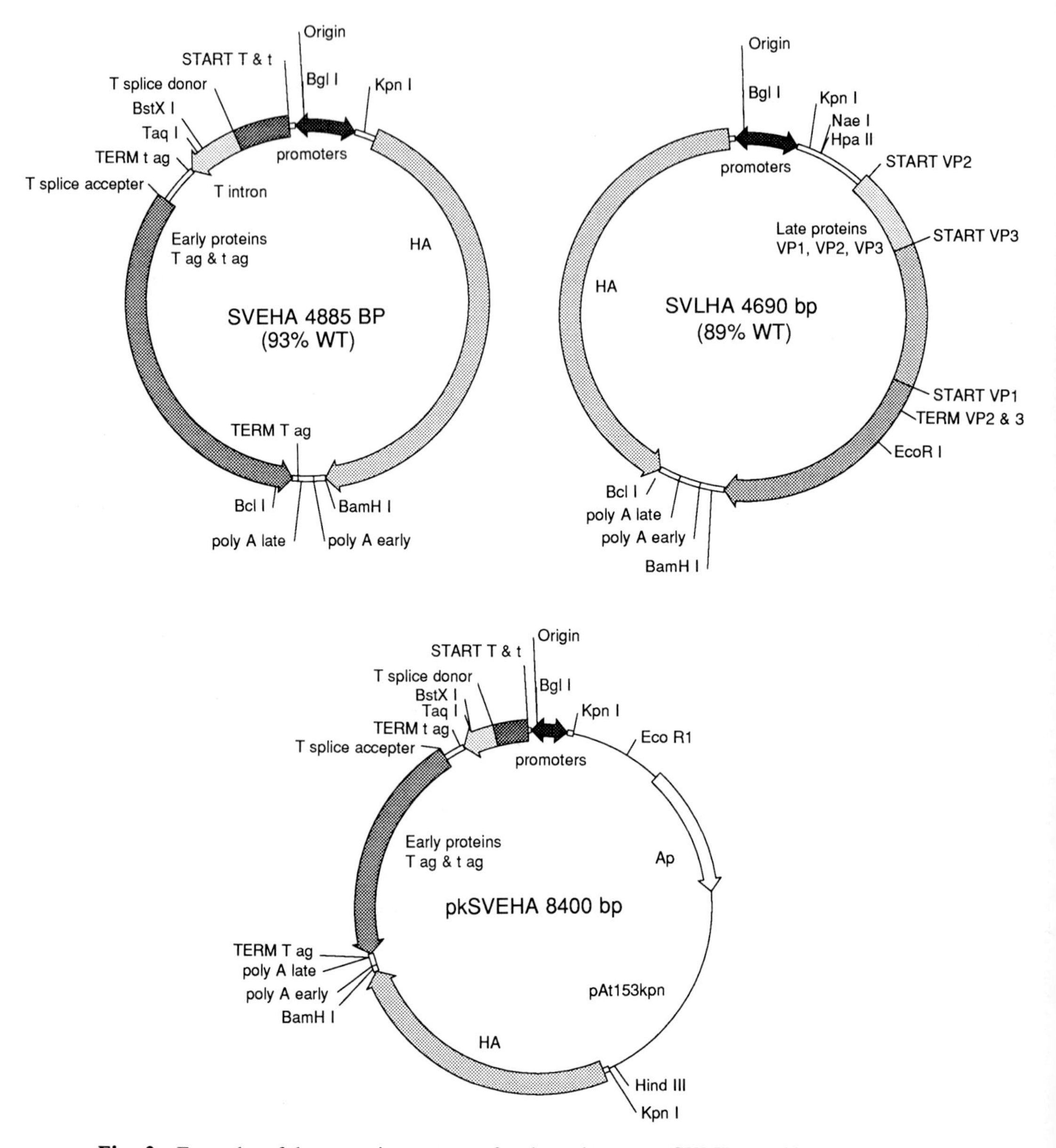


**Fig. 2** Examples of the genomic structure of early-replacement (SVLHA) and late-replacement (SVEHA) recombinant SV40 viruses are shown. In SVLHA, the influenza virus hemagglutinin is expressed from the early promoter, in SVEHA, it is expressed from the late promoter. pkSVEHA is the plasmid shuttle vector in which the virus DNA has been cloned into a unique *Kpn* I site in a derivative of the bacterial plasmid pAT153.

been made by inserting the VSV G protein at the *Kpn* I site (Naim and Roth, 1993), so that advantages of subcloning downstream from the agnoprotein may not be worth the potential complications. A maximum of 2500 bp of the late region of SV40 can be replaced while maintaining the early promoter region and the early polyadenylation site. However, usually the region between the unique *Kpn* I site at 294 or the *Nae* I site at 345 and the *Bam* HI site at 2533 is replaced for this type of vector, allowing a maximum of 2416 or 2366 bp foreign DNA to be inserted. Four sites have been used successfully for inserting bacterial sequences required during vector preparation—the unique *Bst* XI site in the T intron at nucleotide 4759, a modified form of the *Taq* I site at nucleotide 4739, the *Bam* HI site at 2533, and the *Kpn* I site at 294 (Fig. 2).

Late-replacement vectors have been enormously useful in studies of glycoprotein biosynthesis and membrane trafficking. These vectors were the first used to demonstrate that viral glycoproteins were recognized by host cell-sorting processes, which required efficient expression for detection by electron microscopy (Roth *et al.*, 1983). Because of the highly efficient infection and high levels of expression, these vectors allow the very short labeling times required by pulse–chase protocols used for careful kinetic analysis of the effects of site-directed mutagenesis on glycoprotein transport (Gething and Sambrook, 1982; Wills *et al.*, 1984; Doyle *et al.*, 1985; Davis and Hunter, 1987; Perez *et al.*, 1987; Patterson and Lamb, 1987; Roth *et al.*, 1987) and its relationship to protein folding (Copeland *et al.*, 1986; Gething *et al.*, 1986; Heibert and Lamb, 1988; Singh *et al.*, 1990; Lazarovits *et al.*, 1990; Braakman *et al.*, 1991). These vectors have been used to identify the cytoplasmic signals for interaction with coated pits (Roth *et al.*, 1986; Lazarovits and Roth, 1988; Ktistakis *et al.*, 1990; Fire *et al.*, 1991; Naim and Roth, 1993) and for experiments requiring high-level expression of membrane fusion proteins (Doxey *et al.*, 1985,1987; Clague *et al.*, 1991; Schoch and Blumenthal, 1993). We have developed a system in which CV-1 cells infected with SV40 vectors expressing various mutated or wild-type glycoproteins can be superinfected with influenza virus, and the ability of the SV40-produced glycoproteins to be incorporated into the influenza virus envelope can be measured (Naim and Roth, 1993).

## C. Subcloning

If a series of mutant forms of a protein is to be screened in either early- or late-replacement vectors, it is useful to create a subcloning vector first that can aid in the analysis required during vector construction. We have found that it is useful to insert an easily identifiable foreign fragment into the subcloning site of the vector, because the vector with its bacterial sequences is approximately 8.5 kb in size (Fig. 2) and if a region less than 1 kb in size is being removed and replaced, it is difficult to distinguish the completely cut and incompletely cut DNAs by size. This leads to mistakes during fragment purification. Thus, if the DNA being replaced has a unique restriction site and/or differs considerably in size from the

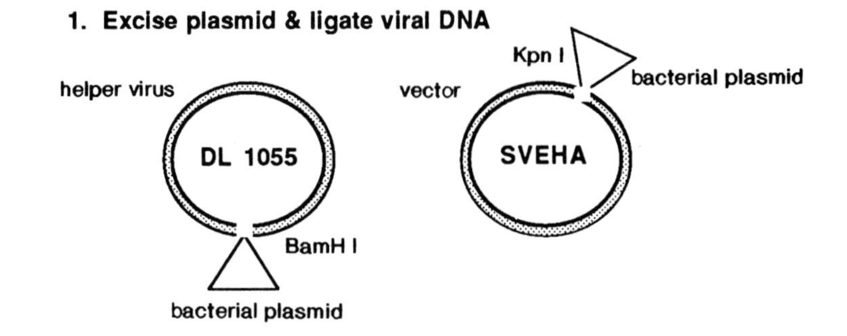


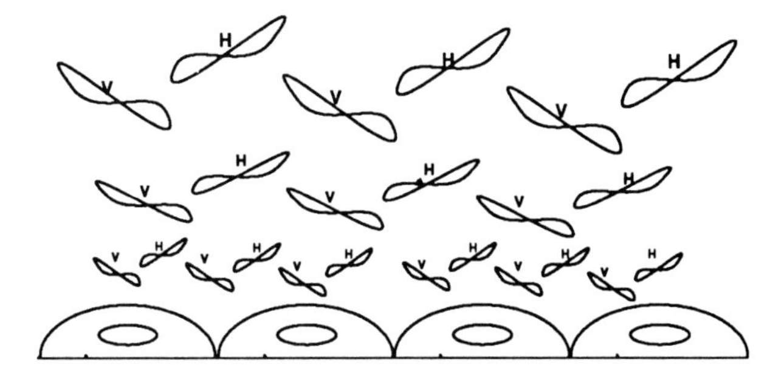


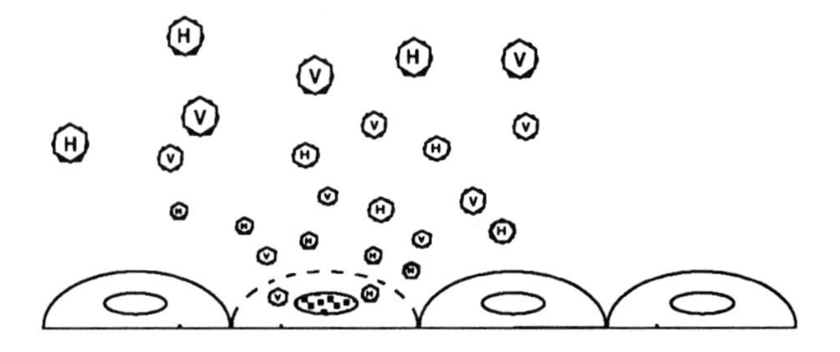


**Fig. 3** Production of SV40 virus stocks. The five steps to virus production described in the text are diagrammed.

fragment being inserted, success of the subcloning reaction can be monitored by a simple digest with the appropriate restriction endonuclease.

## IV. Preparation of High-Titer Recombinant SV40 Virus Stocks

The preparation of recombinant SV40 virus stocks is quite simple (Fig. 3). To prepare high-titer virus stocks, the following factors should be taken into

**4. Infect fresh CV-1 cells with P1 virus**

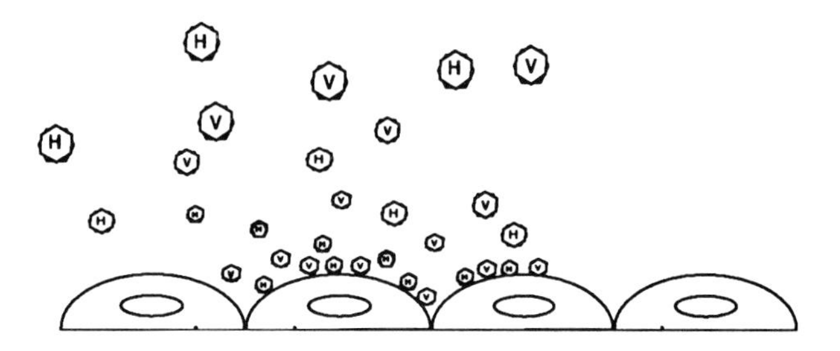

**5. Collect virus from CV-1 cells after 3 - 5 days: P2 virus stock**

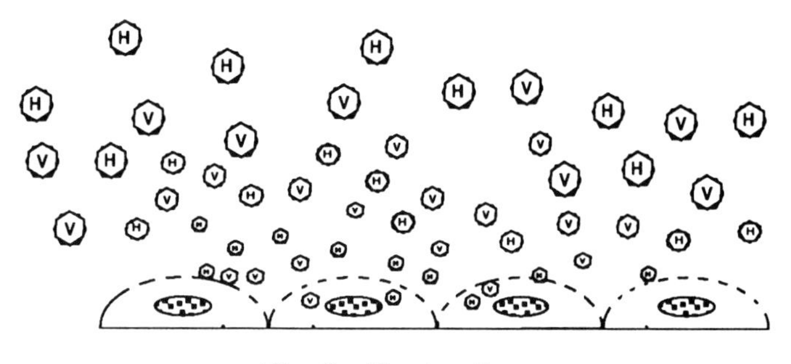

**Fig. 3** (*Continued*)

consideration: (1) the efficiency with which covalently closed viral DNA is formed after bacterial plasmid sequences have been removed, (2) the ratio of the recombinant DNA to the helper virus (Dl1055), and (3) the amount and quality of permissive cells to be transfected. The procedures described here are for late-replacement vectors, but are the same for early-replacement viruses, with the exception that the latter must be transfected into and passaged in COS cells and do not require helper virus DNA.

## A. Dilute Ligation

### 1. Reagents Needed

The source of the reagents, if not listed, is unimportant.

Appropriate restriction enzymes and their reaction buffers

3–4 $\mu$g vector and helper virus DNA that is pure enough to be digested with restriction enzymes

DNA ligase and its reaction buffer

10 $\mu M$ ATP

Reagents and equipment for analysis of DNA by agarose minigels stained with ethidium bromide

Phenol:chloroform prepared according to Sambrook *et al.* (1989)

## 2. Procedure

1. Cut at least 3–4 μg recombinant late-replacement vector (pSVE) or helper virus (Dl1055) DNA with the restriction enzyme (*Kpn*)1 for pSVE and *Bam* HI for the helper) that will release the bacterial plasmid from the SV40 vector (Fig. 3, Step 1).

2. Analyze the restriction enzyme digest by electrophoresis on a 0.8% agarose gel stained with ethidium bromide (Sambrook *et al.*, 1989) to make sure that the digestion is complete. Extract with an equal volume of phenol:chloroform and precipitate the DNA with ethanol by any reliable technique.

3. Resuspend the DNA in 100 μl 10 m*M* Tris, pH 8.0, containing 0.1 m*M* EDTA. Take **≤3 μg** DNA and add distilled water to a final volume of 800 μl; add 100 μl 10× ligase buffer containing 2 units T4 DNA ligase, and 100 μl 10 m*M* ATP. (Although most commercially available ligase buffers contain ATP, this reagent oxidizes over time and supplementing the reaction with extra ATP will improve the efficiency of ligation.) The ligation is performed at a DNA concentration of less than 3 μg/ml to favor a unimolecular ligation in which the vector sequences self-ligate. Save some of the unligated DNA for analysis in Step 5.

4. Incubate 8 hr to overnight at 4°C for complementary ends or 16°C if blunt ends are ligated.

5. It is very important to check the products of ligation on an agarose gel stained with ethidium bromide. The efficiency of the ligation is the single most important determinant of how good a virus stock will become (see *Note*). Run 50 ng unligated restriction digest for comparison. In the ligated sample, bands migrating at the position of the linear bands of the unligated control should be completely absent, having been converted into multiple slower migrating forms. This indicates that the ligation is complete.

*Note:* Poorly ligated DNA can result in a low efficiency for initial infection, but this will ultimately be compensated for by growth of the virus. However, after transfection into CV-1 cells, unligated linear forms are substrates for endonucleases and ligases, producing deleted defective expression vectors. In our experience, virus stocks that show good evidence of virus growth by cytopathic effect, but never express sufficient amounts of protein, result from incomplete ligations.

## B. DEAE–Dextran Transfection and Preparation of P1 Virus Stocks

There are a number of good protocols for transfecting DNA into CV-1 cells for the generation of a virus stock. We use a DEAE–dextran method that is inexpensive, efficient, and reproducible. In our hands, some of the commercially available cationic lipid reagents give a similar efficiency at a much greater cost.

### 1. Reagents Needed

CV-1 cells [American Type Cell Collection (ATCC), Rockville, MD] between passages 16 and 24

Serum-free Dulbecco's modified Eagles medium (SF-DMEM) (GIBCO/BRL, Grand Island, NY). Filter sterilize and store in the dark at 4°C for 1 mo or less.

Trypsin–versene solution (GIBCO/BRL); stores frozen for 6 mo

Fetal calf serum (FCS). Store at −20°C in the dark for 1 yr. (CV-1 cells will also grow in newborn calf serum.)

1 *M* HEPES stock, pH 7.2. Filter sterilize and store at 4°C indefinitely.

50 mg/ml DEAE–dextran (MW 500,000) in water, 200× stock. Sterilize by autoclaving. Store at 4°C indefinitely.

100 m*M* chloroquine in $H_2O$, 1000× stock. Store frozen in the dark indefinitely.

Cell culture incubator set at 37°C, 5% $CO_2$, and 100% humidity

### 2. Procedure

1. Split a fresh plate of cells so they are 50–60% confluent the next day (i.e., the day of transfection). Normally confluent CV-1 cells are split 1:5 or 1:6. It is important that the cells be actively growing.

2. Select dishes that contain evenly distributed cells in cultures that are 50–60% confluent, aspirate the medium, and wash twice with 5 ml SF-DMEM buffered with 10 m*M* (final concentration) HEPES pH 7.2 to remove all traces of serum.

3. Mix 150 ng helper virus (Dl1055) and 200 ng recombinant DNA (for 60-mm dish) or 300 ng helper and 400 ng recombinant DNA (for 100-mm dish) in a final volume of 0.5 ml (for 60-mm dish) or 1–1.5 ml (for 100-mm dish) SF-DMEM buffered with HEPES, pH 7.0, containing 250 μg/ml DEAE–dextran.

4. Aspirate medium from dishes, add the mixture of DNA and DEAE–dextran, distribute evenly over the cells, and incubate at 37°C for 60–90 min. Rock the dishes every 15 min to maintain an even distribution of the DNA over the cells.

5. Remove the mixture of DNA and DEAE–dextran, wash once with SF-DMEM buffered with HEPES, and add 2.5–3 ml (for 60-mm dishes) or 5 ml (for 10-cm dishes) DMEM containing 10% FCS and 100 μ*M* chloroquine. Incubate at 37°C for 4.5–5 hr.

6. Remove the medium containing chloroquine, wash once with DMEM, and add 3–3.5 ml (for 60-mm dishes) or 6 ml (for 10-cm dishes) DMEM containing 10% FCS and incubate at 37°C.

7. At 32 hr after beginning the transfection, add fresh cells to the transfected cells in a ratio 1 : 2 (fresh : transfected); this step will improve the strength of the virus stock by providing actively growing cells for the second round of infection.

8. After 4–5 days, some vacuolated or rounded cells should be visible; there can be considerable cytopathic effects if the initial transfection was very efficient. At this time the first generation of the recombinant virus (P1) must be harvested by freezing and thawing the cells and medium 3 times. Because much of the recombinant virus is contained in or attached to the cells, the cells must be scraped into the medium using a sterile rubber scraper. Add HEPES to a final concentration of 10 m*M*, then aliquot the virus stock into 1- to 2-ml volumes and store at −20°C. The frozen virus remains viable for years and can be thawed multiple times.

At this stage, however, this virus stock is not strong enough for use. Thus a second passage (P2) is needed to generate a high-titer recombinant virus stock (see Fig. 3).

*Note:* After transfection, there will be some cell death the next day due to the effects of DEAE and chloroquine; however, the percentage should be very low. The CV-1 cells used to make virus stocks should be the lowest passage available. These cells are not an established cell line, since they have not been through crisis. We have taken CV-1 cells through crisis and have established continuous cell lines that support SV40 growth, but expression levels are lower in these established cells and we prefer to use low passage cells to make high-titer virus stocks.

## C. Preparation of P2 Virus Stocks

### 1. Reagents Needed

Cell culture reagents and incubator described earlier

Low-speed cell culture centrifuge capable of 500 *g*

### 2. Procedure

1. Remove CV-1 cells from one freshly confluent 10-cm dish with warm trypsin–versene by rinsing the cells twice with 1 ml. Aspirate the cells, leaving them wet, and replace the dish in the cell incubator for about 5 min. Tap the dish and the cells should shake loose. Collect the cells in 5–6 ml DMEM containing 10% FCS. Centrifuge at 500 *g* for 3–5 min.

2. Aspirate the medium from the tube and resuspend the cell pellet with 0.3 ml P1 recombinant virus stock. Incubate on ice for 30–40 min, shaking the cells gently every 10 min.

3. Add 12 ml DMEM containing 10% FCS and dispense 6 ml cell suspension into each of two 10-cm culture dishes; incubate at 37°C in a 5% $CO_2$ atmosphere.

4. When more than 50% of the cells are dead (usually 3–5 days), harvest the virus by freezing and thawing the cells; then aliquot as described earlier.

This virus stock is now the working stock. It consists of approximately equal numbers of helper viruses and recombinants and can be used for the high multiplicity infections required to insure that the kinetics of protein expression follow those of a single cycle of virus infection. An infected CV-1 cells begins to flatten on the second day of infection and no longer proceeds through cell division. With practice, one can assess the progress of infection by observing the culture by light microscopy and by determining the proportion of relatively large, flat cells and smaller, more densely packed cells. The latter are likely to be uninfected. If the infected P2 cells grow into a dense, tightly packed monolayer, it is likely that the transfection to produce the P1 virus stock was inefficient. It is possible to use the P2 stock to make a P3 stock, but this stock may be contaminated with defective viruses if the reason for the poor P1 virus stock was an incomplete ligation. In general, we do not make P3 stocks, unless we have reason to believe that the original transfection used too little DNA, or if the P1 stock being used has been frozen and thawed many times.

## V. Analysis of the Efficiency of Protein Expression

### A. Titering Virus Stocks

We commonly titer recombinant SV40 stocks by immunofluorescence with an antibody specific for the exogenous protein to be expressed. Since the virus stocks are a mixture of two viruses and since SV40 plaques take 10–12 days for form, standard SV40 plaque assays are complicated and slow. In comparison, the immunofluorescence assay is fast and measures the parameter of interest: the expression of the exogenous gene. If a good antibody for the exogenous gene product is unavailable, monoclonal antibodies against SV40 T antigen are commercially available (ATCC). However, this assay will measure SV40 production and not expression of the foreign gene. In the protocol presented here, we describe our assay for SV40 vectors expressing HA.

#### 1. Reagents Needed

Circular 12-mm diameter cover slips, sterilized with ethanol and air dried
Freshly growing CV-1 cells

3% formaldehyde in 100 m$M$ sodium phosphate, pH 7.2. (Add 6 g paraformaldehyde to 100 ml $H_2O$ at 60°C and adjust the pH with NaOH until the formaldehyde dissolves, which usually occurs near pH 9.0. Dilute with concentrated sodium phosphate solution to 200 ml at a final concentration of 100 m$M$ and adjust the pH to 7.2. Store at 4°C. This fixative will remain active for several months, but after 1 wk will begin to permeabilize cells. For experiments that require surface labeling, make fresh formaldehyde.)

DMEM

Protein blocking solution: either 1% bovine serum albumin (BSA) in phosphate-buffered saline (PBS), pH 7.0, or NET/Gel [10 m$M$ Tris-HCl, pH 8.0, 0.5% NP-40, 5 m$M$ EDTA, 150 m$M$ NaCl, 0.25% gelatin (Bloom No. 60)]

Primary antibody and appropriate commercial secondary antibody conjugated to a fluorescent dye

Mounting medium (PolyMount; Polysciences)

## 2. Procedure

1. The virus stock is titered by preparing a series of dilutions between 1 : 1 and 1 : 100. These are used to infect separate cultures of CV-1 cells that will subsequently be processed for immunofluorescence. An uninfected culture should be included as a control for labeling specificity.

2. One can either infect cells in suspension as described for preparation of virus stocks, or after letting them grow 1 day on the cover slip. The latter method makes the cells adhere more tightly to the cover slip so there is less chance of losing cells during washes, but the former produces a titer under the conditions used for most experimental infections. In either case, $10^4$ CV-1 cells, infected or uninfected, should be seeded into wells of a 24-well plate that contain sterile cover slips. A simple way to do this is to suspend a freshly confluent 10-cm dish of CV-1 cells in 10 ml medium in a 10-ml cell culture pipette and dispense 1 drop into each well, taking care not to let the cells settle in the pipette during this process (work quickly).

3. For infection of cells on cover slips, 1 day after the cells were seeded the medium is aspirated and replaced with 200 $\mu$l virus dilution to be titrated. The plate is placed in the cell incubator for 1 hr. Then the inoculum is aspirated and replaced with fresh maintenance medium (DMEM + 10% FCS).

4. At 28 hr postinfection, wash the cells twice with DMEM lacking serum.

5. Fix cells with 3% formaldehyde at room temperature for 20 min.

6. Quench the fixation reaction by washing twice with DMEM lacking serum, each wash lasting 5–10 min.

7a. Dilute the first antibody (polyclonal or monoclonal) appropriately—we find that most of our rabbit antisera require a dilution of 1 : 250—in PBS containing 1% BSA. For surface antigens it is not necessary to permeabilize the cells and the next step is optional.

7b. Incubate the cover slips for 30 min in either PBS containing 1% BSA and 0.1% Triton X-100 or in NET/Gel containing that concentration of the detergent. If cells are permeabilized, use this same solution to dilute antibodies.

8. Add 10 μl diluted antibody to the cover slip and incubate at room temperature for 30 min. To keep the cells from drying, cover the multiwell dishes containing the cover slips with a wet paper towel.

9. Rinse cells twice for 5 min each with PBS–BSA or NET/Gel; perform a third wash for 30 min. The longer incubation time for the last wash allows the unbound antibody to diffuse from the cells and reduces nonspecific staining.

10. Add 10 μl 1 : 200 dilution of fluorescent anti-mouse or anti-rabbit IgG, and incubate for 30 min at room temperature. Fluorescent antibodies are sensitive to light. During the incubation time, dishes must be covered with wet paper towels and aluminum foil.

11. Rinse as in Step 8. Dip the cover slips in distilled water to remove salt, and briefly dry by touching the edge to a paper towel.

12. For each cover slip, apply approximately 2 μl PolyMount (or similar medium) to a microscope slide and mount cell-side down. Usually we mount 6 cover slips/slide.

13. The end point of the titration is the dilution at which 50% of the cells are expressing the protein of interest. The working dilution for experiments will depend on the object of the experiment, but is usually the lowest concentration that expresses the desired protein in 100% of the cells, or a more concentrated inoculum.

CV-1 cells expressing proteins from SV40 vectors are extremely good subjects for immunofluorescence. The cells are quite large and flat, facilitating the identification of intracellular structures. For immunofluorescence, the asynchrony of SV40 infection can be turned to advantage. By using the least amount of virus that infects all the cells and by labeling cells early in the infection, cells that differ greatly in the amount of protein expressed can be analyzed. Under these conditions it is possible to discover by a simple experiment if there are saturable cellular processes that influence either the location or the processing of the protein.

## B. Kinetics of Protein Expression

For biochemical assays such as pulse–chase experiments, the optimum infection is at high multiplicity. Under these conditions the kinetics of expression become more synchronous since every cell receives sufficient virus to produce enough T antigen rapidly to stimulate production of cellular polymerases. A protocol for determining the kinetics of protein expression from SV40 late-replacement vectors is described here and the results from such an experiment are shown in Fig. 4.

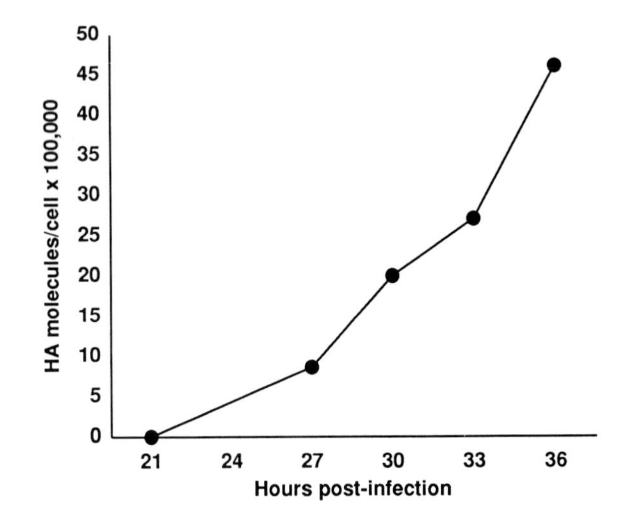


**Fig. 4** The time course of protein expression from the vector SVEHA. HA was recovered from infected cells at different times postinfection and was quantified by densitometry. The data shown assume that HA binds Coomassie Blue to the same extent as BSA does.

## 1. Reagents Needed

Antibody specific for the protein of interest. Rabbit antiserum is stored at −20°C in small aliquots that are thawed only once, then stored undiluted indefinitely at 4°C. Hybridoma cell culture supernatants are stored frozen, thawed once, and used quickly.

Cell culture plates or dishes for preparing 35-mm diameter monolayers of infected cells (for detection of unlabeled proteins) or 18-mm diameter monolayers (for radioactive labeling)

Actively growing CV-1 cells

Protein-A sepharose

Microfuge

Equipment for polyacrylamide gel electrophoresis (PAGE)

Laser scanning densitometer or similar detection device

## 2. Procedure

1. CV-1 cells were infected with the SVE–HA virus in suspension and infected cells were seeded into the wells of 6-well plates at a concentration of $10^5$ cells/well.

2. Cells were lysed at various times (we used 21, 27, 30, 33, and 36 hr postinfection) with 600 $\mu$l lysis buffer [1% NP-40, 0.1% SDS, 50 m*M* Tris-HCl,

pH 8.0, 0.1 unit aprotinin/ml (Boehringer Mannheim Biochemicals, Indianapolis, IN).

3. The lysates were centrifuged at 10,000 *g* at 4°C for 10 min to remove nuclei, mitochondria, and large insoluble elements.

4. HA was immunoprecipitated with sufficient antibody to precipitate 100% of the protein. This was determined beforehand by using different amounts of the antibody and precipitating cells that were late in the infectious cycle. Each lysate was then precipitated again with additional antibody to determine the amount required to precipitate all the HA in a sample of a certain size.

5. BSA was serially diluted to concentrations of 0.5, 1, 2, 4, and 10 $\mu$g/10 $\mu$l and subjected to electrophoresis on an SDS polyacrylamide gel with the HA immunoprecipitates. The gel was stained with Coomassie blue and dried. The optical density of each band was determined and the amount of HA recovered from the infected CV-1 cells at various intervals after infection was compared to the standard curve produced by scanning the BSA samples. These data, converted to molecules of HA/cell, are plotted as a function of time after infection in Fig. 4.

Under the conditions of infection used with this SV40 expression vector, we found that production of influenza virus HA increased rapidly after 24 hr of infection (Fig. 4). By 30 hr postinfection, there were 2 million HA trimers/cell, most of which were at the cell surface; synthesis of HA had reached a maximum, as measured by comparing the amounts of proteins labeled with $^{35}$S-labeled amino acids at various times postinfection. Truly enormous amounts of HA [$6 \times 10^8$/molecules/cell (Gething and Sambrook, 1981)] have been reported to have been produced by late in infection. That amount of HA, if present at the cell surface, would be present at slightly less than the density at which it is found in an influenza virus envelope, in which HA is calculated to occupy approximately one-third of the surface area. When CV-1 or African Green monkey kidney primary cells expressing HA from SV40 vectors are labeled with antibodies conjugated to ferritin and observed by electron microscopy late in infection, the density of labeling does approximate that of virus particles (Roth *et al.,* 1983). This large production of protein can be an advantage in experiments in which cell–cell fusion is studied or can be used for some other purpose. CV-1 cells expressing the vesicular stomatitis virus G protein from a late-replacement vector fuse together into massive syncytia around 48 hr postinfection. This occurs at neutral pH, although G protein has a pH optimum of 6.0 for membrane fusion. Presumably so much G is made that the small percentage of the population active for membrane fusion is sufficient to form polykaryons late in infection. This system has been used for controlled fusion of infected CV-1 cells, which are sensitive to the drug Brefeldin A, to rat kangaroo PtK1 cells, which are drug resistant (Ktistakis *et al.,* 1991). By plating infected CV-1 cells expressing G protein with PtK1 cells and lowering the pH to 6.0 briefly at 36 hr postinfection, heterokaryons were formed in which the effects of the drug could be monitored in a common cytoplasm.

Some cellular processes, such as endocytosis through coated pits, are saturable, so care must be taken with SV40 vectors to perform experiments at the period postinfection during which the amount of protein present is suitable for the mechanism under study. However, this concern does not apply to many of the biosynthetic processes in the secretory pathway. Using pulse–chase protocols such as those described next, we and others have not detected differences in translocation rate, folding, oligomerization, glycosylation, or kinetics of transport of wild-type HA in the period between 26 and 40 hr postinfection. This result suggests that there is a point of control for entry into the secretory pathway or that message levels have reached a steady state at a level that does not saturate secretory processes. Around 48 hr postinfection we have observed an increase in the amount of HA that fails to leave the endoplasmic reticulum. This might be due to cytopathic effects in the portion of the cells that are most advanced in the infectious cycle.

## C. Standard Pulse–Chase Protocol

### 1. Reagents Needed

Antibody specific for the protein of interest

4- or 6-well cell culture plates

Actively growing CV-1 cells and maintenance medium PBS

DMEM lacking cysteine and methionine (DMEM$^{met-cys-}$). We make this in 1-liter batches from a Selectamine kit (GIBCO/BRL) and store it at 4°C in the dark in 100-ml bottles that have the cap wrapped tightly with parafilm. This medium lasts for up to 6 mo.

Source of $^{35}$S-labeled amino acids (e.g., Tran$^{35}$Slabel; ICN, Costa Mesa, CA)

DMEM containing one-third the usual amount of bicarbonate (low-bicarbonate DMEM). We make this in 1-liter batches and store tightly capped in the dark at 4°C for several months.

Circulating waterbath with temperature control to 0.1°C

An aspirator containing bleach in the reservoir flask for virus-infected cell cultures

An aspirator for radioactive medium. Our system contains a reserve flask and activated charcoal filter between the reserve flask and the vacuum source.

Multichannel timers

Lysis buffer (1% NP-40, 0.1% SDS, 50 m*M* Tris-HCl, pH 8.0, 0.1 unit aprotinin/ml (Boehringer Mannheim)

Protein A-Sepharose

Microfuge and tubes

Equipment for PAGE

X-ray film and cassettes and −80°C Freezer, or PhosphorImager (Molecular Dynamics) cassettes

PhosphorImager, laser scanning densitometer, or similar detection device

## 2. Procedure

1. CV-1 cells in suspension are infected with vector virus stocks of equal titers, as described previously, and $10^5$ infected cells are seeded into wells of 6-well plates. The sample size on a 6-well plate is convenient to handle, but 4-well plates or individual 35-mm diameter dishes can be used if only a single time point is required. We regularly compare the kinetics of transport or processing of several different proteins in the same experiment, and it is convenient to use a single multi-well plate for each time point. This insures that all samples are processed identically.

2. At 28 hr after infection, cells are rinsed once with PBS to remove the DMEM and are starved for methionine and cysteine in DMEM$^{met-cys-}$ for 15–30 min. Intracellular pools of these amino acids are depleted rapidly, but since DMEM contains anywhere from $10^5$ to $10^6$ times as much methionine and cysteine as the radioactive label, it is important to rinse out this medium.

*Note:* For a very short pulse followed by brief chases, control of times of labeling and chasing are extremely important. In these cases, each plate is pulsed and chased separately. The following description is for an experiment in which the pulse time is 5 min and the chases are intervals of 0, 5, 15, 30, 60, and 120 min. For a detailed description and analysis of very short labeling times followed by chases on the order of several minutes, refer to Braakman *et al.* (1991).

3. Before labeling, the radioactive medium (DMEM$^{met-cys-}$ containing Tran$^{35}$Slabel at a concentration of 100 $\mu$Ci/ml) is prepared, mixed well, and warmed to the desired temperature, usually 37°C. It is best to prewarm the chase medium at the same location. We use a water bath placed next to the circulating water bath in which labeling will be done. If the chase is to be performed in the circulating water bath (we do this for brief chases or chases at temperatures other than 37°C), the chase medium should be low-bicarbonate DMEM.

4. For labeling, each plate is removed from the $CO_2$ incubator and placed on a flooded tray in a circulating water bath pre-equilibrated to the desired temperature (usually 37°C). The starvation medium is rapidly removed from the cells and replaced with 300 $\mu$l prewarmed labeling mixture. It is convenient to have the labeling mix in a small glass beaker to ease access to the solution. We use a Rainin EDP electronic pipet to increase the speed of dispensing radioactive media, but a pipetman also works well. The plate is only briefly out of the water bath while being tilted during the aspiration and maintains its temperature well.

*Note:* For labeling times of 15 min or longer, the radioactive label can be added to the plate on a bench and the plate returned to the $CO_2$ incubator. For labeling times done in a water bath for longer than 5 min, we add HEPES pH 7.0 to the labeling mix to a final concentration of 10 m*M*.

5. The radioactive label is aspirated from the plate with a separate aspirator reserved for radioactive use. To each well of the plate, 2 ml prewarmed chase medium is added and the plate is returned to the $CO_2$ incubator. However, if the medium is to be collected for analysis, it is convenient to use 1 ml so processing can take place in 1.5-ml microfuge tubes. A timer preset for 120 min is activated. We use multichannel timers for this step.

6. Repeat the labeling procedure for each chase time. Start with the longest chase time. By starting with the samples that receive the longest chase and working quickly, all the samples can be processed in the least amount of time. For the shortest chases, however, leave the plates in the circulating waterbath. This works well for chases of up to 15 min without the use of low-bicarbonate medium, provided that the chase medium contains 10 m*M* HEPES pH 7.0.

7. At the end of each chase, place the plate on ice. If the medium is to be assayed for secreted proteins, collect it. Otherwise, quickly add to each well a volume of ice-cold medium equal to the chase medium, aspirate, and replace with 600 $\mu$l ice-cold lysis buffer.

8. Cells are lysed and the lysates are processed by immunoprecipitation, as described in the previous protocol, with the exception that only half the immunoprecipitate is analyzed by PAGE. Typically we use 10 or 12.5% acrylamide minigels and the Laemmli buffer system (Laemmli, 1972).

9. Depending on the type of experiment, gels are either treated for fluorography by soaking for 30 min in 1 *M* sodium salicylate, 10% glycerol or are directly dried. Gels treated for fluorography are exposed to X-ray film (X-OMAT; Kodak, Rochester, NY), at −80°C. For the experiment described here, the signal is sufficient for autoradiography with X-ray film overnight at room temperature or exposure for a few hours on PhosphorImager plates.

## D. Summary

SV40 late-replacement vectors provide an excellent means for expressing large amounts of proteins from cDNAs of less than 2400 bp. The amount of protein made 28 hr after infection is sufficient for extremely brief pulse–chase protocols. At periods postinfection when cells are still healthy, sufficient protein has been made for techniques in which only a small fraction of the protein is detected, including immunocytochemistry on cryosections. Because of the ease of subcloning DNA fragments into them, the ease of making virus stocks, and the ability to achieve comparable amounts of protein expression reproducibly with different infections, these vectors are superb tools for comparing series of mutants made by site-directed mutagenesis, especially if a normal cellular environment is important for the measurements to be made.

## Acknowledgments

This work was supported by Grant GM37547 from the National Institutes of Health and a grant from the Human Frontier Science Program Organization.

## References


Braakman, I., Hoover-Litty, H., Wagner, K. R., and Helenius, A. (1991). Folding of influenza hemagglutinin in the endoplasmic reticulum. *J. Cell Biol.* **114,** 401–411.

Clague, M. J., Schoch, C., and Blumenthal, R. (1991). Delay time for influenza virus hemagglutinin-induced membrane fusion depends on hemagglutinin surface density. *J. Virol.* **65,** 2402–2407.

Clayson, E. T., Brando, L. V., and Compans, R. W. (1989). Release of simian virus 40 virions from epithelial cells is polarized and occurs without cell lysis. *J. Virol.* **63,** 2278–2288.

Copeland, C. S., Doms, R. W., Bolzau, E. M., Webster, R. G., and Helenius, A. (1986). Assembly of influenza hemagglutinin trimers and its role in intracellular transport. *J. Cell Biol.* **103,** 1179–1191.

Davis, G. L., and Hunter, E. (1987). A charged amino acid substitution within the transmembrane anchor of the Rous sarcoma virus envelope glycoprotein affects surface expression but not intracellular transport. *J. Cell Biol.* **105,** 1191–1203.

Doxsey, S. J., Sambrook, J., Helenius, A., and White, J. (1985). An efficient method for introducing macromolecules into living cells. *J. Cell Biol.* **101,** 19–27.

Doxsey, S. J., Brodsky, F. M., Blank, G. S. and Helenius, A. (1987). Inhibition of endocytosis by anti-clathrin antibodies. *Cell* **50,** 453–463.

Doyle, C., Roth, M. G., Sambrook, J. and Gething, M. J. (1985). Mutations in the cytoplasmic domain of the influenza virus hemagglutinin affect different stages of intracellular transport. *J. Cell Biol.* **100,** 704–714.

Fiers, W., Contreras, R., Haegemann, G., Rogiers, R., Van de Voorde, A., Van Heuverswyn, H., Van Herreweghe, J., Volckaert, G., and Ysebaert, M. (1978). Complete nucleotide sequence of SV40 DNA. *Nature (London)* **273,** 113–120.

Fire, E., Zwart, D. E., Roth, M. G., and Henis, Y. I. (1991). Evidence from lateral mobility studies for dynamic interactions of a mutant influenza hemagglutinin with coated pits. *J. Cell Biol.* **115,** 1585–1594.

Gething, M. J., and Sambrook, J. (1981). Cell surface expression of influenza haemagglutinin from a cloned DNA copy of the RNA gene. *Nature (London)* **293,** 620–625.

Gething, M. J., and Sambrook, J. (1982). Construction of influenza haemagglutinin genes that code for intracellular and secreted forms of the protein. *Nature (London)* **300,** 598–603.

Gething, M. J., McCammon, K., and Sambrook, J. (1986). Expression of wild-type and mutant forms of influenza hemagglutinin: The role of folding in intracellular transport. *Cell* **46,** 939–950.

Gluzman, Y. (1981). SV40-Transformed simian cells support the replication of early SV40 mutants. *Cell* **23,** 175–182.

Hartman, J. R., Nayak, D. P., and Fareed, G. C. (1982). Human influenza virus hemagglutinin is expressed in monkey cells using simian virus 40 vectors. *Proc. Natl. Acad. Sci. USA* **79,** 233–237.

Hiebert, S. W., and Lamb, R. A. (1988). Cell surface expression of glycosylated, nonglycosylated, and truncated forms of a cytoplasmic protein pyruvate kinase. *J. Cell Biol.* **107,** 865–876.

Jay, G., Nomura, S., Anderson, C. W., and Khoury, G. (1981). Identification of the SV40 agnogene product: a DNA binding protein. *Nature (London)* **291,** 346–349.

Ktistakis, N. T., Thomas, D., and Roth, M. G. (1990). Characteristics of the tyrosine recognition signal for internalization of transmembrane surface glycoproteins. *J. Cell Biol.* **111,** 1393–1407.

Ktistakis, N. T., Roth, M. G., and Bloom, G. S. (1991). PtK1 cells contain a nondiffusible, dominant factor that makes the Golgi apparatus resistant to brefeldin A. *J. Cell Biol.* **113,** 1009–1023.

Laemmli, U. K. (1972). Cleavage of structural proteins during assembly of the head of bacteriophage T4. *Nature (London)* **227,** 680–685.

Lazarovits, J., and Roth, M. (1988). A single amino acid change in the cytoplasmic domain allows the influenza virus hemagglutinin to be endocytosed through coated pits. *Cell* **53,** 743–752.

Lazarovits, J., Shia, S. P., Ktistakis, N., Lee, M. S., Bird, C., and Roth, M. G. (1990). The effects of foreign transmembrane domains on the biosynthesis of the influenza virus hemagglutinin. *J. Biol. Chem.* **265,** 4760–4767.

Naim, H. Y., and Roth, M. G. (1993). Basis for selective incorporation of glycoproteins into the influenza virus envelope. *J. Virol.* **67,** 4831–4841.

O'Banion, M. K., Winn, V. D., Settleman, J., and Young D. A. (1993). Genetic definition of a new bovine papillomavirus type 1 open reading frame, E5B, that encodes a hydrophobic protein involved in altering host-cell protein processing. *J. Virol.* **67,** 3427–3434.

Paterson, R. G., and Lamb, R. A. (1987). Ability of the hydrophobic fusion-related external domain of a paramyxovirus F protein to act as a membrane anchor. *Cell* **48,** 441–452.

Perez, L. G., Davis, G. L., and Hunter, E. (1987a). Mutants of the Rous sarcoma virus envelope glycoprotein that lack the transmembrane anchor and cytoplasmic domains: Analysis of intracellular transport and assembly into virions. *J. Virol.* **61,** 2981–2988.

Perez, L., Wills, J. W., and Hunter, E. (1987b). Expression of the Rous sarcoma virus env gene from a simian virus 40 late-region replacement vector: Effects of upstream initiation codons. *J. Virol.* **61,** 1276–1281.

Pipas, J. M., Adler, S. P., Peden, K. W. C., and Nathans, D. (1979). Deletion mutants of SV40 that affect the structure of viral tumor antigens. *Cold Spring Harbor Symp. Quant. Biol.* **44,** 285–291.

Roth, M. G., Compans, R. W., Giusti, L., Davis, A. R., Nayak, D. P., Gething, M. J., and Sambrook J. (1983). Influenza virus hemagglutinin expression is polarized in cells infected with recombinant SV40 viruses carrying cloned hemagglutinin DNA. *Cell* **33,** 435–443.

Roth, M. G., Doyle, C., Sambrook, J., and Gething, M. J. (1986). Heterologous transmembrane and cytoplasmic domains direct functional chimeric influenza virus hemagglutinins into the endocytic pathway. *J. Cell Biol.* **102,** 1271–1283.

Roth, M. G., Gundersen, D., Patil, N., and Rodriguez-Boulan, E. (1987). The large external domain is sufficient for the correct sorting of secreted or chimeric influenza virus hemagglutinins in polarized monkey kidney cells. *J. Cell Biol.* **104,** 769–782.

Sambrook, J., Hanahan, D., Rodgers, L., and Gething, M. J. (1986). Expression of human tissue-type plasminogen activator from lytic viral vectors and in established cell lines. *Mol. Biol. Med.* **3,** 459–481.

Sambrook, J., Fritsch, E., and Maniatis, T. (1989). "Molecular Cloning: A Laboratory Manual," 2d Ed. Cold Spring Harbor Laboratory Press, Cold Spring Harbor, New York.

Schoch, C., and Blumenthal, R. (1993). Role of the fusion peptide sequence in initial stages of influenza hemagglutinin-induced cell fusion. *J. Biol. Chem.* **268,** 9267–9274.

Settleman, J., and DiMaio, D. (1988). Efficient transactivation and morphologic transformation by bovine papillomavirus genes expressed from a bovine papillomavirus/simian virus 40 recombinant virus. *Proc. Natl. Acad. Sci. USA* **85,** 9007–9011.

Settleman, J., Fazeli, A., Malicki, J., Horwitz, B. H., and DiMaio, D. (1989). Genetic evidence that acute morphologic transformation, induction of cellular DNA synthesis, and focus formation are mediated by a single activity of the bovine papilloma virus E5 protein. *Mol. Cell. Biol.* **9,** 5563–5572.

Singh, I., Doms, R. W., Wagner, K. R., and Helenius, A. (1990). Intracellular transport of soluble and membrane-bound glycoproteins: Folding, assembly and secretion of anchor-free influenza hemagglutinin. *EMBO J.* **9,** 631–639.

Sladek, T. L., and Jacobberger, J. W. (1992). Simian virus 40 large T-antigen expression decreases the G1 and increases the G2 + M cell cycle phase durations in exponentially growing cells. *J. Virol.* **66,** 1059–1065.

Sveda, M. M., and Lai, C. (1981). Functional expression in primate cells of cloned DNA coding for the hemagglutinin surface glycoprotein of influenza virus. *Pro. Natl. Acad. Sci. USA* **78,** 5488–5492.

Tooze, J. (1981). "DNA Tumor Viruses," 2d Ed. Cold Spring Harbor Laboratory Press, Cold Spring Harbor, New York.

Valcarcel, J., Fortes, P., and Ortin, J. (1993). Splicing of influenza virus matrix protein mRNA expressed from a simian virus 40 recombinant. *J. Gen. Virol.* **74,** 1317–1326.

Wills, J. W., Srinivas, R. V., and Hunter, E. (1984). Mutations of the Rous saroma virus *env* gene that affect the transport and subcellular location of the glycoprotein products. *J. Cell Biol.* **99,** 2011–2023.

# CHAPTER 7

# Use of Recombinant Vaccinia Virus Vectors for Cell Biology

**Ora A. Weisz and Carolyn E. Machamer**

Department of Cell Biology and Anatomy
Johns Hopkins University School of Medicine
Baltimore, Maryland 21205



## I. Introduction

Vaccinia virus has been exploited as a powerful and convenient tool for transient expression of proteins in mammalian cells. This chapter describes the use of several of the recombinant vaccinia expression systems pioneered by Bernard Moss and colleagues. We will focus on the systems that are most useful for cell biologists, and discuss their advantages and limitations. Vaccinia-mediated expression can be used for assessing cellular localization, post-translational modifications, oligomerization, and transport and turnover rates. The system provides a rapid method for screening mutant proteins for expression and targeting. It is an excellent way of quickly deciding which mutant proteins might be worth further study using stable expression systems. Some uses of vaccinia vectors will not be discussed here, including methods for large-scale production of proteins. For detailed methods for large-scale production of proteins from cDNAs, or other vaccinia methods not discussed here, see Earl and Moss (1991), Moss *et al.* (1990), and Moss (1991).

## A. Biology of Vaccinia Virus

Vaccinia virus is the best-studied member of the Poxviridae, the largest and most complex of the animal viruses. Widespread vaccination with vaccinia virus (probably derived from cowpox virus) resulted in the worldwide eradication of smallpox. The double-stranded linear genome of vaccinia virus is nearly 200 kb, and there are over 250 potential genes (Goebel *et al.*, 1990; Moss, 1991). The virions are enveloped and are approximately 200 $\times$ 300 nm in size, with a characteristic brick shape (Fig. 1). Several features of the vaccinia life cycle make it unique as a eukaryotic expression vector. At least 25 kb of DNA can be inserted into the vaccinia genome without detrimental effects on viral replication or assembly. Vaccinia replicates completely within the cytoplasm of the host cell, and thus imports or directs the synthesis of its own polymerases and transcription factors. In addition, the virus has a wide host range, so most cultured mammalian cell lines are susceptible to infection. The virus is easy to grow and purify in large quantities, and is relatively safe to work with.

Considering the size of vaccinia virus, the complexity of replication and assembly is not surprising (Figs. 1 and 2). After binding to the host cell, the viral envelope is believed to fuse directly with the plasma membrane (Doms *et al.*, 1990), releasing the core virion into the cytoplasm. Little is known about the uncoating steps, but transcription of early genes begins within 1 hr of infection. The virion core contains all the enzymes it needs for transcription of the early genes. Transcribed RNAs are capped, methylated, and polyadenylated in the cytosol by viral enzymes. DNA replication begins with 3–4 hr, and occurs in discrete juxtanuclear regions of the cytosol termed "viral factories" near the Golgi complex. After transcription of intermediate and late viral genes (by different viral RNA polymerases), viral assembly begins. This complex process appears to involve two separate membrane "enwrapping" events, each of which results in a double lipid bilayer surrounding the core particle. The first double membrane is derived from the "intermediate compartment" between the endoplasmic reticulum (ER) and Golgi (Sodeik *et al.*, 1993). The viral DNA is

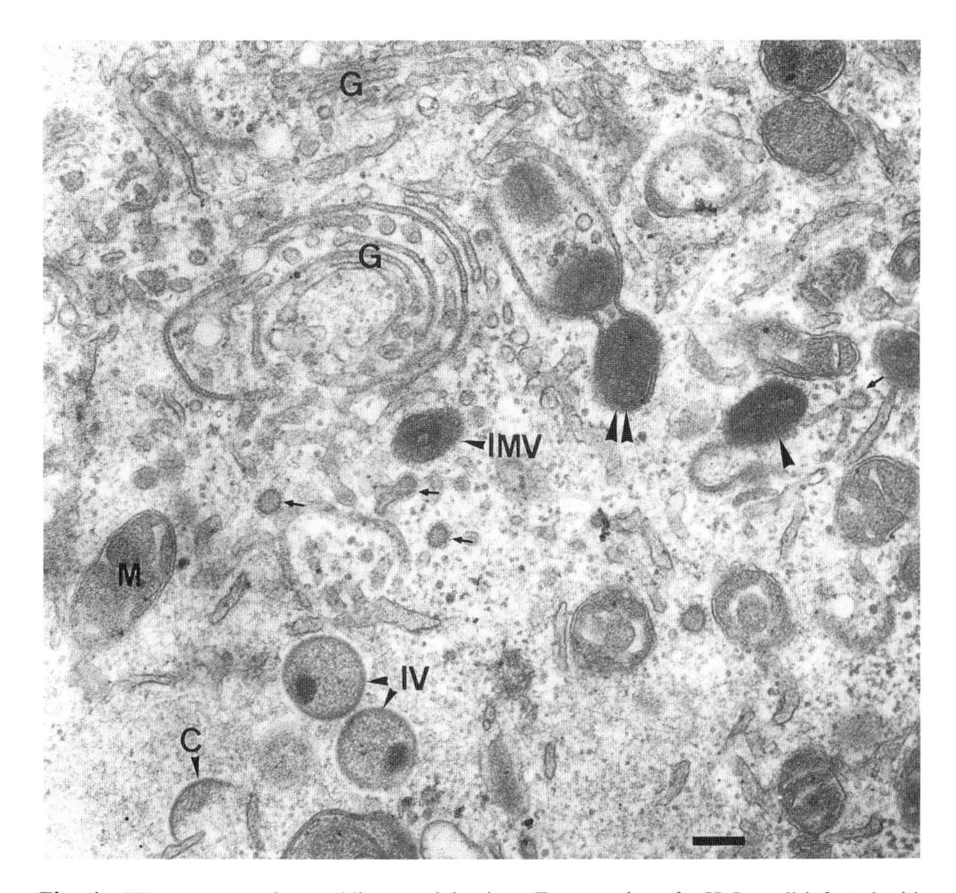


**Fig. 1** Ultrastructure of assembling vaccinia virus. Epon section of a HeLa cell infected with vaccinia virus for 8 hr. The viral factories (where viral DNA replication and early assembly events occur) can be seen near the Golgi complex, as can several forms of assembling virions. C, Crescent; IV, immature virus, IMV, intracellular mature virus (first infectious form). The single arrowhead indicates attachment of an IMV to a cisterna of the *trans*-Golgi network (arrows indicate putative clathrin buds and vesicles). The double arrowhead shows wrapping of an IMV by *trans*-Golgi network membrane. All the membranes surrounding the virion are clearly visualized in ultrathin cryosections (Sodeik *et al.*, 1993). Bar, 200 nm. (Micrograph courtesy of Drs. B. Sodeik and G. Griffiths.)

inserted into a crescent-shaped region of the intermediate compartment membrane which then fuses around it, generating intracellular mature virus (IMV), formerly called intracellular "naked" virus. The second enwrapping membranes are derived from the *trans*-Golgi network (Schmelz *et al.*, 1994), resulting in the intracellular enveloped virus (IEV) form. Both membrane enwrapping events are thought to depend on certain viral membrane-associated proteins that localize specifically to the intermediate compartment or the *trans*-Golgi network. IEV with four membranes (doubly enwrapped) can fuse with the

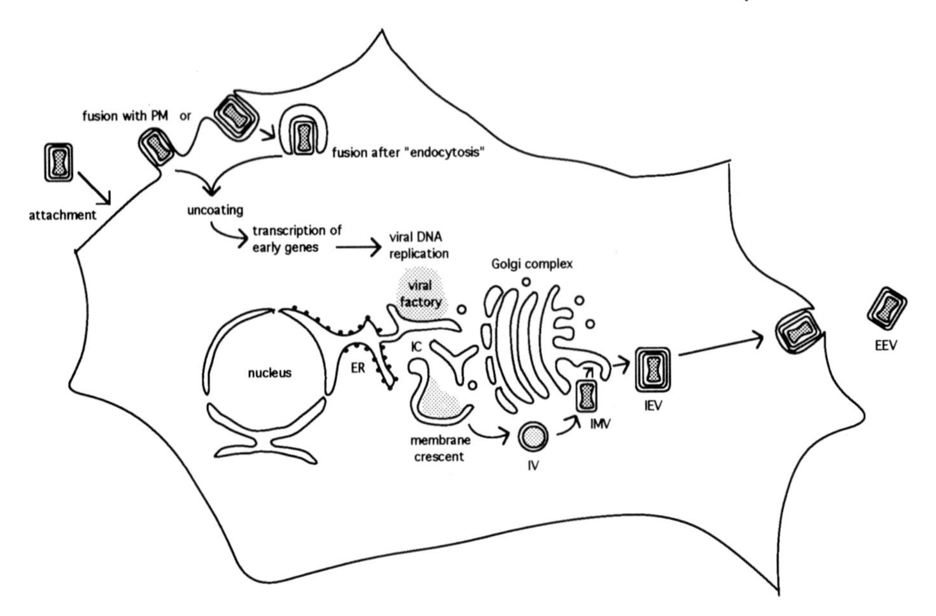


**Fig. 2** Infection cycle of vaccinia virus. After virus entry, uncoating, transcription of early genes, and DNA replication, virus assembly occurs (Griffiths and Rottier, 1992). Two membrane enwrapping events result in intracellular mature virus (IMV) and intracellular enveloped virus (IEV). Fusion of IEV with the plasma membrane (an inefficient process for many vaccinia strains) results in release of virions with three membranes (EEV, extracellular enveloped virus). Both IMV and IEV forms are infectious. ER, Endoplasmic reticulum; IC, intermediate compartment; IV, immature virus.

plasma membrane, releasing virions with three membranes into the extracellular space. However, some strains of vaccinia are inefficiently released from cells. Approximately 95% of the commonly used WR strain remains cell associated. Other strains (e.g., IHD-J) produce substantial amounts of extracellular virus. Since both forms of intracellular virus (singly and doubly enwrapped) are infectious, virus is usually purified from cell homogenates rather than from the culture medium.

### B. Systems for Protein Expression Using Vaccinia Virus

Expression of foreign genes using vaccinia virus is based on recombinant viruses constructed by insertion of cDNA into the nonessential thymidine kinase (TK) gene. Both direct and indirect methods of expression are possible (Fig. 3). The foreign gene can be inserted into the vaccinia genome by homologous recombination using a plasmid with flanking regions of vaccinia DNA. The recombinant virus is selected, expanded, and used to infect cells, which then express high levels of the foreign protein. In an alternative (''indirect'') approach, a recombinant vaccinia virus encoding bacteriophage T7 RNA polymer-

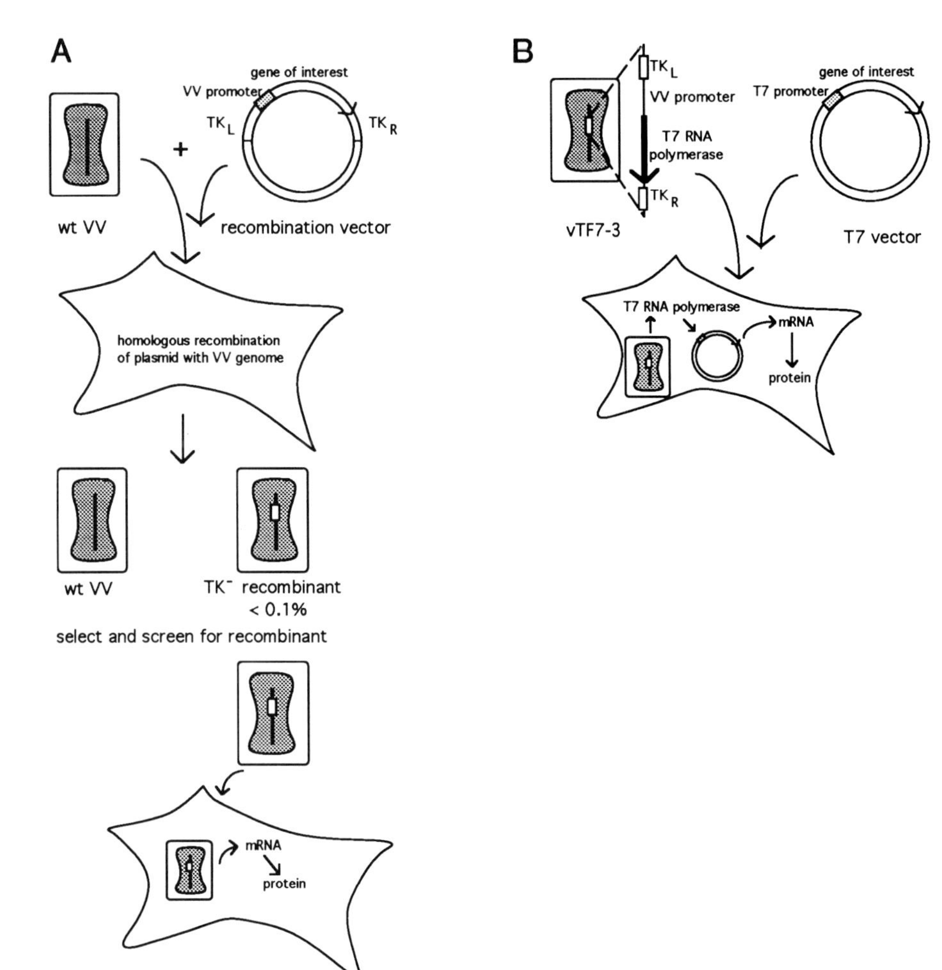


**Fig. 3** Comparison of direct expression using a recombinant vaccinia virus and the vaccinia/T7 RNA polymerase hybrid system. (A) Expression using the direct system requires production of a recombinant virus encoding the foreign gene as shown. (B) In the hybrid system, cells are infected with the recombinant virus vTF7-3 encoding T7 RNA polymerase, then transfected with plasmids encoding the foreign gene behind a T7 promoter.

ase (vTF7-3) is used to infect cells (Fuerst *et al.*, 1986). The T7 RNA polymerase is expressed efficiently in the cytoplasm early after infection. Transfection of cells immediately after infection with a vector containing the gene of interest cloned behind a T7 promoter results in rapid and efficient expression of the encoded protein. This system is convenient, since recombinant virus production is unnecessary. Commonly used vectors such as pBluescript (Stratagene, La Jolla, CA) can be used.

## II. Generating a Recombinant Vaccinia Virus

Recombinant vaccinia viruses are generated by subcloning the foreign gene into a plasmid transfer vector so it is flanked by DNA from the vaccinia (TK) gene, which is nonessential for growth of the virus in tissue culture. This plasmid is then transfected into vaccinia-infected cells. Homologous recombination of the plasmid and the vaccinia genome generates a recombinant virus with an inactive TK gene. A lysate from the infected cells is used to infect a thymidine kinase negative ($TK^-$) cell line, usually human osteosarcoma 143B cells (Mackett *et al.*, 1984). Recombinant viruses are enriched by growing the cells in the presence of bromodeoxyuridine (BrdU), since only recombinant viruses (and TK mutants) are able to replicate. If the plasmid transfer vector used for recombination contains the $\beta$-galactosidase gene (see subsequent discussion), recombination viruses can be distinguished from TK mutants by color screening using 5-bromo-4-chloro-3-indolyl-$\beta$-D-galactoside (X-gal; Chakrabarti *et al.*, 1985).

### A. Vectors and Subcloning

Various vectors for recombination into the vaccinia genome are available (Moss *et al.*, 1990; Earl and Moss, 1991; Moss, 1992). The use of an early vaccinia promoter to drive the foreign gene is essential for cell biology applications, since the protein will be expressed before most of the cytopathic effects of the virus infection become evident. The vector we use for production of recombinant virus is pSC11ss (Fig. 4A) or its derivative pSC65. pSC11ss contains the promoter $P_{7.5}$ that is transcribed both at early and late times after infection; pSC65 contains a synthetic early/late promoter. Following the promoter is a short polylinker into which the foreign gene is cloned. Another vaccinia promoter, $P_{11}$ (a late promoter), drives expression of the bacterial $\beta$-galactosidase gene and allows color screening of plaques to help identify recombinants. Both expression cassettes are flanked by portions of the vaccinia TK gene.

For subcloning the gene of interest into the recombination vector, several things must be kept in mind. Since splicing of the subcloned DNA will not occur, cDNAs must be used. The 5′ and 3′ untranslated regions should be kept short. Also, it is important to check for a transcription termination sequence that is recognized by the early RNA polymerase. The sequence TTTTTNT ($T_5NT$, where N can be any nucleotide) will cause termination of transcripts about 50 bp downstream (Yuen and Moss, 1987). Thus, silent mutations that eliminate any $T_5NT$ sequences should be introduced by site-directed mutagenesis prior to subcloning. Although proteins can still be expressed from recombinant viruses made with pSC11ss or pSC65 when their cDNAs contain $T_5NT$ sequences, expression will only occur during the late phase of virus infection

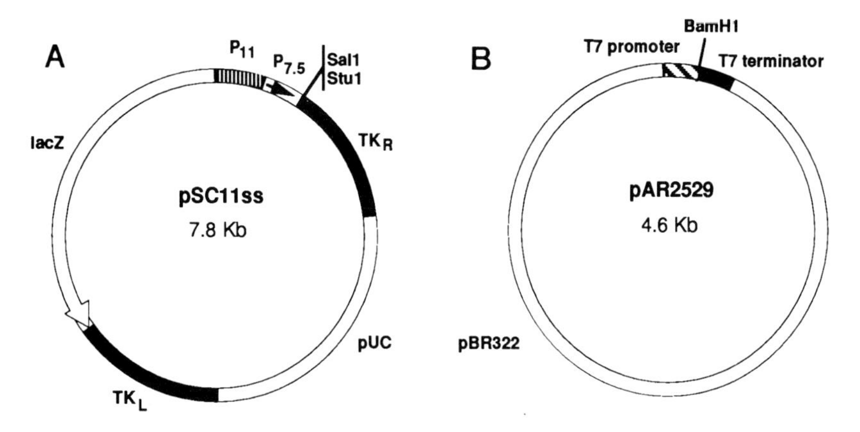


**Fig. 4** Examples of vectors used for recombination and for T7 polymerase-mediated expression. (A) pSC11ss is used for producing vaccinia recombinants. The foreign gene (with its own start and stop codons) is subcloned into the *Stu* I or *Sal* I sites behind the vaccinia $P_{7.5}$ promoter. (B) pAR2529 is used for T7 polymerase-mediated expression after infection with vTF7-3 (Section IV,A). Other plasmids such as pBluescript (Stratagene) can also be used; however, the absence of the T7 polymerase terminator sequence and the presence of the G-C rich region in the Bluescript polylinker upstream of the foreign gene can reduce mRNA production and consequently decrease protein expression.

(>6 hr) when the late RNA polymerase (which does not terminate transcripts at $T_5NT$) is active.

After the subcloning step is completed, purified plasmid DNA should be prepared in CsCl gradients or by the Qiagen method (QIAGEN, Inc., Chatsworth, CA). Nonpurified miniprep DNA does not work when the $Ca_3(PO_4)_2$ method of transfection is used.

## B. Recombination

Work with vaccinia virus should be performed under standard Biosafety Level 2 (BL-2) conditions, including the use of class I or II biological safety cabinets. National Institutes of Health (NIH) and Centers for Disease Control (CDC) guidelines recommend that workers be vaccinated every 3 years, although each institution sets its own requirements for vaccination and physical containment. Regulations should be obtained from institutional biosafety offices before initiating projects involving vaccinia virus. We use class II biological safety cabinets, autoclave disposable items that have been in contact with virus, and inactivate used solutions containing infectious virus with bleach. A convenient way of inactivating virus-containing solutions is to aspirate them into the reservoir flask in the hood and, immediately afterward (using the same pasteur pipet), aspirate bleach through the line. Additional precautions for using large quantities of vaccinia virus (for example, when growing and purifying large stocks) are discussed in Section III,A.

### 1. Production of Recombinant Viruses

1. HeLa or CV-1 cells are usually used for production of recombinant viruses. They are grown in Dulbecco's modified Eagle's medium (DMEM) containing 5% fetal calf serum (FCS). Plate approximately $5 \times 10^5$ cells in a 3.5-cm dish for cells to be about 80% confluent the next day.

2a. Rinse the cells once in serum-free DMEM, and add 0.25 ml serum-free medium containing 0.05–0.1 plaque-forming units (pfu)/cell of the WR strain of vaccinia virus (we use $10^5$ pfu total).

2b. Some investigators employ a trypsinization step prior to infection, which is required when the virus stock is a cell lysate instead of a purified preparation. Add an equal volume of 0.25 mg/ml trypsin to the amount of lysate that will be required and incubated at 37°C for 30 min, with occasional vortexing. This helps dissociate aggregates and releases infectious virus from cell debris. If trypsinization is used, DMEM containing 2.5% FCS should be used to dilute and adsorb the virus. FCS inhibits the trypsin.

3. After addition of the virus inoculum, cells are returned to the incubator and rocked every 5–10 min for 45 min (or 2 hr if trypsinized virus in serum-containing medium is used).

Toward the end of the infection period, prepare a $Ca_3(PO_4)_2$ precipitate of the DNA for transfection.

4. Add 5 $\mu$g DNA to 125 $\mu$l 2× HEPES buffer (50 m*M* HEPES, pH 7.1, 0.28 *M* NaCl, 1.5 m*M* sodium phosphate). Add an equal volume of 0.25 *M* $CaCl_2$ dropwise with continuous vortexing, and allow a precipitate to form by leaving the mixture at ambient temperature for 20–30 min. The precipitate should be very fine, turning the solution slightly opaque.

5. After the adsorption period, remove the virus inoculum and add 2.25 ml DMEM/5% FCS to the dish. Allow the pH to equilibrate in the $CO_2$ incubator for 15 min.

6. Add the precipitated DNA solution dropwise while gently swirling the dish, and return the cells to the incubator.

7. Replace DNA-containing medium with fresh growth medium the following morning.

8. Prepare a cell lysate 2 days after infection by scraping the cells into their medium and centrifuging for 5 min at 650 *g* at 4°C.

9. Resuspend the cell pellet in 0.25 ml complete medium and lyse the cells by freezing (in dry ice/ethanol) and thawing (at 37°C) three times.

10. The lysate is then stored at −80°C until selection and screening is performed.

## C. Selection and Screening of Recombinants

1. Prepare a BrdU stock solution (5 mg/ml) in water, sterilize by filtering, and store at −20°C.

2. Grow human $TK^-$ 143B cells in DMEM containing 10% FCS and 25 $\mu$g/ml BrdU. Plate 143B cells in a 6-well tissue culture dish at $5 \times 10^5$ cells/well in 2 ml complete medium and grow to near confluence (usually overnight).

3. Thaw an aliquot (100 $\mu$l) of transfected cell lysate, add an equal volume of 0.25 mg/ml trypsin, and incubate the mixture at 37°C for 30 min.

4. Make 10-fold serial dilutions (to $10^{-4}$) of trypsinized lysate in DMEM with 2.5% FCS.

5. Aspirate the medium from the cells and infect cells in the wells in duplicate with 1 ml $10^{-2}$–$10^{-4}$ dilutions for 2 hr with intermittent rocking.

Toward the end of the infection period, the agarose overlay is prepared. Although most of the virus remains cell-associated, an overlay helps insure that a plaque is derived from a single virion.

6. Melt a previously autoclaved solution of 2% low melting point agarose (No. 5517; GIBCO/BRL, Grand Island, NY) in water, aliquot enough for 1 ml per well into a tube, and equilibrate to 50–55°C. [Some lots of Difco (Detroit, MI) agar work well, but some are toxic to cells and/or plaque formation.]

7. Warm 2-fold concentrated DMEM containing 10% FCS and 50 $\mu$g/ml BrdU (1 ml/well) to 37°C.

8. At the end of the infection period, aspirate the inoculum from the cells, mix the agarose and 2× medium, and overlay each well with 2 ml. After the overlay has solidified at room temperature, return the dishes to the $CO_2$ incubator.

9. After 2 days, the infected cells will be rounded and dead and can be seen as clear areas (plaques) in the cell lawn. To see these plaques more clearly and to identify recombinants, prepare a second overlay by mixing an equal volume of melted agarose with 2× DMEM containing 0.1 mg/ml neutral red (from a 10 mg/ml stock) and 0.6 mg/ml X-gal (from a 60 mg/ml stock in dimethyl sulfoxide or dimethyl formamide).

10. Overlay each well with 2 ml, allow the agarose to solidify, and incubate the dishes overnight.

11. The following morning, pick any blue plaques by inserting a pasteur pipet or a yellow pipet tip through the agarose until it contacts the cell layer; scrape the cell monolayer gently and discharge the agarose plug into a sterile microfuge tube containing 0.4 ml DMEM/5% FCS.

12. These plaque isolates should be vortexed vigorously, then frozen and thawed three times, and stored at −80°C until further purification. Typically 6 plaques are picked and taken through two additional rounds of plaque purification, as described subsequently, to ascertain that the recombinant virus is clonally derived and devoid of nonrecombinant virus.

For each round of plaque purification, a 6-well dish of 143B cells is plated as just described for each isolated plaque. The frozen and thawed cell lysate is diluted serially from $10^{-1}$ to $10^{-3}$ and duplicate wells of the confluent monolayers are infected with 1 ml of each dilution of virus for 2 hr. After infection, the wells are overlaid first with BrdU–agarose for 2 days, and then with a second overlay containing X-gal, as described.

The plaque-purified recombinants can be tested by several different methods (Northern or Southern blotting, immunofluorescence, dot blotting, or immunoprecipitation) to insure that they contain the gene of interest inserted into the vaccinia TK gene. We commonly use immunofluorescence after 2–3 plaque purifications. A portion of an isolated plaque is used to infect cells on a cover slip, and cells are fixed and stained 1–2 days later (see protocol in Section IV,C,3). Cells surrounding the plaques should express the foreign protein.

For other methods of screening, plaques should be expanded by successively infecting larger and larger numbers of cells. This is usually done by infecting cells in one well of a 12-well dish with half of an isolated plaque; after 2 days a cell lysate is made (as described) and is used to infect cells on a 35-mm dish. Before growing a large-scale preparation of the recombinant, we also test that the expressed protein is the correct size by immunoprecipitation from radiolabeled infected cells, followed by electrophoresis in SDS polyacrylamide gels (see Section IV,C,1).

## D. Potential Problems and Practical Considerations

After subcloning into the recombination vector, approximately 3 wk will be required for generating and purifying the vaccinia recombinant. If all goes well, and if cells for successive plaque purifications are plated to be ready the same day plaques are picked, the time required may be somewhat shorter.

The recombination frequency should be about 1 in 1000, which is relatively high. However, occasionally only a few blue and many clear plaques are obtained. One problem might be that the vaccinia virus used to infect the cells for the recombination contains a large population of $TK^-$ virus. The solution is to plaque purify the wild-type vaccinia, and test several plaques for their ability to grow in 143B cells in the presence of BrdU. Select a plaque that *fails* to grow (i.e., $TK^+$) for future use.

Another problem might be that expression of the foreign protein is incompatible with vaccinia replication or assembly. We have had this problem with several proteins that accumulate in the intermediate compartment and *cis*-Golgi network. Either the recombinant is never obtained, or DNA encoding the foreign gene is lost, giving the virus a growth advantage over virus that still contains the foreign gene. In this case, one alternative is to use a double infection system (Moss *et al.*, 1990; Earl and Moss, 1991). The cDNA encoding the foreign gene is cloned into a recombination vector with the T7 RNA promoter instead of the $P_{7.5}$ promoter (pTM-1), and a recombinant virus is produced as described already. Since the protein is not expressed, it will not interfere with recombinant virus production. Expression is mediated by co-infecting the cells with another vaccinia virus encoding T7 RNA polymerase (vTF7-3). Alternatively, the infection/transfection system can be used (Section IV,B).

## III. Large-Scale Growth and Purification of Vaccinia Virus Stocks

### A. Growth of Vaccinia Virus

1. Grow HeLa cells in DMEM with 5% FCS and expand into four 15-cm dishes. Grow the cells until they are approximately 80% confluent.

2a. If the virus inoculum is a purified preparation, rinse the dishes with phosphate-buffered saline (PBS) and infect with vaccinia virus (or a vaccinia recombinant) at 0.05–0.1 pfu/cell in 2 ml serum-free DMEM per dish (a confluent dish contains approximately $2 \times 10^7$ cells). Return the infected cells to the incubator and rock every 5–10 min for 30 min.

2b. If the virus inoculum is a crude cell lysate, the volume required to give 0.05–0.1 pfu/cell should be trypsinized as described in Section II,C and diluted in medium with 2.5% serum for infection. Rock the infected cells every 15 min for 2 hr.

3. At the end of the infection period, add 20 ml DMEM/5% FCS to each dish.

4. Purify the virus 2–3 days after infection, at which time most or all of the cells should appear rounded, but should remain attached to the dish.

### B. Purification of Vaccinia Virus

Although a frozen and thawed preparation of infected cells can be used for expression work, purified virus stocks are recommended. Sonication and trypsinization of cell lysates is the only way to disrupt virus aggregates generated by freezing in medium, and must be performed before infecting cells. Purified virus is stored at high pH in a buffer lacking salt, which minimizes aggregation. Thus trypsinization and sonication are unnecessary when infecting with purified preparations.

1. Virus purification should be performed in a laminar flow hood and gloves should be worn. Pipets and any glassware used should be soaked in bleach or autoclaved after use to destroy infectious virus.

2. Scrape the infected HeLa cells into their medium using a rubber policeman; then combine and transfer to 50-ml disposable sterile tubes and centrifuge for 5 min at 200 *g* (1000 rpm in a tabletop centrifuge) at 4°C.

3. Resuspend the pellets in a total of 8 ml of 10 m*M* Tris HCl, pH 9.0, and homogenize (in two batches) on ice in a 7-ml dounce with 40–60 strokes using the tight pestle.

4. Centrifuge the homogenate in two 15-ml disposable tubes for 5 min at 200 *g* at 4°C to remove nuclei. Collect the supernatant and save on ice while the pellets are washed.

5. Resuspend the pellets in a small volume of 10 m*M* Tris-HCl, pH 9.0, dilute to 10 ml with the same solution, and recentrifuge as in step 4.

6. Transfer the supernatants along with those from the previous spin into 50-ml tubes and centrifuge at 650 *g* (2000 rpm) for 10 min at 4°C to remove any remaining debris.

7. Sonicate the supernatants for 2–5 min in a water bath sonicator; then distribute into two SW28 tubes and underlay with an equal volume of 36% (w/v) sucrose in 10 m*M* Tris-HCl, pH 9.0, to fill the tubes.

8. Centrifuge the tubes at 13500 rpm (33000 *g*) in a SW28 rotor for 80 min at 4°C.

9. After centrifugation, remove the supernatant, treat with bleach, and discard. Resuspend the pellets in 2 ml of 1 m*M* Tris-HCl, pH 9.0, per tube (they should be easy to resuspend but can be sonicated if necessary). Combine and dispense into small (50–100 $\mu$l) aliquots.

10. Purified virus is stored at −80°C. Usually, virus purified with this protocol yields 2–8 $\times$ $10^9$ pfu/ml. After an aliquot is thawed for use, any remainder can be refrozen at −80°C and used once or twice more without a substantial loss of titer.

## C. Titering a Vaccinia Virus Preparation

Titering the virus is performed by infecting cells (typically HeLa, CV-1, or BSC-1) with various dilutions of virus and determining the number of infectious particles per milliliter by counting the number of plaques that form. We usually use HeLa cells, but plaques formed on BSC-1 monolayers are easier to distinguish and quantify. The protocol for titering the virus is basically the same as that described for selection of vaccinia recombinants (Section II,C). However, the purified virus does not require treatment with trypsin prior to plaquing. Also, an agarose overlay is not necessary, since most of the virus remains cell-associated and thus does not spread readily through the culture medium. An agarose overlay containing neutral red stain and/or X-gal added to the cells just prior to counting the plaques can aid in visualizing them.

1. Grow cells in a 6-well dish until they are nearly confluent (usually 4–5 $\times$ $10^5$ cells/well plated the day before).

2. To titer the virus, thaw an aliquot of purified virus and dilute serially into serum-free DMEM to $10^{-8}$. Include a negative control (no virus) and a positive control (purified vaccinia virus of known titer diluted to a concentration designed to give 50–100 plaques).

3. After rinsing the cells in PBS, add 100 $\mu$l serum-free DMEM to each well; then add 100 $\mu$l diluted virus (use dilutions ranging from $10^{-5}$ to $10^{-8}$).

4. Place the dish in the incubator and rock every 5–10 min.

5. After 30 min, replace the inoculum with 2 ml complete medium.

6. After 3 days, aspirate the medium and stain the cells with 0.1% crystal violet in 20% ethanol for 5 min. Aspirate the stain and allow the well to dry before counting the plaques. Alternatively, the wells can be overlaid with 3 ml agarose containing neutral red and/or X-gal. The overlay procedure is described in Section II,C. Count the plaques in each well (where possible) and determine the titer (pfu/ml) based on the dilution of virus used to infect the cells.

## IV. Vaccinia Virus-Mediated Protein Expression

### A. Vectors for Use with the T7 RNA Polymerase System

As mentioned in the introduction, the T7 RNA polymerase method of vaccinia-mediated expression is quite convenient since only one recombinant virus is needed (vTF7-3). In addition, the gene to be expressed can be cloned into commonly used vectors such as pBluescript. However, specially designed vectors containing a slightly extended version of the T7 promoter and the T7 terminator sequence give better expression (Fig. 4B). For very high levels of expression, vectors have been constructed that contain an untranslated leader region from encephalomyocarditis virus as well as the T7 promoter and terminator. The message produced by T7 polymerase is inefficiently capped by the vaccinia capping enzyme, and the leader region from encephalomyocarditis virus allows cap-independent translation (Elroy-Stein *et al.,* 1989). However the 5- to 10-fold increase in protein expression obtained with this system may not be practical for cell biological studies (see Section V,B).

### B. Infection and Infection/Transfection

The protocols for the direct and the indirect methods of expression are essentially the same, with the exception of an additional transfection step following the infection for the T7 polymerase-mediated system. We discuss both protocols for protein expression together in this section, with additional steps for the T7 system described in the appropriate places.

A wide variety of cell lines can be used for expression of the protein of interest. Chinese hamster ovary (CHO) cells are an exception, since they are nonpermissive for vaccinia infection. In addition, polarized Madin–Darby canine kidney (MDCK) and Caco-2 cells are poorly infected.

#### 1. Protein Expression

1. To examine protein expression using metabolic labeling, immunoblotting, or immunofluorescence, cells are plated the day before infection to reach

40–70% confluence (approximately 2 × $10^5$ cells/35-mm dish). Larger dishes may be used for some applications; however, expression is usually high enough that this is not required.

2. Cells are rinsed once in PBS, and then infected with vTF7-3 or another recombinant virus. The virus (10–20 pfu/cell or $10^7$ pfu/35-mm dish) is added in 0.3 ml serum-free DMEM per dish.

3. After addition of the virus, dishes are returned to the incubator and rocked every 5–10 min for 30 min. If the protein of interest is expressed directly by the recombinant virus, the inoculum is replaced after infection with 2 ml regular growth medium containing serum and Step 4 is omitted.

*Note:* If expression is mediated via T7 polymerase (vTF7-3), the infection period can be used to prepare the DNA for transfection. Transfection can be performed using either cationic lipid or calcium phosphate as a carrier.

4a. For lipid-mediated transfection, 2–5 μg plasmid DNA (CsCl-purified or Qiagen-purified works best, but miniprep DNA can also be used) is added to 0.75 ml serum-free DMEM per 35-mm dish of cells. After mixing, 10 μl LipofectACE (No. 18301; GIBCO/BRL) is added to the side of the tube and immediately vortexed for 5 sec. [Lipofectin (No. 18292; GIBCO/BRL) also works well, but is more expensive.] The mixture is incubated at ambient temperature for up to 30 min. At the end of the viral infection period, the inoculum is replaced with the DNA–TransfectACE mixture.

4b. For calcium phosphate-mediated transfection, a precipitate containing 5 μg purified plasmid DNA (encoding the foreign gene behind the T7 promoter) is prepared and added to cells exactly as described in Section II,B. We no longer use the $Ca_3(PO_4)_2$ method of transfection because it is much less reproducible than the cationic lipid method. After transfection, dishes are returned to the incubator. The incubation time varies depending on the type of analysis that will be performed (see the next section).

## C. Analysis of Protein Expression

### 1. Metabolic Labeling

Metabolic labeling can be initiated as early as 3 hr postinfection. Cells should be starved for 15 min in medium lacking the amino acid that will be used to radiolabel cells if a short pulse label will be used. Because of the high expression level, pulse periods can be short (5 min in 100 μCi/ml [$^{35}$S]methionine typically gives an overnight exposure). After the appropriate pulse–chase period, cells are lysed in nonionic detergent. We use a detergent solution containing 50 m*M* Tris-HCl, pH 8.0, 62.5 m*M* EDTA, 0.4% deoxycholate, 1% NP-40, and 0.04 TIU aprotinin/ml. Other typical lysis buffers contain 1% Triton X-100 in Tris-buffered saline with protease inhibitors. After spinning out nuclei and debris (1–15 min in a microfuge), the protein of interest is immunoprecipitated and electrophoresed as desired. Solubilization and immunoprecipitation conditions must be determined empirically for each antibody–antigen combination.

*a. Cell lysis and immunoprecipitation*

1. We typically lyse cells from a 35-mm dish in 0.5 ml detergent solution (described earlier). All polyclonal antibodies and many monoclonal antibodies seem to work well in this mixture.

2. After 10 min on ice, lysates are transferred to a microfuge tube and nuclei and debris are removed by spinning for 1 min.

3. The supernatants are transferred to a fresh tube, and 10% SDS is added to a final concentration of 0.2%. The SDS helps reduce background binding. For immunoprecipitation with monoclonal antibodies, we generally omit the SDS.

4. Antibody is added to the tubes, and they are rotated (2 hr to overnight) at 4°C.

5. We use fixed *Staphylococcus aureus* cells (No. 507861; Calbiochem, San Diego, CA) to collect the antigen–antibody complexes (20 min at 4°C); protein A–Sepharose can also be used.

6. The immunoprecipitates are washed 3 times in RIPA buffer (0.15 *M* NaCl, 10 m*M* Tris-HCl, pH 7.4, 1% NP-40, 1% deoxycholate, 0.1% SDS).

7. The samples are usually eluted directly in SDS sample buffer containing reducing agent and are electrophoresed in SDS–polyacrylamide gels. Alternatively, the samples can be eluted and treated with endo- or exoglycosidases to analyze carbohydrate modifications. Our protocol for digestion with endoglycosidase H is given in the legend to Fig. 7.

## 2. Immunoblotting

1. If the expressed protein is to be detected by immunoblotting, it is advisable to wait until 6–8 hr postinfection before lysing the cells to allow chemical amounts of protein to accumulate. Cells are infected (and transfected if necessary) as described already.

2. After the appropriate incubation time, the medium is removed (and trichloroacetic acid precipitated if the protein is secreted) and the dishes are rinsed once in cold PBS.

3. After removing the last trace of PBS, cells are lysed on the dish by adding 50 μl Laemmli sample buffer containing a reducing agent and swirling with a pipet tip to draw the lysate together (the DNA from the lysed nuclei makes a viscous solution).

4. The samples are transferred to microfuge tubes, heated at 100°C for 3 min (or longer if required to fully shear DNA), and loaded onto SDS–polyacrylamide gels.

5. After electrophoresis, the proteins are transferred to nitrocellulose (or other suitable membrane) and incubated with primary antibodies.

6. Proteins are detected using colorimetric or chemiluminescence methods after incubation with an appropriately conjugated second antibody. An example of immunoblotting used to follow the time course of protein expression is shown in Fig. 5.

## 3. Immunofluorescence

Significant levels of expressed proteins can usually be detected by immunofluorescence staining between 4 and 8 hr after infection, depending on the cell type used and the subcellular localization of the protein. Proteins that become concentrated in subcellular compartments (e.g., nuclear, Golgi, or lysosomal proteins) may be detectable at shorter times after infection than proteins with diffuse localizations (e.g., plasma membrane or cytoplasmic proteins). At long times after infection (>7–8 hr), localization by immunofluorescence becomes

**Fig. 5** Time course of VSV G protein expression after vTF7-3-mediated expression. BHK cells were infected with vTF7-3, then transfected as described with 5 $\mu$g per dish of pAR/G, which encodes the G protein of vesicular stomatitis virus (VSV). At each time point, individual dishes were rinsed once with PBS, then solubilized in 50 $\mu$l Laemmli sample buffer containing 5% $\beta$-mercaptoethanol. Samples were heated to 100°C for 3 min prior to electrophoresis in a 10% polyacrylamide gel. Proteins were transferred to nitrocellulose and incubated overnight with anti-VSV polyclonal antibody after blocking in Tris-buffered saline containing 0.05% Tween-20 and 5% nonfat dry milk. Primary antibody was detected with horseradish-conjugated goat anti-rabbit IgG and enhanced chemiluminescence (ECL; Amersham, Arlington Heights, IL).

increasingly difficult as the cells become rounded. The change in cell shape makes it difficult to focus on certain intracellular compartments. If long infection times must be used, confocal microscopy is recommended.

1. When proteins are to be detected by immunofluorescence, cells are plated on glass cover slips in 35-mm dishes the day before infection. For many cell types, plain cover slips can be used (after sterilization by autoclaving). However, some cells attach better if the glass is pretreated by acid washing, or by coating with poly-L-lysine (1 mg/ml) or extracellular matrix components such as fibronectin (1 $\mu$g/ml).

2. Vaccinia infection and subsequent DNA transfections are performed as described in Section IV,B.

3. After the desired incubation period, cells are fixed in formaldehyde or methanol–acetone. Methanol–acetone fixation may be required if components of the cytoskeleton are being analyzed, and usually consists of 5–10 min in methanol at −20°C, followed by 30 sec to 1 min in acetone at −20°C. No further permeabilization is required.

*Note:* We find that formaldehyde fixation works well for all the antibodies we use (see subsequent discussion). As for other methods of analysis, optimal antibody dilutions and other parameters should be determined for each protein. As a guideline, polyclonal sera are usually diluted 1 : 200–1 : 1000, monoclonal antibodies from tissue culture supernatant are used straight or diluted up to 1:200, and affinity-purified antibodies or purified IgGs are used at 2–10 $\mu$g/ml. Commercially prepared secondary antibodies conjugated to fluorescein, rhodamine, or Texas red are available from many suppliers. We have had excellent success with affinity-purified preparations from Jackson Immunoresearch (West Grove, PA).

4. Our protocol includes fixation in 3% paraformaldehyde in PBS for 20–30 min at room temperature.

5. The fixative is quenched using PBS–gly (PBS with 10 m*M* glycine).

6. If an intracellular epitope is being labeled, the cells are permeabilized for 4 min with 0.5% Triton X-100 in PBS–gly.

7. After rinsing once with PBS–gly, we typically add a 5 min incubation in blocking solution (0.25% ovalbumin in PBS–gly.)

8. After removing aggregates by centrifuging antibodies in a microfuge for about 30 sec, both primary and secondary antibodies are diluted into the blocking solution. Throughout the staining procedure, it is important that the cover slips never dry out.

9. The cover slips are lifted with forceps, a corner is touched to a Kimwipe® to drain as much liquid as possible, and the slip is placed cell-side down onto a drop (50 $\mu$l) of primary antibody on a sheet of parafilm and incubated 20 min at room temperature.

10. The cover slips are then returned (cell-side up) to the dishes, washed for 10 min in several changes of PBS–gly, and incubated in 50 $\mu$l fluorochrome-conjugated secondary antibody, as described, for 20 min in the dark.

11. After washing in PBS–gly for 20–30 min, the backs of the cover slips are gently wiped and they are inverted onto a small drop of glycerol on a microscope slide. We use glycerol containing 100 m*M* *n*-propyl gallate to reduce photobleaching; glycerol containing phenylaminediamine or a commercial "mounting medium" can also be used.

12. The excess mounting medium is aspirated (and the cover slip may be sealed to the slide with nail polish) before viewing in a microscope equipped with epifluorescence and the appropriate barrier filters. An example of indirect immunofluorescence detection of proteins expressed using the T7 polymerase–vaccinia hybrid system is shown in Fig. 6.

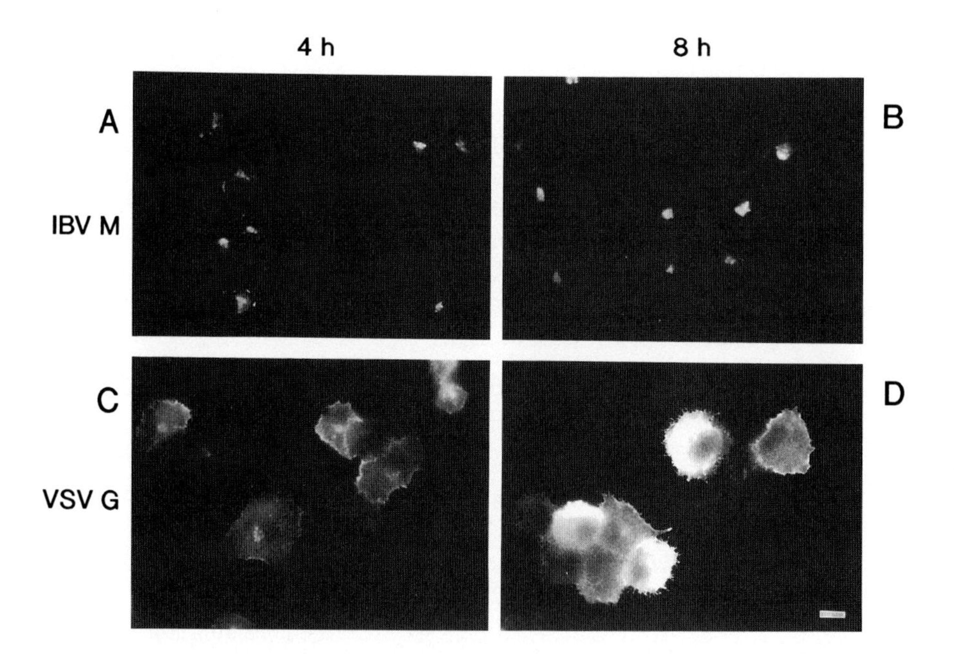


**Fig. 6** Indirect immunofluorescence staining of vaccinia-infected BHK cells. BHK cells were infected with vTF7-3 followed by LipofectACE-mediated transfection with pAR2529 encoding either the Golgi-resident M protein of avian infectious bronchitis virus (A,B) or the cell surface VSV G protein (C,D). At 4 (A,C) or 8 (B,D) hr postinfection, cells were rinsed with PBS, fixed, and labeled using appropriate antibodies followed by Texas red-conjugated secondary antibody. Note that cell rounding at 8 hr postinfection makes it difficult to focus on the Golgi complex (compare A and B), whereas it is easier to detect VSV G at the plasma membrane at the later time point. Bar, 10 $\mu$m.

# V. Discussion

This chapter has discussed the preparation and use of recombinant vaccinia viruses to express proteins in mammalian cells. This method offers many advantages over other expression systems for the expression of some proteins (see below). However, the limitations of this method should be taken into account before choosing this technique.

## A. Advantages of Vaccinia Virus-Mediated Protein Expression

1. Vaccinia virus-mediated expression is rapid and efficient: experiments can be performed in 1 day. Furthermore, most of the cells in a dish can be infected, resulting in a much higher efficiency of expression than other in transient expression systems.

2. Vaccinia has a wide host range, so most mammalian tissue culture cell lines are susceptible to infection. One exception is the CHO cell line, which is not susceptible to infection with the virus. However, cowpox virus does infect CHO cells, and a gene that allows this infection has recently been characterized (Spehner *et al.*, 1988). It is possible that construction of a vaccinia virus recombinant containing this cowpox gene will allow productive infection of CHO cells.

3. Because of the high efficiency of expression, coexpression of two or more proteins is simple using this technique. The expression level of each protein can be varied with the amount of virus or DNA used, depending on the method of expression (e.g., Zagouras *et al.*, 1991).

4. Because this expression system operates cytoplasmically, it can be used to express genes from RNA viruses that contain cryptic splice sites that are used when the cDNA is inserted into a vector that must be transcribed in the nucleus (Machamer and Rose, 1987).

5. Infection with recombinant vaccinia viruses results in high expression levels in >90% of cells on a dish soon after infection. This method is therefore ideal for subcellular localization of expressed proteins by immunoelectron microscopy (Machamer *et al.*, 1990; Krijnse Locker *et al.*, 1992). Early times postinfection (<6 hr) should be analyzed to avoid complications from viral cytopathic effects. Localization of proteins expressed via the vaccinia–T7 polymerase hybrid system is more difficult since the cellular morphology can be altered by the DNA carriers used for transfection, especially cationic lipid.

6. Vaccinia virus-mediated expression can be useful for production of biologically active proteins. Because proteins are expressed in mammalian cell lines, glycosylation and other post-translational modifications that may be necessary for activity are preserved.

7. An advantage specific to the T7 polymerase hybrid expression system is that it can be used to screen rapidly for expression of newly cloned cDNAs

and of mutant proteins generated by site-directed mutagenesis. The same vector can be used for mutagenesis, sequencing, *in vitro* transcription/translation, and vaccinia virus-mediated expression. In addition, a high percentage of cells is transfected because the DNA only needs to reach the cytoplasm and not the nucleus to be expressed.

## B. Disadvantages and Limitations

1. A serious disadvantage of the vaccinia virus system is that the expression level may be so high that it overwhelms cellular translation and translocation machinery. This can result in decreased efficiency of membrane translocation, post-translational modifications, and transport through the secretory pathway. At very high expression levels, newly synthesized proteins that enter the secretory pathway may accumulate in the ER. We have observed reduced rates of transport of the vesicular stomatitis virus (VSV) G protein to the cell surface, compared with VSV-infected cells, when the protein is expressed with the vaccinia–T7 hybrid system (Fig. 7). Whereas VSV G becomes resistant to endoglycosidase H (as it moves through the Golgi complex) with a half-time of $<25$ min in VSV-infected HeLa cells, the half-time is considerably slower ($\sim40$ min) in vTF7-3-infected and -transfected cells. The difference in transport kinetics may be less dramatic in other cell types.

2. The cytopathic effects of vaccinia infection can be a problem for certain types of analysis. For example, an early effect of the infection is rounding of cells (probably due to changes induced in the cytoskeleton), which can make conventional immunofluorescence difficult to interpret (Fig. 6). In addition, host cell protein synthesis is inhibited in infected cells. Vaccinia encodes homologs of cellular proteins including superoxide dismutase, epidermal growth factor, and profilin (Goebel *et al.*, 1990), expression of which could affect cell morphology and behavior in unpredictable ways. Although viral DNA replication and early assembly steps can be blocked (by treating infected cells with hydroxyurea or rifampicin, respectively), early cytopathic effects such as cell rounding are unfortunately not prevented.

3. Recombinant vaccinia viruses can be difficult and time-consuming to produce. Because they inhibit transcription from early promoters, any $T_5NT$ sequences in the gene of interest must be mutagenized prior to subcloning into the recombination vector. Even then, some recombinant viruses are impossible to produce, perhaps because expression of the foreign protein is incompatible with vaccinia replication or assembly.

4. A final consideration prior to using the vaccinia virus-mediated protein expression system is the safety of laboratory workers. Institutional guidelines for use of vaccinia virus recombinants must be followed. Some institutions require a vaccination against smallpox by all vaccinia users once every 3 years (as recommended by the NIH and CDC; however, see Grist, 1989; Wenzel and

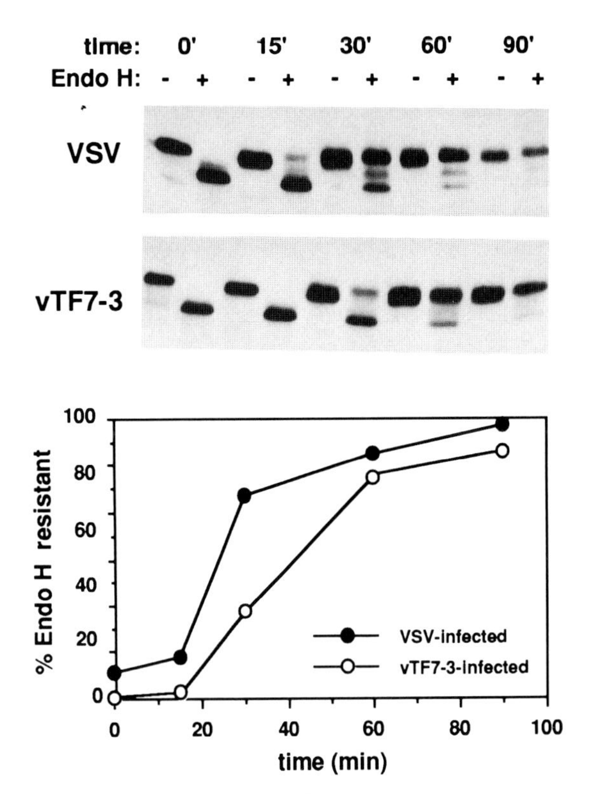


**Fig. 7** Comparison of intracellular transit kinetics of VSV G protein in VSV-infected versus vTF7-3-infected/transfected cells. HeLa cells grown in 35-mm dishes were infected with vTF7-3 followed by LipofectACE-mediated transfection with 2 $\mu$g pAR/G per dish, as described in Section IV,B, or with VSV (20 pfu/cell). At 3.75 hr postinfection, cells were rinsed once in PBS, starved for 15 min in methionine-free DMEM, and pulse-labeled for 5 min with 50 $\mu$C $^{35}$S *in vivo*-labeling mix (Amersham) in 0.5 ml methionine-free DMEM per dish. Cells were chased in growth medium. At the indicated chase times, individual dishes were lysed in detergent solution and immunoprecipitated using a polyclonal anti-VSV antibody, as described in Section IV,C,1. Samples were eluted in 20 $\mu$l 50 m*M* Tris pH 6.8, 1% SDS for 3 min at 100°C. Eluents were divided in half, and either 10 $\mu$l 0.15 *M* citrate, pH 5.5, or the same amount of buffer containing 0.3 mU endoglycosidase H (endo H) was added. After overnight incubation at 37°C, concentrated Laemmli sample buffer with $\beta$-mercaptoethanol was added; the samples were heated to 100°C for 3 min and electrophoresed on Laemmli SDS–polyacrylamide gels. The percentage of VSV G protein that was resistant to endo H at each time point was quantitated by densitometric scanning of fluorographed gels.

Nettleman, 1989). BL-2 restrictions also apply. If available, a virus-only laminar flow hood is recommended to avoid accidental infection of cell lines during routine tissue culture, although we have not had problems with accidental infection when the ultraviolet light is left on for an adequate time following work with infected cells.

## C. Availability of Vaccinia Virus Stocks and Plasmids

A small aliquot of vaccinia virus (both wild-type and the recombinant vTF7-3) and vectors for producing recombinant viruses and for T7 polymerase-mediated expression can be obtained by writing to Dr. Bernard Moss (Laboratory of Viral Diseases, NIAID, NIH, 9000 Rockville Pike, Bldg 4-Rm 229, Bethesda, MD 20892). New users must demonstrate that their facilities meet biosafety requirements, and that they have followed the institutional guidelines regarding vaccination. In addition, a material transfer agreement from the NIH must be completed and signed by your institution prior to receipt of these stocks.

## References


Chakrabarti, S., Brechling, K., and Moss, B. (1985). Vaccinia virus expression vector: Coexpression of β-galactosidase provides visual screening of recombinant virus plaques. *Mol. Cell. Biol.* **5,** 3403–3409.

Doms, R. W., Blumenthal, R., and Moss, B. (1990). Fusion of intra- and extracellular forms of vaccinia virus with the cell membrane. *J. Virol.* **64,** 4884–4892.

Earl, P. L., and Moss, B. (1991). Generation of recombinant vaccinia viruses. *In* "Current Protocols in Molecular Biology" (F. M. Ausubel, R. Brent, R. E. Kingston, D. D. Moore, J. G. Seidman, J. A. Smith, K. Struhl, eds.), 16.17.1–16.17.16. Greene Publishing Associates and Wiley Interscience, New York.

Elroy-Stein, O., Fuerst, T. R., and Moss, B. (1989). Cap-independent translation of mRNA conferred by encephalomyocarditis virus 5′ sequence improves the performance of the vaccinia virus/bacteriophage T7 hybrid expression system. *Proc. Natl. Acad. Sci. USA* **86,** 6126–6130.

Fuerst, T. R., Niles, E. G., Studier, W., and Moss, B. (1986). Eukaryotic transient-expressions system based on recombinant vaccinia virus that synthesizes bacteriophage T7 RNA polymerase. *Proc. Natl. Acad. Sci. USA* **83,** 8122–8126.

Goebel, S. J., Johnson, G. P., Perkus, M. E., Davis, S. W., Winslow, J. P., and Paoletti, E. (1990). The complete DNA sequence of vaccinia virus. *Virology* **179,** 247–266.

Griffiths, G., and Rottier, P. (1992). Cell biology of viruses that assemble along the biosynthetic pathway. *Sem. Cell Biol.* **3,** 367–381.

Grist, N. R. (1989). Smallpox vaccination for investigators. *Lancet* **ii,** 919.

Krijnse Locker, J., Griffiths, G., Horzinek, M. C., and Rottier, P. J. M. (1992). O-glycosylation of the coronavirus M protein. *J. Biol. Chem.* **267,** 14094–14101.

Machamer, C. E., and Rose, J. K. (1987). A specific transmembrane domain of a coronavirus E1 glycoprotein is required for its retention in the Golgi region. *J. Cell Biol.* **105,** 1205–1214.

Machamer, C. E., Mentone, S. A., Rose, J. K., and Farquhar, M. G. (1990). The E1 glycoprotein of an avian coronavirus is targeted to the Golgi complex. *Proc. Natl. Acad. Sci. USA* **87,** 6944–6948.

Mackett, M., Smith, G. L., and Moss, B. (1984). General method for production and selection of infectious vaccinia virus recombinants expressing foreign genes. *J. Virol.* **49,** 857–864.

Moss, B. (1991). Vaccinia virus: A tool for research and vaccine development. *Science* **252,** 1662–1667.

Moss, B. (1992). Poxvirus expression vectors. *Curr. Top. Microbiol. Immunol.* **158,** 25–38.

Moss, B., Elroy-Stein, O., Mizukami, T., Alexander, W. A., and Fuerst, T. R. (1990). New mammalian expression vectors. *Nature (London)* **348,** 91–92.

Schmelz, M., Sodeik, B., Ericsson, M., Wolffe, E., Shida, H., Hiller, G., and Griffiths, G. (1994). Assembly of vaccinia virus: The second wrapping cisterna is derived from the *trans* Golgi network. *J. Virol.* **68,** 130–147.

Sodeik, B., Doms, R. W., Ericsson, M., Hiller, G., Machamer, C. E., van't Hoff, W., van Meer, G., Moss, B., and Griffiths, G. (1993). Assembly of vaccinia virus: Role of the intermediate compartment between the endoplasmic reticulum and the Golgi stacks. *J. Cell Biol.* **121,** 521–541.

Spehner, D., Gillard, S., Drillien, R., and Kirn, A. (1988). A cowpox virus gene required for multiplication in Chinese hamster ovary cells. *J. Virol.* **62,** 1297–1304.

Wenzel, R. P., and Nettleman, M. D. (1989). Smallpox vaccination for investigators using vaccinia recombinants. *Lancet* **ii,** 630–631.

Yuen, L., and Moss, B. (1987). Oligonucleotide sequence signaling transcriptional termination of vaccinia virus early genes. *Proc. Natl. Acad. Sci. USA* **84,** 6417–6421.

Zagouras, P., Ruusala, A., and Rose, J. K. (1991). Dissociation and reassociation of oligomeric viral glycoprotein subunits in the endoplasmic reticulum. *J. Virol.* **65,** 1976–1984.

# CHAPTER 8

# Use of Recombinant Adenovirus for Metabolic Engineering of Mammalian Cells

**Thomas C. Becker, Richard J. Noel, Ward S. Coats, Anna M. Gómez-Foix,* Tausif Alam, Robert D. Gerard, and Christopher B. Newgard**

Departments of Biochemistry and Internal Medicine
and Gifford Laboratories for Diabetes Research
University of Texas Southwestern Medical Center
Dallas, Texas 75235



* Current Address: Departament de Bioquímica i Fisiologia, Facultat de Química, Universitat de Barcelona, 08028 Barcelona, Spain.

The ability to transfer DNA rapidly and efficiently into bacteria and yeast via physical techniques has allowed researchers to generate insights into the roles of particular genes in metabolic regulation and has also led to exploitation of certain recombinant strains for industrial purposes (Kispal *et al.*, 1989; Bailey, 1991; Liao and Butow, 1993). In contrast, alteration of dynamic metabolic processes in mammalian cells by introduction of particular genes has traditionally been hindered by the relative inefficiency of available techniques for gene transfer. These problems have also resulted in reduced enthusiasm for the prospects for gene therapy for inherited metabolic diseases. Recently, however, optimism has been rekindled by the rapid development of a number of gene transfer vectors and techniques based on the properties of DNA viruses such as adenovirus, herpesvirus, vaccinia virus and adeno-associated virus (AAV) (Berkner, 1988; Graham and Prevec, 1991; Miller, 1992; Morgan and French Anderson, 1993; Nienhuis *et al.*, 1993). In this chapter, we will describe the utility of these viral systems for transfer of genes involved in metabolic regulation into mammalian cells, with particular reference to primary cell types with low replicative activity such as hepatocytes and cells of the islets of Langerhans. Since most of the work in this area to date has been performed with recombinant adenoviruses, we will focus on this system.

## I. Historical Perspective on the Development and Application of Viral Vectors for DNA Transfer

Researchers have long recognized that viruses exist by virtue of their capacity to transfer genetic information into host cells. With the advent of recombinant DNA technology, it became apparent that this essential property of viruses might be exploited for the creation of efficient gene transfer vectors. The viral genome can consist of RNA, in the case of retroviruses, or of DNA. Research into the life cycle and biological properties of RNA and DNA viruses proceeded in a parallel fashion, beginning with the discovery of these distinct classes several decades ago (Berkner, 1988; Varmus, 1988; Graham and Prevec, 1991; Miller, 1992; Morgan and French Anderson, 1993; Nienhuis *et al.*, 1993). Despite the near-simultaneous development of gene transfer vectors derived from DNA and RNA viruses in the early 1980s, most of the studies on the utility of viral vectors for gene transfer into mammalian cells were initially focused on retroviruses. The reasons for the bias toward retroviruses are not entirely clear, but may have been related to the fact that retroviruses are capable of reverse transcribing their RNA genome to proviral DNA that then integrates into the genomic DNA of mammalian cells. Although integration of the recombinant gene into the cellular genomic DNA insures transfer of genetic information during cell division, it also enhances the risk for cellular transformation caused by insertional mutagenesis or activiation of oncogenes. Furthermore, the retroviral genome must be integrated for the transferred gene to be expressed, and integration only occurs in dividing cells, apparently because the process requires the participation of cellular factors present only during cell division (Varmus, 1988). Nevertheless, the intense focus on retroviruses has led to significant improvements in vector design, especially in the area of production of helper viruses that prevent the production of potentially dangerous replication-competent recombinant strains (for review, see Varmus, 1988; Miller, 1992; Morgan and French Anderson, 1993; Nienhuis *et al.*, 1993). In fact, nearly all currently approved gene therapy protocols employ disabled murine retroviruses (Miller, 1992; Morgan and French Anderson, 1993). Unfortunately for investigators interested in alteration of metabolic regulation in cell types involved in whole-animal fuel homeostasis such as hepatocytes or cells of the islets of Langherhans, adipose tissue, or muscle, the generally low mitotic activity of these cells precludes the use of retroviruses for efficient gene transfer. Furthermore, retrovirus-mediated gene therapy for metabolic disorders can only be efficiently applied to tissues that can be surgically resected, manipulated *in vitro,* and replaced; these procedures may be viable for liver (Wilson *et al.*, 1990; Ponder *et al.*, 1991), but certainly not for inaccessible tissues such as the islets of Langerhans.

DNA viruses have emerged as potentially attractive gene transfer vectors in recent years (Berkner, 1988; Graham and Prevec, 1991; Miller, 1992; Morgan and French Anderson, 1993; Nienhuis *et al.*, 1993). Each virus under study has some distinct advantages and disadvantages. For example, vaccinia viruses are capable of accommodating extremely large inserts (25 kb or more), but the large size of the viral genome (190 kb) dictates that recombinant viruses can only be produced by homologous recombination in virally infected host cells. This process is so inefficient that it requires genetic methods for selection of recombinant viruses (Smith, 1991). AAV appears to integrate in a site-specific manner in human chromosome 19 (19q13.3qter) (Kotin *et al.*, 1990,1991). This site-specific integration appears to be lost, however, in recombinant AAV virions. Furthermore, the small size of the AAV genome (4680 bp) limits the size of recombinant inserts. AAV is actually a parvovirus that requires the activities of adenovirus genes for replication. The AAV vectors also have essential portions of the genome deleted, requiring the propagation of recombinant AAV virions in adenovirus-infected packaging cell lines that provide activities encoded by the deleted genes in *trans*. Given these complications, it is not surprising that viral stocks with titers approaching those of other DNA viruses have not yet been produced.

Of the available DNA virus systems, adenovirus is the best studied, and probably the most accessible and generally useful, although its applicability for long-term gene therapy is still very much unclear. Positive features of adenovirus include the capacity for expression of relatively large DNA inserts (up to 7 kb in currently available helper-independent viral genomes), the ability to propagate high-titer viral stocks, an extremely broad range of infectivity of mammalian cell types, and the availability of an ever-increasing number of vectors containing different promoters. Although early clinical trials on the use of recombinant adenovirus for delivery of the cystic fibrosis transmembrane conductance regulator (CFTR) are underway, the long-term prospects for this approach may be hindered by the fact that adenovirus integrates into genomic DNA with very low efficiency and exists predominantly in an episomal mode, suggesting that duration of expression of the introduced gene may be limited. Activation of the immune system by the initial administration of virus will likely preclude the use of multiple injections as a strategy to circumvent the lack of permanent expression. An interesting problem for the near future will be to determine if the positive features embodied in each of the DNA viruses can be incorporated in combination into a "second generation" of DNA viral vectors by genetic engineering.

## II. Examples of the Utility of Recombinant Adenovirus for Studies on Metabolic Regulation

Although the future of adenoviral vectors for human gene therapy is uncertain, their utility for rapid and efficient gene delivery to mammalian cells, including nonreplicating primary cells, has been clearly demonstrated. This property of the adenoviral vectors opens a new range of experimental possibilities. Before providing details about the specific properties of adenoviral vectors and procedures for their use, a short review is in order of successful experiments performed to date with this system, particularly those in the area of metabolic regulation.

### A. Pioneering Studies

The first adenovirus vectors began to appear in the early 1980s (Solnick, 1981; Thummel *et al.*, 1981). Initially, they were mainly utilized for transfer of genes that could induce transformation, for example, SV40 or polyomavirus T antigens (Solnick, 1981; Thummel *et al.*, 1981; Van Doren and Gluzman, 1984; Berkner *et al.*, 1987; Davidson and Hassell, 1987; Sen *et al.*, 1988). Some of the early vectors had antibiotic selection markers, and were used to show that adenovirus has a rather low efficiency of transformation (0.4 to 0.75%), probably because of its low propensity for integration into genomic DNA (Van Doren *et al.*, 1984). Adenovirus was initially touted as a system for high-level expression of recombinant proteins, following reports that cloning of genes next to the major late promoter of adenovirus could lead to their expression at levels representing 10–20% of newly synthesized cellular protein (Stillman *et al.*, 1985; Yamada *et al.*, 1985; Berkner *et al.*, 1987; Davidson and Hassell, 1987; Alkhatib and Briedis, 1988; Johnson *et al.*, 1988). Dihydrofolate reductase (Berkner *et al.*, 1987) and thymidine kinase (Yamada *et al.*, 1985) were the first enzymes expressed in cells with recombinant adenovirus, but these studies were motivated more by evaluation of the system for selectability and overexpression than by interest in impacting metabolic pathways.

The first use of adenovirus in the context of metabolic regluation was described by Stratford-Perricaudet *et al.* (1990). These investigators constructed a recombinant adenovirus containing the ornithine transcarbamoylase (OTC) cDNA expressed from the adenoviral major late promoter. OTC catalyzes the condensation of ornithine and carbamoyl phosphate to yield citrulline in the mammalian urea cycle. Mice with a partial deficiency in OTC (Spf-ash strain) exhibit elevated levels of ammonia and orotic acid in the circulation and a phenotype of growth retardation and sparse fur (Doolittle *et al.*, 1974). Injection of a single bolus of $2–4 \times 10^7$ plaque-forming units (pfu) of the recombinant adenovirus into newborn mice resulted in high levels of OTC expression sufficient to reverse the sparse fur phenotype in 2 of 8 mice studied 1 mo after virus injection, 2 of 7 mice studied 2 mo after injection, and 1 of 2 mice after 15 mo. Among those mice that did not display a full reversal of phenotype, several had levels of hepatic OTC enzymatic activity that were higher than those observed in control animals injected with adenovirus lacking the OTC insert. Finally, the authors were able to demonstrate a beneficial metabolic impact of adenovirus-mediated gene transfer because orotic aciduria was significantly reduced in 2 mice for periods of up to 13 mo. Despite the lack of consistency in the efficiency of gene delivery among different animals, this study provided the first suggestion that adenovirus-mediated gene transfer could be used to deliver genes direcly to important metabolic tissues *in vivo*.

## B. Adenovirus-Mediated Gene Transfer to Mammalian Cells in Culture

### 1. Hepatocytes

Recombinant adenoviruses also allow manipulation of metabolic regulation in isolated primary cells in tissue culture that would normally be refractory to gene transfer by physical methods or retroviral vectors because of their poor replicative capacity. Of particular interest in our laboratory are studies on metabolic function in liver and the islets of Langerhans, the two tissues that are most intimately involved in the control of fuel homeostasis in mammals. An example of the potential of the adenoviral technique for alteration of metabolic events in cells is provided in the study by Gómez-Foix *et al.* (1992). We constructed a recombinant adenovirus containing the muscle glycogen phosphorylase cDNA linked to the cytomegalovirus (CMV) immediate early promoter (AdCMV-MGP) and used this system to transduce primary rat hepatocytes in culture. The study was undertaken to evaluate whether regulation of hepatic glycogen metabolism could be altered by overexpression of the muscle isozyme of glycogen phosphorylase which, unlike the naturally expressed liver phosphorylase isozyme, is potently activated by the allosteric ligand AMP (Newgard *et al.*, 1989). AMP-activatable glycogen phosphorylase activity was increased

46-fold after infection of the liver cells with AdCMV-MGP. *In situ* hybridization with an antisense cRNA probe specific for muscle phosphorylase revealed that 86% of the cells expressed muscle phosphorylase mRNA. Despite large increases in phosphorylase activity, glycogen levels were only slightly reduced in AdCMV-MGP-infected liver cells relative to uninfected cells or cells infected with wild-type adenovirus. The lack of correlation of phosphorylase activity and glycogen content suggested that the liver cell environment inhibited the muscle phosphorylase isozyme. This inhibition could be overcome, however, by addition of carbonyl cynanide *m*-chlorophenylhydrazone (CCCP), which increased AMP levels 30-fold and caused a much larger decrease in glycogen levels in AdCMV-MGP-infected cells than in control hepatocytes. Introduction of muscle phosphorylase into hepatocytes therefore enables a glycogenolytic response to external effectors that is not provided by the endogenous liver phosphorylase isozyme.

The remarkable efficiency of adenovirus-mediated gene transfer to hepatocytes both *in vitro* (Stratford-Perricaudet *et al.*, 1990; Gómez-Foix *et al.*, 1992) and *in vivo* (Stratford-Perricaudet *et al.*, 1990; see subsequent discussion) suggests that the technique may be applicable for studies on gene therapy for glycogen storage disorders, at least in animal models. An example of a suitable model is the gsd/gsd rat, which appears to lack hepatic phosphorylase kinase activity (Malthus *et al.*, 1980). The obvious strategy of replacement of phosphorylase kinase activity by adenovirus-mediated gene transfer is complicated by the fact that phosphorylase kinase exists as a heterotetramer and that the missing subunit in the gsd/gsd animals has not been defined. An alternative to replacement of phosphorylase kinase might be to overexpress muscle phosphorylase or engineered liver phosphorylase proteins that gain some of the regulatory features of muscle phosphorylase (Coats *et al.*, 1991) in gsd/gsd hepatocytes. Such maneuvers may allow the block in glycogenolysis to be circumvented by alternative modes of phosphorylase activation. Other studies indicate that adenovirus vectors may also be applicable to gene therapy of McArdle's disease, a genetic deficiency in expression of glycogen phosphorylase in skeletal muscle. Baque *et al.* (1993) have shown that muscle phosphorylase cDNA is transferred to muscle cells in culture with the AdCMV-MGP recombinant virus, and with greater efficiency to differentiated myotubes than to myoblasts. These data suggest that deficient muscle phosphorylase activity may be replaceable in patients with McArdle's disease via adenovirus or derivative vector systems.

## 2. Pancreatic Islets of Langerhans

Adenovirus also has utility for metabolic engineering of pancreatic islets. Using a recombinant adenovirus containing a nuclear-localizing variant of the *Escherichia coli* $\beta$-galactosidase gene inserted adjacent to the CMV promoter (AdCMV-$\beta$Gal), Becker *et al.* (1993) demonstrated highly efficient gene transfer into freshly isolated rat islets of Langerhans. Infection of isolated islets with the AdCMV-$\beta$Gal virus resulted in expression of $\beta$-galactosidase in islets for at least 21 days postinfection. Analysis of multiple islet sections showed that the recombinant virus transferred the $\beta$-galactosidase gene into 60–70% of the islet cells compared to 10–20% transfer efficiencies with more traditional methods. Importantly, perifusion studies demonstrated that the normal insulin secretory responses to glucose or glucose + arginine were not affected by infection with the AdCMV-$\beta$Gal virus. Since glucose-stimulated insulin secretion is the signature functional attribute of the normal islet and is a process that is dependent on active glucose metabolism (Meglasson and Matchinsky, 1986; Newgard, 1992), we concluded that recombinant adenovirus is an attractive and highly efficient system for studying the impact of overexpression of specific genes in the islets.

Changes in the rate of glucose metabolism in islet $\beta$ cells in response to changes in the external glucose concentration are thought to be mediated by the GLUT-2 facilitated glucose transporter and the glucose phosphorylating enzyme glucokinase; the latter protein is likely to play the true rate-limiting role because of its relatively low level of expression in islets (for review see Meglasson and Matchinsky, 1986; Newgard, 1992). Since the rate of glucose metabolism in $\beta$ cells appears to be proportional to the magnitude of the glucose-stimualted insulin secretion response, glucokinase and GLUT-2 are thought of as essential components of the "glucose sensing apparatus" of the $\beta$ cells (Newgard, 1992). Although the circumstantial evidence supporting this contention is compelling, support from molecular studies has been slowly forthcoming because of the difficulties inherent in achieving high-efficiency gene transfer into islet cells.

Previous gene transfer studies on isolated islets have employed physical techniques such as lipofection or electroporation (German *et al.*, 1990; Welsh *et al.*, 1990; German, 1993). In one of these studies, immunofluoresence analysis revealed that the gene being transferred, chloramphenicol acetyl transferase (CAT), was expressed in approximately 11% of the islet cells (German *et al.*, 1990). Such efficiencies allow meaningful studies on expression of promoter–reporter chimeric gene constructs (German *et al.*, 1990; German, 1993). In fact, co-transfection of fetal islets by electroporation with plasmids containing hexokinase I and the rat insulin I promoter linked to CAT resulted in a reduced glucose concentration threshold for activation of the insulin promoter (German, 1993). This effect is presumably due to the enhancement in glucose metabolism occurring at low glucose concentration, since hexokinase has a $K_m$ for glucose of 50 $\mu M$, whereas glucokinase has a $K_m$ for glucose (8 m$M$) that falls within the physiological range (4–10 m$M$). The important knowledge gained about regulation of insulin expression in the foregoing study is reliant on the high incidence of co-transfection of two plasmids into a single cell. Clearly, however, an overall transfection efficiency of only 10–20% will not provide meaningful insights into the impact of overexpression of genes on acute regulation of insulin release, because most of the cells will not have received the gene of interest. To circumvent this problem we have prepared recombinant adenoviruses containing various isoforms of glucokinase (AdCMV-GK) and hexokinase I (AdCMV-HKI) and have used these constructs for gene transfer into normal rat islets (Becker *et al.*, 1994). Expression of these proteins was increased approximately 10–30-fold 4 days after infection of islets with AdCMV-GK or AdCMV-HKI. Islets overexpressing glucokinase exhibited an enhanced glucose-stimulated insulin secretion response relative to control islets that were either untreated or treated with the AdCMV-βGal virus. In contrast, islets overexpressing hexokinase I exhibited no increase in the magnitude of response to stimulatory glucose, but showed a doubling in basal insulin release in the presence of the nonstimulatory glucose concentration. These results validate recombinant adenoviruses as an effective means of altering metabolic regulation in isolated pancreatic islets, and open the door for studies on a number of other potential regulatory loci, including phosphofructokinase and enzymes controlling islet lipid metabolism.

### 3. Clonal Cell Lines

Much of the early work with recombinant adenovirus was performed with clonal cell lines. Abundant evidence suggests that the virus efficiently transfers genes to a wide range of cell types (Solnick, 1981; Van Doren and Gluzman, 1984; Van Doren *et al.*, 1984; Stillman *et al.*, 1985; Yamada *et al.*, 1985; Berkner *et al.*, 1987; Davidson and Hassell, 1987; Alkhatib and Briedis, 1988; Johnson *et al.*, 1988; Sen *et al.*, 1988; Stratford-Perricaudet *et al.*, 1990). This property of adenovirus virions can be used to advantage for rapid evaluation of the effects of introduced genes in clonal cells. For example, our laboratory is engaged in studies on engineering of glucose-stimulated insulin secretion in insulin-secreting cell lines that are normally insensitive to the sugar (reviewed by Newgard, 1992; Newgard *et al.*, 1993). These studies involve transfer of genes such as GLUT-2 and glucokinase that have been implicated in the "glucose sensing" function of the islets of Langerhans. Until recently, these studies relied on stable transfection and selection of transfected clones by antibiotic resistance. Although such approaches do produce clonal lines with stable expression of the genes of interest, the transfection and selection of clones is time consuming, and it is necessary to evaluate multiple clones to be certain that phenotypic changes are due to the transferred gene rather than to spurious effects of clonal selection, while adenovirus DNA does not integrate efficiently and thus is lost over time as clonal cells replicate; the system can be extremely valuable for rapid testing of concepts. For example, we have recently demonstrated that rat insulinoma (RIN) cells gain a glucose-stimulated insulin secretion response on stable transfection with a plasmid containing GLUT-2, but the response is found to be maximal at subphysiological glucose concentrations (Ferber *et al.*, 1994). We hypothesized that the hypersensitive response to glucose is due to high levels of expression of hexokinase in RIN cells, a contention supported by studies with the hexokinase inhibitor 2-deoxyglucose, which caused the insulin secretion response to occur at higher glucose concentrations (Ferber *et al.*, 1994). We have now prepared a recombinant adenovirus with the hexokinase I cDNA in antisense orientation (H. BeltrandelRio, T. C. Becker, and C. B. Newgard, unpublished work), which will allow a rapid test of the concept that reduction in expression of hexokinase I will provide the correct glucose-sensing threshold.

### C. *In Vivo* Studies

Recombinant adenovirus also appear to have great applicability for gene delivery *in vivo* (Gerard and Meidell, 1993). Although the impetus for such work is often to evaluate adenovirus as a means of replacing genes that are absent in genetic disorders, the technique also has fascinating potential for elucidating the effects of perturbation of metabolic pathways in whole animals. After the pioneering study involving gene transfer of OTC (Stratford, Perricaudet *et al.*, 1990) the efficacy of adenovirus for *in vivo* gene delivery was confirmed and elaborated by several groups. Initial efforts were directed at delivery of genes to the airway epithelia of lung, since these cells can be specifically targeted by intratracheal instillation (Rosenfeld *et al.*, 1991, 1992). Recombinant adenoviruses have been used successfully to deliver $\alpha$1-antitrypsin (Rosenfeld *et al.*, 1991) and the CFTR gene (Rosenfeld *et al.*, 1992) to the airway epithelium of cotton rats; expression of the introduced human proteins persisted for a period of weeks. These promising results have led rapdily to multiple clinical trials of recombinant adenovirus for delivery of the CFTR to patients, with as yet unknown results. More recently, Herz and Gerard (1993) and co-workers, (Ishibashi *et al.*, 1993) have elegantly demonstrated the utility of recombinant adenovirus for whole animal metabolic studies. In an initial study (Herz and Gerard, 1993) AdCMV-$\beta$Gal virus (approximately $10^9$ pfu) was injected into the jugular vein of normal mice resulting in gene transfer to approximately 90% of the hepatic parenchymal cells. Similar efficiencies of gene transfer to the liver were measured by immunofluorescent staining following injection of a recombinant virus containing the cDNA encoding the human low density lipoprotein (LDL) receptor (AdCMV-LDLR). After confirming hepatic overexpression of the LDL receptor by Western blotting, the authors demonstrated as much as a 10-fold increase in the rate of $^{125}$I-labeled LDL clearance, with rates being roughly proportional to the amount of AdCMV-LDLR virus administered (Herz and Gerard, 1993). Among tissues studied (liver, kidney, lung, spleen, and heart), only the liver exhibited enhanced accumulation of labeled LDL, indicating that the increased rate of clearance was due to hepatic expression of the LDL receptor. In a subsequent study, Ishibashi *et al.* (1993) employed homologous recombination to create mice lacking a functional LDL receptor gene. These mice exhibited a doubling of total plasma cholesterol that was due to a 7- to 9-fold increase in intermediate density lipoproteins (IDL) and LDL, with little change in high density lipoproteins (HDL). Injection of the AdCMV-LDLR virus into such mice resulted in a return of the IDL/LDL levels to normal, and mice with restored LDL receptor levels also cleared $^{125}$I-labeled very low density lipoproteins (VLDL) much more rapidly than LDL receptor-deficient mice injected with the AdCMV-$\beta$Gal virus as a control (Ishibashi *et al.*, 1993).

The utility of recombinant adenovirus for gene delivery to tissues other than lung or liver is not yet clear. Injection of a recombinant virus containing the firefly luciferase gene (AdCMV-Luc) resulted in accumulation of the vast majority of luciferase activity in liver, exceeding the activities found in lung, skeletal muscle, spleen, kidney, or heart by several orders of magnitude (Herz and Gerard, 1993). Other investigators have also observed highly efficient gene transfer to the liver by systemic delivery of recombinant adenovirus (Stratford-Perricaudet *et al.*, 1990, 1992; W. Coats and C. B. Newgard, unpublished observations), but claims of efficient systemic delivery to tissues such as skeletal muscle and heart (Stratford-Perricaudet *et al.*, 1992), or brain (Le Gal Le Salle *et al.*, 1993) remain to be confirmed. We have infused the AdCMV-$\beta$Gal virus for several days at a rate of $1 \times 10^8$ pfu/hr and found preferential expression of $\beta$-galactosidase in the islets of Langerhans among pancreatic cells (W. Coats and C. B. Newgard, unpublished observations). The result has been confirmed by isolation of the islets after virus infusion, but reliable estimates of gene transfer efficiency will require further investigation. Preferential targeting of recombinant adenovirus to tissues such as liver and the islets of Langerhans may be due to the fact that blood can easily come in direct contact with such cells, via the sinusoidal vasculature of the liver or through the fenestrations that exist in the intense vasculature of the islets (Henderson and Moss, 1985). Targeting tissues that are not as well served by the systemic circulation might ultimately be achieved by direct injection or inoculation with recombinant virus, as has been demonstrated for the hippocampus and the substantia nigra of the brain (Le Gal Le Salle *et al.*, 1993) and for skeletal muscle fibers (Quantin *et al.*, 1992; Ragot *et al.*, 1993). Further studies in these areas will be required.

## III. Adenovirus Biology

Wild-type adenovirus exists as an icosahedral particle approximately 75 nm in diameter with a dense core containing the DNA genome (Horwitz, 1990). The virus was originally isolated from human adenoidal tissue in 1953 (Rowe *et al.*, 1953). It has been associated with a number of human respiratory illnesses and was the first human virus demonstrated to cause malignant tumors in animals, although, interestingly, no human malignancies have yet been described that are due to adenovirus transformation. Human adenovirus serotypes 2 (Ad2) and 5 (Ad5) are the most extensively studied, and their DNA genome has a size of 36 kb. The interested reader is referred to several excellent reviews on adenovirus biology that provide detailed information about the viral genome (Berkner, 1988; Horwitz, 1990; Graham and Prevec, 1991). The lytic life cycle of the wild-type virions can be divided into early and late phases, which are defined as occurring before and after the onset of viral DNA replication, respectively. Genes expressed during the early phase are noncontiguous in the genome and include the E1 regions E1A and E1B, the E2 regions E2A and E2B, E3, and E4 (see Figure 1). After the onset of DNA replication triggered by expression of the early genes, the major late promoter (MLP) is activated and is responsible for most of the transcriptional activity in the late phase. MLP-driven transcripts contain multiple open reading frames that are alternatively spliced to yield mRNAs containing identical 5′ untranslated regions known as the tripartite leader (Berkner, 1988; Horwitz, 1990; Graham and Prevec, 1991). Late in infection, 20–40% of total cellular RNA is viral and cell lysis begins to occur 48–72 hr after infection.

Current strategies for adenovirus vector design focus on deleting portions of the genome to allow packaging of relatively large DNA inserts. Wild-type adenovirus can only accommodate 2 kb of foreign DNA, but deletion of one or more of the early genes of the virus can allow recombinants with inserts up to 7 kb in size. Before describing some of the currently available vectors in detail, it seems useful to describe the function of those genes that have been targeted for deletion. The E1 genes are required for viral transformation. E1A is particularly important because it appears to be required for activation of all other early genes (E1B, E2, E3, and E4). Although the mechanism(s) responsible for global activation by E1A is not yet understood, an intriguing structural relationship between certain E1A and *myc* and *myb* gene products has been noted (Ralston and Bishop, 1983; McLachlan and Boswell, 1985). An E1A protein has also been shown to be similar to SV40 T antigen and the human papillomavirus-16 E7 transforming proteins; all bind to the retinoblastoma (RB) growth suppressor gene (DeCaprio *et al.*, 1988; Whyte *et al.*, 1988), suggesting a common mechanism of action for these transforming genes. E1A and E1B

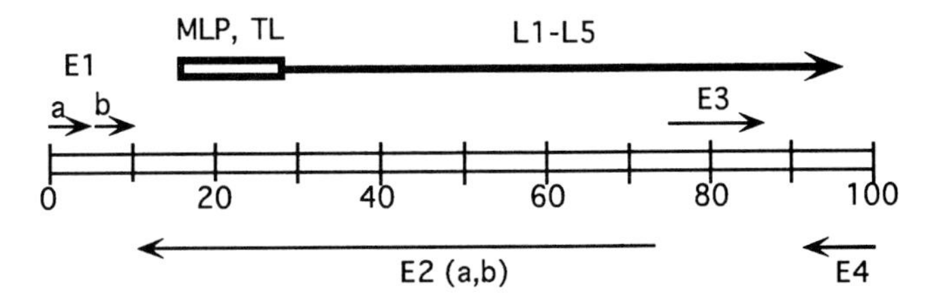


**Fig. 1** Transcriptional map of adenovirus genome. The 36-kb adenovirus genome is a linear molecule that can be divided into 100 map units of 360 bp each. Early gene transcripts (E1–E4) are represented by thin arrows. Late gene transcripts (L1–L5) are initiated at the major late promoter (MLP) and all contain the tripartite leader (TL) sequence.

appear to work together in mediating cellular transformation by adenovirus. E2A and E2B encode DNA binding and DNA polymerase activities, respectively, and are essential for DNA replication. In contrast, E3 appears to have no essential function for adenovirus growth or DNA replication, and may instead be involved in modulating host responsiveness to adenovirus infection. A 19-kDa protein encoded by the E3 region has been shown to bind to major histocompatibility complex (MHC) polypeptides in the endoplasmic reticulum, thereby inhibiting appearance of MHC peptides at the cell surface (Persson *et al.*, 1979; Paabo *et al.*, 1986). The E4 region is currently not well understood, although a 34-kDa protein encoded by this region is thought to interact with an E1B 55-kDa protein in the nucleus. Apparently six of the seven open reading frames within E4 can be deleted without affecting viral replication or infectivity (Halbert *et al.*, 1985).

## IV. Development of Adenovirus Vectors

The strategy employed for construction of a particular recombinant adenovirus depends in part on the regions of the adenovirus genome that are deleted to accommodate DNA inserts. To date, the most common strategies have been to delete portions of the E1 or the E3 regions of the viral genome, or both. Wild-type Ad5 can accommodate only 2 kb of insert DNA. Deletion of E3 with two naturally occurring *Xba* I restriction enzyme sites removes 1.9 kb from the viral genome, thereby accommodating inserts approaching 4 kb in length (Berkner, 1988; Graham and Previc, 1991). Up to 3 kb can be deleted from the E1 region, allowing a maximum insert size in doubly deleted viral genomes of approximately 7 kb. Theoretically, it should also be possible to delete nonessential regions of E4, allowing inserts of up to 10 kb, but such vectors are not yet available. Recombinant viruses can be generated either by direct ligation of insert sequences into the deleted E1 or E3 regions or, more commonly, by homologous recombination of overlapping fragments of the viral genome in cell lines. If the recombinant genome is deleted only in E3, the virus is described as nonconditional since E3 is not essential for viral replication and growth. In such constructs, recombinant virions can be grown in a wide array of cell lines, including HeLa or KB cells. In the case of deletion of the essential E1 gene, viral propagation is conditional because the function of the E1 gene must be provided in *trans*. This is achieved by growing E1-deleted recombinant virus in the human 293 cell line, which was originally transformed with Ad5 and contains the left 14% of the adenovirus genome integrated into cellular DNA, including the E1 region (Graham *et al.*, 1977).

Since the adenovirus genome is very large, the chances of finding unique restriction sites that allow direct ligation into the E1 or E3 regions are small. One strategy that has been used successfully is to remove the far-left 2.6 map units of Ad5 by digestion at a unique *Cla* I site. For example, recombinant adenovirus containing the OTC cDNA was prepared by direct ligation of a cassette consisting of the adenovirus MLP, the tripartite leader, the OTC cDNA, and SV40 3′ sequences via a 3′ *Acc* I site, which can ligate to *Cla* I-restricted DNA (Stratford-Perricaudet *et al.*, 1990). This DNA construct is then tranfected into 293 cells for viral propagation. A current focus of investigators in the field is to attempt to prepare new viral vectors that add or eliminate restriction sites to allow more flexibility in direct ligation strategies. The discovery of restriction endonucleases that recognize 8-bp palindromic sequences will facilitate this process. In the interim, most recombinant viruses are generated by homologous recombination. A particularly useful strategy in our hands has been adapted from procedures devised by Graham and co-workers (1991). An important discovery that has contributed to current methods was that adenovirus DNA circularizes in infected cells, allowing isolation of the viral DNA as infectious plasmids (Ruben *et al.*, 1983; Graham, 1984). Recombinant virions can therefore be produced by co-transfection with the circular viral genome and a plasmid containing a fragment of the viral genome and the recombinant gene of interest. A problem with this approach is that significant contamination with wild-type virus can occur; in fact, the wild-type virus will often have a growth advantage over the recombinant. This problem has been circumvented by preparation of a new plasmid called pJM17, which contains a DNA segment of 4.3 kb inserted at 3.7 map units that includes the ampicillin and tetracyline resistance genes and a bacterial origin of replication (McGrory *et al.*, 1988; see Fig. 2 for map). The size of this insert exceeds the packaging limit of wild-type virus, thus preventing propagation of wild-type virions and selecting for recombinants.

More recently, new vectors have been designed that will allow the use of a variety of promoter–enhancer elements, including those with tissue-specific properties such as the insulin promoter. A very useful vector, pACCMVpLpA, has been developed by modification of the pAC vector (Gluzman *et al.*, 1982) by R. Gerard, by replacement of a region of the adenovirus genome between map units 1.3 and 9.1 with the CMV early promoter (Fig. 3). A convenient cloning cassette has also been inserted immediately downstream from the CMV promoter, which in turn is followed by a fragment of the SV40 genome that includes the small t antigen intron and the polyadenylation signal (see map, Fig. 1). A large number of cDNAs including muscle glycogen phosphorylase (Gómez-Foix *et al.*, 1992), the LDL receptor (Herz and Gerard, 1993), glucokinase and hexokinase I (Becker *et al.*, 1994), the GLUT-2 facilitated glucose transporter (Ferber *et al.*, 1994), and apolipoprotein A1 (R. S. Meidell and R. D. Gerard, unpublished work) have been cloned into pACCMVpLpA and have been used to generate recombinant adenoviruses by recombination with pJM17, as shown in Fig 4. In our hands, generation of virions by co-transfection of 293 cells with purified pAC constructs and pJM17 requires 2–4 wk, allowing for the relatively rare recombination events to occur (see also subsequent discussion). Future development of this vector system will focus on two features. (1) The CMV promoter will be substituted with other nonviral promoters

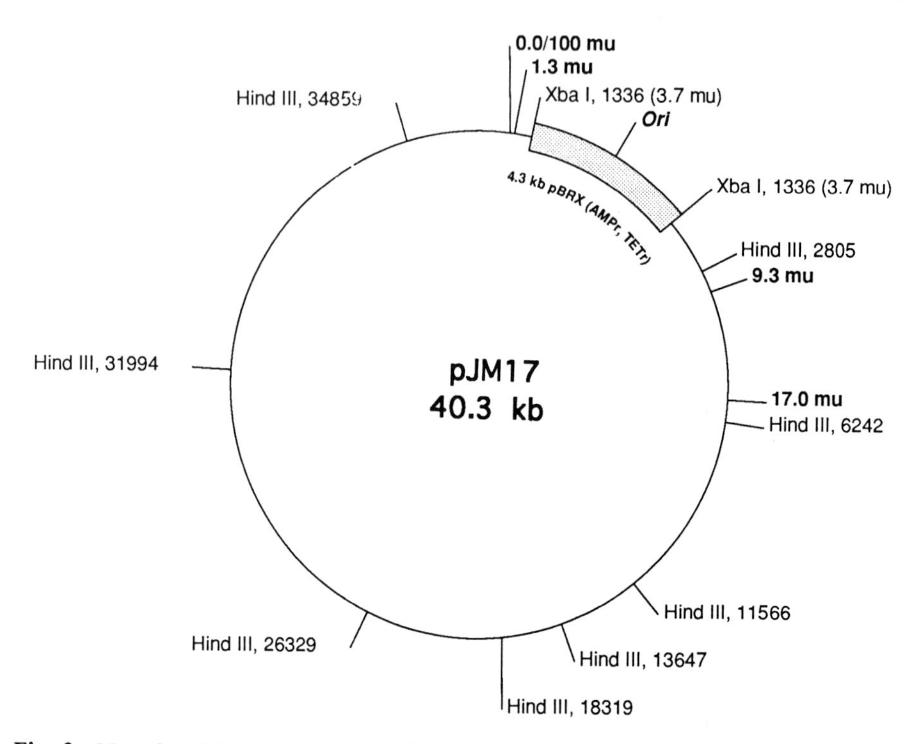


**Fig. 2** Map of pJM17 plasmid. The plasmid was prepared by McGrory *et al.* (1988) and consists of a modified Ad5 genome containing a 4.3-kb pBRX insert at 3.7 map units (mu). Also shown are the positions of the *Hin*d III restriction endonuclease sites; the numbers adjacent to each *Hin*d III position refer to the distance in base pairs from the 0/100 mu position. Regions of overlap with the pACCMV.pLpA plasmid are highlighted (0.0–1.3 and 9.3–17.0 mu).

in the pAC plasmid. For example, we have recently succeeded in replacing the CMV promoter with the human insulin promoter and are beginning to test the efficacy of this virus for tissue-specific expression in insulin-producing islet $\beta$-cells (L. Moss, R. Noel, T. Becker, P. Antinozzi, and C. B. Newgard, unpublished observations). (2) Restriction site engineering must occur so the DNA insert of interest can be cloned into one of the series of pAC plasmids, which can then be linearized and ligated directly to the remainder of the adenovirus genome, thus obviating the need for the homologous recombination step.

## V. Specific Procedures for Preparation of Recombinant Viruses

For simplicity, we will focus in this section on construction of recombinant adenovirus by co-transfection of 293 cells with the plasmids pJM17 and

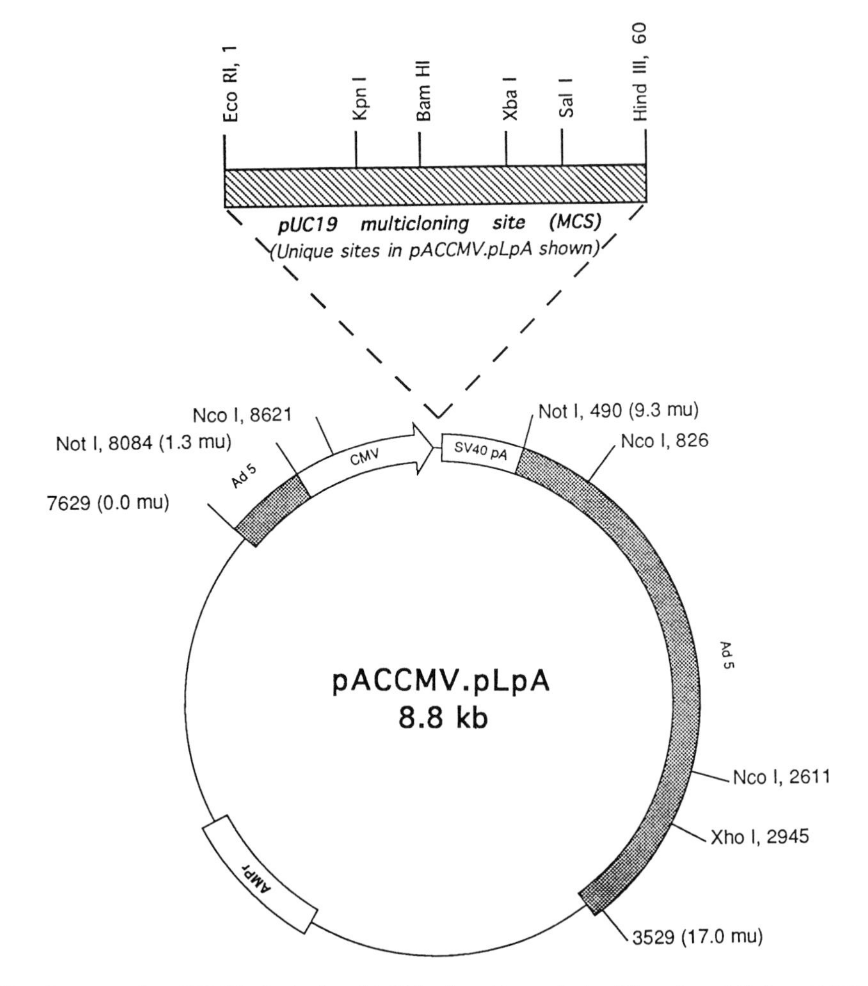


**Fig. 3** Map of pACCMV.pLpA plasmid. This plasmid was derived from the pAC plasmid (Gluzman *et al.*, 1982) by insertion of the CMV promoter/enhancer element, a pUC19 cloning cassette, and the SV40 polyadenylation signal. The plasmid consists of Ad5 sequence (shown as the shaded bars) and bacterial plasmid sequence that includes the ampicillin resistance gene.

pACCMVpLpA. Viruses derived from these particular reagents lack the E1 but not the E3 region of the virus, so there is an inherent limitation on the size of the cDNA that can be inserted. Note that the pJM17 plasmid (McGrory *et al.*, 1988) was derived from Ad5 containing insertions and deletions in the E3 gene, so the E3 gene is largely intact but nonfunctional (Jones and Shenk, 1978). Using pACCMVpLpA and pJM17, we have been able to prepare a recombinant virus containing the 3.6-kb hexokinase I cDNA (Becker *et al.*, 1994), close to the theoretical upper limit of 3.8 kb. This system of preparing recombinant adenoviruses requires a rare recombination event to occur between

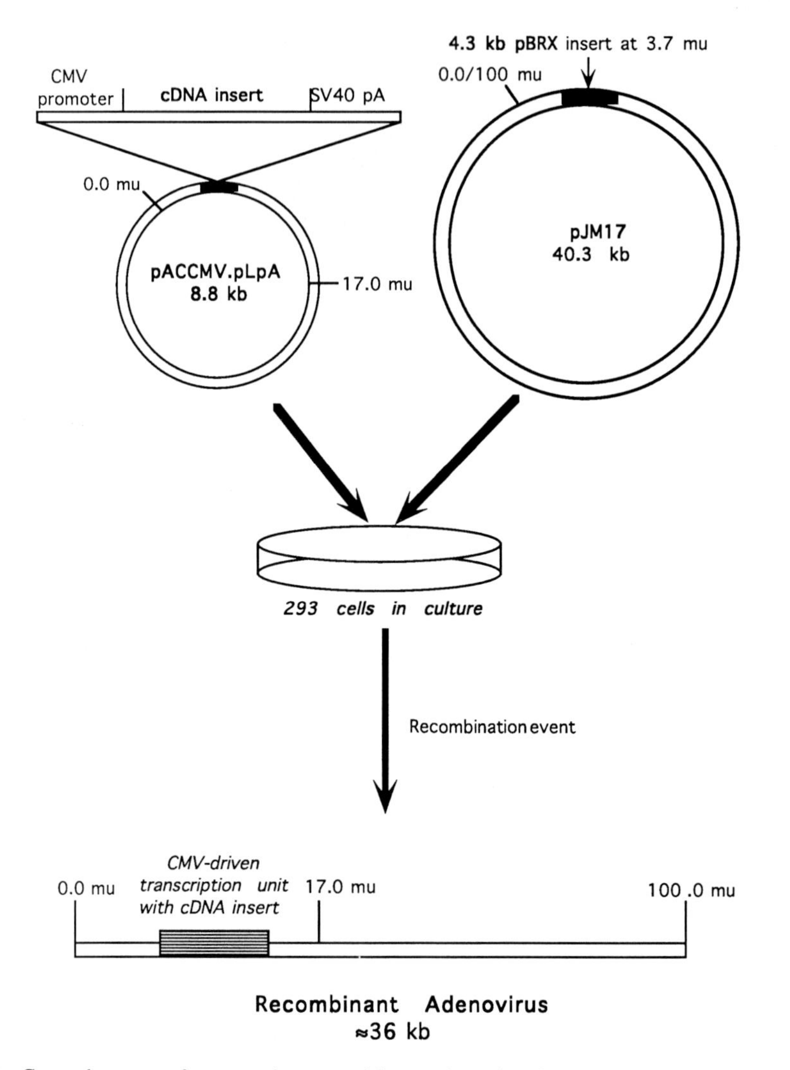


**Fig. 4** General strategy for preparing recombinant adenovirus by homologous recombination. The cDNA of interest is cloned into the pACCMVpLpA vector. The recombinant pAC and pJM17 are purified and co-transfected into 293 cells as described in detail in the text. Since 293 cells were originally produced by adenovirus transformation, the missing E1 gene function of pJM17 is provided in trans. Virus is isolated, titered, and amplified as described in the text. The final product is a recombinant adenovirus that is replication defective (at least in cells lacking the E1 region of adenovirus) but fully infectious.

two plasmids co-transfected at a low efficiency into 293 cells. In spite of this additional limitation, we have successfully prepared a large number of recombinant viruses using the following protocols. Note that production of a recombinant virus may require as little as a single co-transfection, or may require several co-transfections. A flow chart providing an overview of the procedure used to generate recombinant virions is provided in Fig. 5.

## A. Purification of Plasmid DNA

Obtaining high quality DNA for co-transfections is a prerequisite for preparing recombinant viruses. Preparation of the large (40 kb) pJM17 plasmid poses

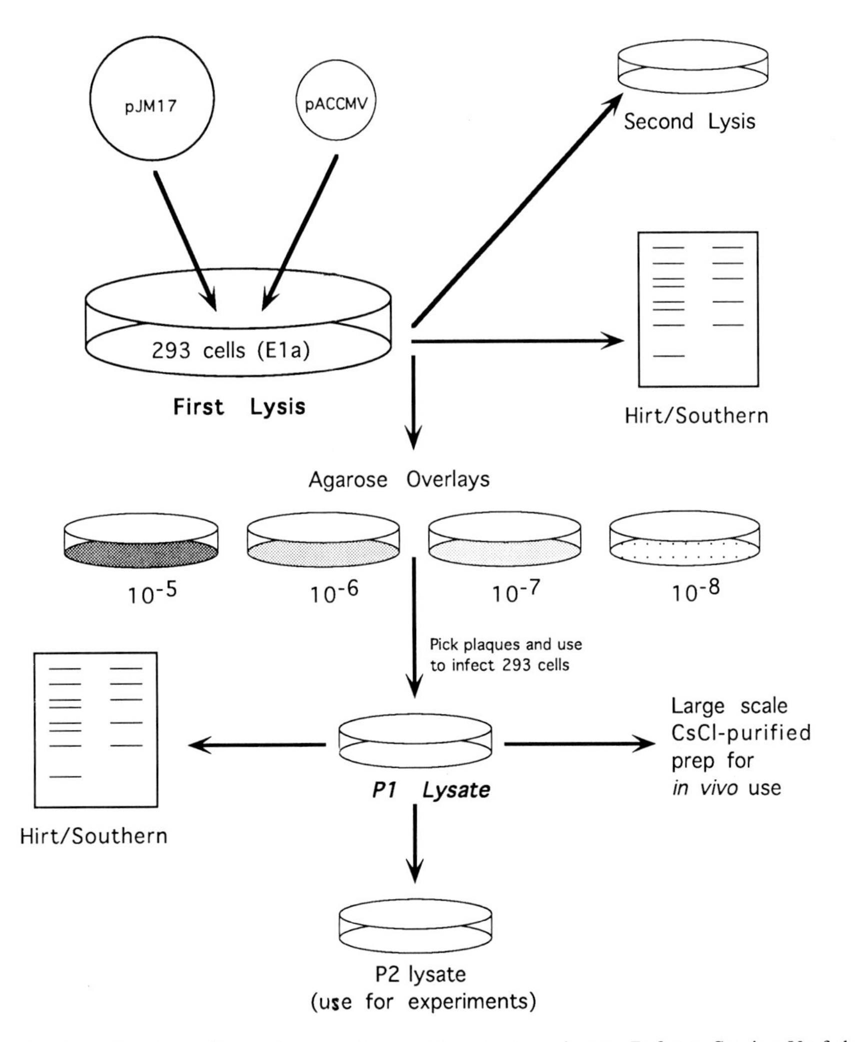


**Fig. 5** Overall scheme for production of recombinant adenoviruses. Refer to Section V of the text for details.

particular problems because of its propensity for shearing during handling. Bacteria containing pJM17 or pAC-derived plasmids are grown on LB plates or in LB or TB media (Sambrook *et al.*, 1989) containing 100 $\mu$g/ml ampicillin. For large-scale plasmid preparations, 1-liter cultures of LB or TB with ampicillin are inoculated and grown overnight. Plasmid DNA is isolated using the QIAGEN MaxiPrep system (QIAGEN, Chatsworth, CA). Using this method, we obtain 300–500 $\mu$g pACCMVpLpA DNA per 75 ml TB/ampicillin and 150–200 $\mu$g pJM17 per liter LB/ampicillin. Note that the yield of pJM17 DNA is typically lower than the yield of pACCMVpLpA.

To assess the quality of the purified pJM17 DNA, an aliquot of the DNA is typically digested with Hind III (see Fig. 1). If the DNA was purified with a minimum of shearing, this digest will resolve as a distinct "ladder" of 7–8 bands between 2 and 8 kb on a 0.8% agarose gel. Uncut DNA resolves as several high MW bands, presumably corresponding to various supercoiled forms of the DNA. Some pJM17 preparations may include a 4-kb band corresponding to a pBRX sequence that is spontaneously excised from pJM17 and that has the capacity to replicate autonomously because of retention of the bacterial origin of replication and the antibiotic resistance genes. Although recombinant virions can be produced from pJM17 preparations containing this contaminant, preparations in which it represents the major DNA species should be discarded.

## B. Co-transfection of 293 Cells

The method of choice for co-transfection is calcium phosphate co-precipitation. This method is chosen because it allows for relatively efficient co-transfection of two plasmids, even though other methods such as electroporation or lipofection may be more efficient for single plasmids (Sambrook *et al.*, 1989). Although the importance of passage number has yet to be firmly established, we routinely co-transfect 293 cells before they reach passage 30. The following specific procedure is followed for co-transfection in our laboratories.

1. The E1-tranformed 293 cell line is maintained in 60-mm plates at 37°C in an atmosphere of 5% $CO_2$ in high-glucose (4.5 g/liter) Dulbecco's modified Eagle's medium (DMEM) containing 10% fetal bovine serum (FBS) and 100 U/ml penicillin/streptomycin.
2. The 293 cells should be co-transfected when they are subconfluent, typically at approximately 80% confluence.
3. For each co-transfection, mix 500 $\mu$l sterile 2× HEPES-buffered saline (2× HBS; 280 m*M* NaCl, 42 m*M* HEPES, 10 m*M* KCl, 1.4 m*M* $Na_2HPO_4 \cdot 7H_2O$, 0.2% dextrose, pH 7.05–7.15), 10 $\mu$g pACCMVpLpA DNA containing the cloned cDNA fragment, 10 $\mu$g pJM17 DNA, and sterile water to a volume of 950 $\mu$l. To avoid shearing the pJM17 DNA, mix by inversion rather than using a vortex mixer.

4. Slowly add 50 $\mu$l 2.5 *M* $CaCl_2$ and mix by inversion. Allow the DNA/calcium phosphate precipitate to form for 45–60 min at room temperature. The precipitate should appear as a very fine, slightly turbid suspension.

5. Add 500 $\mu$l precipitate to each of two 60-mm plates of semiconfluent 293 cells containing 5 ml complete DMEM. Incubate at 37°C and 5% $CO_2$ for 4–6 hr.

6. Aspirate the medium and precipitate and wash once with phosphate-buffered saline (PBS).

7. Since a brief treatment of cells with glycerol is believed to increase the efficiency of transfection (Halbert *et al.*, 1985), the cells are treated for 1–2 min with 15% glycerol in serum-free DMEM, followed by an additional PBS wash and replacement of complete medium.

8. Cells are maintained by replacing the majority (but not all) of the medium on a weekly basis. We and others have noted a transient "yellowing" or acidification of the medium in some plates following refeeding. We have found this to be neither harmful to the cells nor a predictor of virus formation.

9. When a recombination event occurs in a co-transfected 293 cell, the resulting recombinant virus completes the life cycle in the permissive host, resulting in cell lysis and the formation of plaques of dead cells. Plaque formation may occur as early as 2 wk and as late as 4 wk after co-transfection and first appears as a visible "hole" in the monolayer. This initial plaque will enlarge as the viral infection progresses, so by 3–5 days after the appearance of the plaque, the monolayer should be completely lysed.

10. After the monolayer lyses, collect the medium and lyse any remaining intact cells by several freeze/thawing cycles. Pellet the cell debris by centrifugation and store the viral supernatant at −20°C.

11. After isolating viral DNA from infected 293 cells, the presence of the cDNA insert can be confirmed by preparation of viral DNA (see subsequent section) and Southern blotting. In addition, a single clone must be isolated and amplified to be used as the working viral stock.

12. Alternatively, single clones can be isolated from the original transfection by replacement of the medium (see Step 7) with 0.65% Noble agar in DMEM with 2.5% FBS.

## C. Extraction of Viral DNA from 293 Cells

The procedure for isolating viral DNA from 293 cells is adapted from an established method (Hirt, 1967).

1. On Day 1, infect a subconfluent 60-mm plate of 293 cells with recombinant virus. For initial lysates obtained after co-transfection, infections are typically done using 10–100 $\mu$l viral lysate in a final volume of 2 ml complete medium for 1 hr at 37°C, followed by a PBS wash and replacement of complete medium.

When the titer of the viral stock is known, a multiplicity of infection of 10 pfu/cell is recommended.

2. Approximately 36 hr later, the cells should appear rounded and refractile, yet should still be adherent to the plate. At this point, aspirate the medium and gently wash twice with PBS.

3. Aspirate the PBS wash, add 800 μl freshly prepared lysis buffer (0.6% SDS, 10 m*M* EDTA, 100 μg/ml proteinase K), and incubate at 37°C for 1 hr.

4. Add 200 μl 5 *M* NaCl dropwise while swirling the plate to uniformly mix the lysate; then incubate on ice for 1 hr.

5. Transfer the viscous lysate to a 1.5-ml eppendorf tube and centrifuge at ≥12,000 *g* for 30–45 min at 4°C.

6. The "pellet" at this point is likely to be very loose and undefined, and should be removed with an inoculating loop, leaving behind the supernatant.

7. Remove proteinase K activity from the supernatant by extracting once with an equal volume of phenol : chloroform. To avoid shearing the viral DNA, mix by inversion rather than using a vortex mixer. After a brief centrifugation (≥12,000 *g*), recover the upper phase and precipitate the viral DNA at −20°C in the presence of 0.25 *M* sodium acetate, pH 5.2, and 2 volumes of ethanol.

8. Pellet the viral DNA by centrifugation (≥12,000 *g*), wash the pellet with 70% ethanol, and gently resuspend the viral DNA in 25–50 μl TE containing 10 μg/ml RNase A. The DNA is now ready for restriction enzyme digestion and Southern blot analysis.

## D. Isolation and Amplification of Viral Clones

One of the great advantages of recombinant adenovirus is that high-titer viral stocks (≥$10^9$ pfu/ml) can easily be obtained. The initial lysate obtained after co-transfection, however, typically has a titer one or more orders of magnitude lower and, in addition, is not necessarily a clonal stock. For these reasons, it is desirable to isolate single discrete plaques and to use them for amplification to generate pure high-titer recombinant stocks. A convenient technique for titering viral stocks and for isolating individual viral plaques is plating by agar overlay.

1. Prepare serial dilutions of the viral stock in 2 ml complete medium. Typically, dilutions between $10^{-3}$ and $10^{-9}$ are prepared.

2. Infect subconfluent 293 cells with the 2 ml containing the various viral dilutions for 1 hr at 37°C.

3. Aspirate the virus-containing media and wash the monolayers twice with PBS.

4. In a separate tube, premix equal volumes of melted 1.3% agarose in water with 2× DMEM (prepared from DMEM powder; GIBCO/BRL, Grand Island, NY) containing 4% PBS, 200 U/ml penicillin/streptomycin + 0.5 μg/ml Fungizone, and 30 mg/liter of phenol red. It is important that the melted agarose

be neither too hot nor too cool. For this reason, the melted agarose should be incubated in a 56°C water bath prior to use.

5. After aspirating the final PBS wash from the infected 293 cells, slowly add 6 ml of the warm agarose/DMEM mixture and allow the mixture to solidify at room temperature for several minutes before returning the plates to the incubator.

6. Viral plaques should form 7–10 days after plating.

To amplify a clonal viral stock, single plaques are isolated by taking a core with a pipet tip and transferring to a tube containing 2 ml complete medium. The following procedure is then used to amplify the pure virus.

1. Elute the virus from the agarose by several freeze/thaw cycles of the medium containing the viral plug.

2. Infect a plate of subconfluent 293 cells with the entire 2 ml of relatively dilute viral sample for 1 hr at 37°C, followed by a PBS wash and addition of complete medium.

3. On cell lysis, which can be anticipated approximately 1 wk after infection, the lysate is collected, cell debris is removed by centrifugation, and the supernatant is collected as the initial or "P1" viral stock. This stock, which will generally have a titer in the range of $10^7$ pfu/ml, should be checked by Southern blotting as described in Section V,C.

4. To increase the titer of the clonal stock further, infect one or more 10-cm plates of subconfluent 293 cells with a minimal volume (5–25 $\mu$l) of the clonal stock in a final volume of 3 ml complete medium for 1 hr at 37°C.

5. Wash with PBS and replace complete medium.

6. After the monolayer lyses, collect the media in a sterile 50-ml polypropylene tube and lyse any remaining intact cells with several rounds of freezing and thawing. Pellet cell debris by centrifugation and store the viral supernatant at −20°C. The titer of this high-titer stock can be determined as described in Section V,E.

In some instances, particularly when planning experiments involving chronic infusion of recombinant adenoviruses, it is desirable to obtain large quantities of purified recombinant virus.

1. Inoculate 30 dishes (150 × 25 mm) of 293 cells at 80% confluence with enough adenovirus to achieve a multiplicity of infection of at least 10 pfu/cell. Allow approximately 36–48 hrs for cells to lyse completely.

2. Harvest lysate and add 0.5% Nonidet P–40. Place on orbital shaker at room temperature for 10 min to ensure thorough mixing and then clear debris by centrifugation at 20,000 *g*.

3. Recover supernatant and add 0.5 volume of 20% polyethylene glycol 8000/2.5 *M* NaCl. Incubate on ice 30 min to overnight with frequent shaking. Centrifuge 10 min at 20,000 *g* to pellet adenovirus. Discard supernatant into bleach.

4. Resuspend adenovirus pellets in 3–6 mL phosphate-buffered saline and centifuge to clear debris (10 min at 4000 *g* in a table-top centrifuge). Recover supernatant and add solid CsCl until 1 mL of the solution weighs 1.32–1.34 *g* (approximately 0.5 g CsCl for each mL of supernatant). Load solution into 13 × 32 mm sealable ultracentrifuge tubes and spin 3 hrs at 361,000 *g* (90,000 rpm in a Beckman TLA-100 rotor).

5. Extract white adenovirus band from tubes with needle and syringe in a total volume of 0.5–1.0 mL and load onto a Pharmacia PD-10 Sephadex column equilibrated with 137 m*M* NaCl, 5 m*M* KCl, 10m*M* Tris-HCl pH 7.4, and 1 m*M* $MgCl_2$. Collect 8–10 drop fractions and determine adenovirus peak by measuring absorbance at 260 nm (use 1/50 dilutions). Pool fractions and determine titer by plaque assay or using the formula 1 $OD_{260} \approx 10^{12}$ pfu/mL. Add 0.1% BSA and store in small aliquots at −80 degrees centigrade.

## E. Transduction of Mammalian Cells with Recombinant Virions *in Vitro*

Recombinant adenovirus can be used for transduction of isolated primary cells such as hepatocytes, small cell clusters such as the islets of Langerhans, or clonal cell lines. In all cases, the basic procedure is the same. For cells that adhere to tissue culture plates, the recombinant viral stock is added directly to the plates in a minimal volume of complete medium (usually 3 ml per 10-cm plate). Generally, a multiplicity of infection of 5–10 pfu/cell applied for 1 hr insures high-efficiency gene transfer (≥80%). Use of higher ratios (as high as 500:1) can enhance the level of expression of the recombinant gene, since each cell can be multiply infected. The potential for overexpression is limited, however, by eventual toxic effects of the virus; the upper limit appears to be different for different cell types. For each new cell type that is considered for adenoviral gene transfer, we recommend a "functional titration" to determine the range of expression that is possible.

Isolated islets are generally maintained intact in suspension culture. Transduction with recombinant virus is achieved by collection of the islets by mild centrifugation, aspiration of their media, and replacement with the viral stock for 1 hr. The virus is then removed by aspiration, the cells are washed with PBS, and the islets are returned to suspension culture in complete media.

Although detailed time-course studies have not been performed, high levels of expression of recombinant genes are evident 48 hr after exposure to recombinant adenovirus that persists for at least 6 days in hepatocytes and 21 days in islets (Gómez-Foix *et al.*, 1992; Becker *et al.*, 1994).

## F. Delivery of Recombinant Virions *in Vivo*

As detailed earlier, recombinant adenoviruses have been useful for delivering genes to whole animals. In general, viruses have been administered as single injections via accessible blood vessels such as the external jugular vein (Stratford-Perricauclet *et al.*, 1990; Rosenfeld *et al.*, 1992; Herz and Gerard, 1993). High efficiency gene delivery to hepatocytes has been reported with a protocol in which a single injection containing approximately $2 \times 10^9$ pfu was administered systemically into mice weighing 20–40 g (Rosenfeld *et al.*, 1992; Herz and Gerard, 1993). Other groups have reported widespread gene delivery to tissues of neonatal animals with a comparable dose (Ishibashi *et al.*, 1993). An alternative method for delivery of large numbers of virions is to infuse the virus slowly over time. We have achieved this in rats by chronic cannulation of the left lateral ventricle with silastic tubing (Dow Corning, Midland, Michigan) under sodium pentothal anesthesia (W. Coats and C. B. Newgard, unpublished observations). The tubing is exteriorized between the shoulders and held in place and protected by a device consisting of a swivel, a harness, and a spring tether, as previously described (Komiya *et al.*, 1990). This configuration allowed the animals to be housed in an unrestrained state with free access to food and water. AdCMV-βGal virus containing approximately $1 \times 10^{10}$ pfu/ml was prepared by CsCl gradient ultracentrifugation of a crude viral stock. The purified stock was diluted to a concentration of $2.5 \times 10^8$ pfu/ml in a solution of 0.9% NaCl, 10% glycerol and was infused continuously via the ventricular cannula for several days at a rate of 0.4 ml/hr. This protocol allowed gene delivery to the islets of Langerhans of the pancreas, whereas a single bolus injection did not. Whether viral infusion allows enhanced gene delivery to other tissues as well remains to be determined.

## VI. Future Directions

As indicated earlier, adenovirus is one of several DNA viruses currently being exploited as a gene transfer vector for mammalian cells, each with its own advantages and disadvantages. In closing, we will speculate on future directions for this rapidly expanding area of research, including in our discussion the applicability of new physical gene transfer methods, some of which have been combined with viral technologies to create "hybrid" approaches.

As summarized earlier, recombinant adenovirus provides distinct advantages over most physical DNA transformation techniques such as $Ca_3(PO_4)_2$ co-precipitation, lipofection, or electroporation in terms of efficiency of gene transfer (generally 10–20% with the physical methods versus 60–100% with recombinant adenovirus). Adenovirus-mediated DNA delivery *in vivo* is also clearly possible, albeit with varying efficiency in different tissues. Recent developments have suggested other alternative or potentially complementary approaches. One approach is to prepare the DNA to be transferred as a chemically linked conjugate with ligands such as galactose-bearing asialo-orosomucoid (Wu and Wu, 1987,1988) or transferrin (Wagner *et al.*, 1990) that targets the complex to the receptor-mediated endocytosis pathway of cells. Two distinct advantages of the DNA-conjugate approach over the recombinant adenovirus approach are (1) the absence of theoretical size limitations for the DNA to be transferred and (2) the tissue-specific delivery of the gene of interest to those tissues that express receptors recognizing the coupled ligand. An apparent disadvantage of this approach is that it appears to work with limited efficiency, probably because a substantial percentage of the DNA that is internalized via receptor-mediated endocytosis is degraded within lysosomes.

Recently, combination viral/physical transformation protocols have been described that show considerable promise. Replication-defective adenovirus greatly enhances the transfer of conjugated DNA by virtue of its ability to lyse endosomes, thereby releasing the internalized DNA before it is degraded (Curiel *et al.*, 1991). The efficacy of simple co-infection of cells that are transfected with conjugated DNA is limited by the relatively low probability of internalization of the conjugate and a virus into the same endosome. To circumvent this problem, more recent approaches have emphasized physical coupling of the adenovirus to the DNA conjugate. This has been achieved either by chemically linking an anti-adenovirus antibody to the DNA conjugate (Curiel *et al.*, 1992) or, more simply, by coupling adenovirus to poly-L-lysine through the action of transglutaminase or the use of biotin/streptavidin (Wagner *et al.*, 1992). The poly-L-lysine-conjugated adenovirus is then combined with the DNA to be transferred and the poly-L-lysine-conjugated ligand (i.e., transferrin) to form a virus–DNA–ligand complex. Such complexes enhance the efficiency of gene transfer into cell lines and primary cells by several orders of magnitude compared with conjugated DNA lacking adenovirus; efficiencies of 100% have been reported for HeLa cells (Wagner *et al.*, 1992) and rodent hepatocytes (Wagner *et al.*, 1992; Christiano *et al.*, 1993). Although adenovirus conjugation overcomes the problem of low efficiency of gene transfer with DNA conjugates, it removes tissue specificity since adenovirus receptors are found on most mammalian cells. A technique for overcoming this final hurdle has been presented recently, involving treatment of a transferrin–DNA–adenovirus complex with a monoclonal antibody against the adenoviral fiber protein (Michael *et al.*, 1993). Such treatment was found to block the permissive effect of free adenovirus on conjugate-mediated DNA uptake into HeLa cells but had no effect on complexes with physically coupled adenovirus. Whether these multifunctional molecular conjugate vectors will retain their efficacy and specificity of gene delivery in whole animals remains to be determined.

Another issue is whether even simpler methods may ultimately prevail. Thus, it has been appreciated for some time that efficient DNA transfer can be achieved *in vitro* with lipofection, a technique in which DNA is packaged into cationic liposomes consisting of the synthetic lipid *N*-[1-(2,3-dioleyloxy)propyl]-*N*,*N*,*N*-trimethylammonium chloride (DOTMA) Felgner *et al.*, 1987). Optimization of the DNA:liposome ratio has allowed efficient DNA delivery to a wide array of tissues in rodents injected with a single dose of the conjugate (Zhu *et al.*, 1993). Expression was found to persist in many tissues for as long as 9 wk, although the status of the plasmid DNA (integrated verus episomal) was not clearly established. Importantly, re-administration of the conjugate caused a second peak of transgene expression, indicating that the liposomes may be different from recombinant adenoviruses in the degree to which they are tolerated by the immune system. Further studies will be required to evaluate the safety and consistency of this exciting new method.

## VII. Conclusions

Rapid progress has been made in recent years in the area of techniques of gene transfer to mammalian cells. In this chapter we have selected recombinant adenovirus as one of the best-studied examples of the new technology, and have attempted to provide the rationale and precedent for its use as an exciting new tool for studying metabolic regulation. We have also provided methods and procedures for constructing and propagating new recombinant virions in the laboratory. Adenovirus in its current form is unlikely to represent the ultimate transfer vector for human gene therapy, given problems such as the lack of integration of the viral genome and the potential for immunological response to injected virus. It is therefore likely that new research initiatives will focus on attempting to engineer "second generation" virions that combine functional features of different viruses, that is, the site-specific integration function of AAV with the growth and infectivity characteristics of adenovirus. In the interim, however, recombinant adenovirus greatly enhances our power to investigate the impact of gene transfer in normal animals as well as in animal models of disease, thereby providing the knowledge base that will allow us to take maximal advantage of the safer and more efficacious gene therapy strategies that are likely to emerge in the future.

### Acknowledgments

Studies from the authors' laboratory relevant to this article have been supported by NIH grants R29-DK40734 and R01-DK46492 (to C. B. N.). R. D. G. is a recipient of an Established Investigatorship from the American Heart Association-Genentech Incorporated.

### References

Alkhatib, G., and Briedis, D. J. (1988). High-level expression of biologically active measles virus hemagglutinin by using an adenovirus type 5 helper-free vector system. *J. Virol.* **62,** 2718–2727.

Baque, S., Newgard, C. B., Gerard, R. D., Guinovart, J. J., and Gómez-Foix, A. M. (1994). Adenovirus-mediated delivery into myocytes of muscle glycogen phosphorylase, the enzyme deficient in glycogen storage disease type v. *Biochem. J.,* in press.

Bailey, J. E. (1991). Toward a science of metabolic engineering. *Science* **252,** 1668–1675.

Becker, T. C., Noel, R. J., Johnson, J. H., Quaade, C., Meidell, R. S., Gerard, R. D., and Newgard, C. B. (1993). Use of recombinant adenovirus vectors for high-efficiency gene transfer into the islets of Langerhans. *Diabetes* **42,** Suppl. 1:11A.

Becker, T. C., Noel, R., Johnson, J. H., and Newgard, C. B. (1994). Divergent effects of glucokinase, hexokinase, and glucokinase mutants associated with MODY on glucose-induced insulin release. *J. Cell. Biochem.* Suppl. 18A:133.

Berkner, K. L. (1988). Development of adenovirus vectors for the expression of heterologous genes. *BioTechniques* **6,** 616–629.

Berkner, K. L., Schaffhausen, B. S., Roberts, T. M., and Sharp, P. A. (1987). Abundant expression of polyomavirus middle T antigen and dihydrofolate reductase in an adenovirus recombinant. *J. Virol.* **61,** 1213–1220.

Coats, W. S., Browner, M. F., Fletterick, R. J., and Newgard, C. B. (1991). An engineered liver glycogen phosphorylase with AMP allosteric activation. *J. Biol. Chem.* **266,** 16113–16119.

Cristiano, R. J., Smith, L. C., and Woo, S. L. C. (1993). Hepatic gene therapy: Adenovirus enhancement of receptor-mediated gene delivery and expression in primary hepatocytes. *Proc. Natl. Acad. Sci. USA* **90,** 2122–2126.

Curiel, D. T., Agarwal, S., Wagner, E., and Cotten, M. (1991). Adenovirus enhancement of transferrin-polylysine mediated gene delivery. *Proc. Natl. Acad. Sci. USA* **88,** 8850–8854.

Curiel, D. T., Wagner, E., Cotten, M., Birnsteil, M. L., Agarwal, S., Li, C., Loechel, S., and Hu, P. (1992). High-efficiency gene transfer mediated by adenovirus coupled to DNA-polylysine complexes. *Hum. Gene Ther.* **3,** 147–154.

Davidson, D., and Hassell, J. A. (1987). Overproduction of polyomavirus middle T antigen in mammalian cells through the use of an adenovirus vector. *J. Virol.* **61,** 1226–1239.

DeCaprio, J. A., Ludlow, J. W., Figge, J., Stein, J. Y., Huang, C. M., Lee, W. H., Marsilio, E., Paucha, E., and Livingstone, D. M. (1988). SV40 large tumor antigen forms a specific complex with the product of the retinoblastoma susceptibility gene. *Cell* **54,** 275–283.

Doolittle, D. P., Hulbert, L. L., and Cordy, C. A. (1974). A new allele of the sparse fur gene of the mouse. *J. Hered.* **65,** 194–195.

Felgner, P. L., Gadek, T. R., Holm, M., Roman, R., Chan, H. W., Wenz, M., Northrop, J. P., Ringold, G. M., and Danielsen, M. (1987). Lipofection: A highly efficient, lipid-mediated DNA-transfection procedure. *Proc. Natl. Acad. Sci. USA* **84,** 7413–7417.

Ferber, S., BeltrandelRio, H., Noel, R., Becker, T. C., Johnson, J. H., Cassidy, L., Clark, S., Hughes, S. D., and Newgard, C. B. (1994). GLUT-2 gene transfer into insulinoma cells confers both low and high affinity glucose-stimulated insulin release: Relationship to glucokinase activity. *J. Biol. Chem.* **269,** 11523–11529.

Gerard, R. D., and Meidell, R. S. (1993). Adenovirus-mediated gene transfer. *Trends Cardiovasc. Med.* **3,** 171–177.

German, M. S. (1993). Glucose sensing in pancreatic islet beta cells: The key role of glucokinase and the glycolytic intermediates. *Proc. Natl. Acad. Sci. USA* **90,** 1781–1785.

German, M. S., Moss, L. G., and Rutter, W. J. (1990). Regulation of insulin gene expression by glucose and calcium in transfected primary islet cultures. *J. Biol. Chem.* **265,** 22063–22066.

Gluzman, Y., Reichl, H., and Solnick D. (1982). Helper-free adenovirus type-5 vectors. *In* "Eucaryotic Viral Vectors" (Y. Gluzman, ed.), pp. 187–192. Cold Spring Harbor Laboratory Press, Cold Spring Harbor, New York.

Gómez-Foix, A. M., Coats, W. S., Baque, S., Alam, T., Gerard, R. D., and Newgard, C. B. (1992). Adenovirus-mediated transfer of the muscle glycogen phosphorylase gene into hepatocytes confers altered regulation of glycogen metabolism. *J. Biol. Chem.* **267,** 25129–25134.

Graham, F. L. (1984). Covalently closed circles of human adenovirus DNA are infectious. *EMBO J.* **3,** 2917–2922.

Graham, F. L., and Prevec, L. (1991). Manipulation of viral vectors. *Meth. Mol. Biol.* **7,** 109–128.

Graham, F. L., Smiley, J., Russell, W. C., and Nairn, R. (1977). Characteristics of a human cell line transformed by DNA from human adenovirus type 5. *J. Gen. Virol.* **36,** 59–72.

Halbert, D. N., Cutt, J. R., and Shenk, T. (1985). Adenovirus early region 4 encodes functions required for efficient DNA replication, gene expression and host cell shutoff. *J. Virol.* **56,** 250–257.

Henderson, J. R., and Moss, M. C. (1985). A morphometric study of the endocrine and exocrine capallaries of the pancreas. *Q. J. Exp. Physiol.* **70,** 347–356.

Herz, J., and Gerard, R. D. (1993). Adenovirus-mediated transfer of low density lipoprotein receptor gene acutely accelerates cholesterol clearance in normal mice. *Proc. Natl. Acad. Sci. USA* **90,** 2812–2816.

Hirt, B. (1967). Selective extraction of polyoma DNA from infected mouse cell cultures. *J. Mol. Biol.* **26,** 365–369.

Horwitz, M. S. (1990). Adenoviridae and their replication. *In* "Virology" (B. N. Fields, ed.), 2nd Ed., pp. 1679–1721. Raven Press, New York.

Ishibashi, S., Brown, M. S., Goldstein, J. L., Gerard, R. D., Hammer, R. E., and Herz, J. (1993). Hypercholesterolemia in low density lipoprotein receptor knockout mice and its reversal by adenovirus-mediated gene delivery. *J. Clin. Invest.* **92,** 883–893.

Johnson, D. C., Ghosh-Choudhury, G., Smiley, J. R., Fallis, L., and Graham, F. L. (1988). Abundant expression of Herpes simplex virus glycoprotein gB using an adenovirus vector. *Virology* **164,** 1–14.

Jones, N., and Shenk, T. (1978). Isolation of deletion and substitution mutants of adenovirus type 5. *Cell* **13,** 181–188.

Kispal, G., Evans, C. T., Malloy, C., and Srere, P. A. (1989). Metabolic studies on citrate synthase mutants of yeast. *J. Biol. Chem.* **264,** 11204–11210.

Komiya, I., Baetens, D., Kuwajima, M., Orci, L., and Unger, R. H. (1990). Compensatory capabilities of islets of BB/Wor rats exposed to sustained hyperglycemia. *Metabolism* **39,** 614–618.

Kotin, R. M., Siniscalco, M., Samulski, R. J., *et al.* (1990). Site-specific integration by adeno-associated virus (AAV). *Proc. Natl. Acad. Sci. USA* **87,** 2211–2215.

Kotin, R. M., Menninger, J. C., Ward, D. C., and Berns, K. I. (1991). Mapping and direct visualization of a region-specific viral DNA integration site on chromosome 19q13-qter. *Genomics* **10,** 831–834.

Le Gal Le Salle, G., Robert, J. J., Berrard, S., Ridoux, V., Stratford-Perricaudet, L. D., Perricaudet, M., and Mallet, J. (1993). An adenovirus vector for gene transfer into neurons and glia in the brain. *Science* **259,** 988–990.

Liao, X., and Butow, R. A. (1993). RTG1 and RTG2: Two yeast genes required for a novel path of communication from mitochondria to the nucleus. *Cell* **72,** 61–71.

Malthus, R., Clark, D. G., Watts, C., and Sneyd, J. G. T. (1980). Glycogen storage disease in rats, a genetically determined deficiency in liver phosphorylase kinase. *Biochem. J.* **188,** 99–104.

McGrory, J., Bautista, D., and Graham, F. L. (1988). A simple technique for the rescue of early region I mutations into infectious human adenovirus type 5. *Virology* **163,** 614–617.

McLachlan, A. D., and Boswell, D. R. (1985). Confidence limits for homology in protein or gene sequences. The c-*myc* oncogene and adenovirus E1A protein. *J. Mol. Biol.* **185,** 39–49.

Meglasson, M. D., and Matschinsky, F. M. (1986). Pancreatic islet glucose metabolism and regulation of insulin secretion. *Diabetes/Metab. Rev.* **2,** 163–214.

Michael, S. I., Huang, C., Romer, M. U., Wagner, E., Hu, P., and Curiel, D. T. (1993). Binding-incompetent adenovirus facilitates molecular conjugate-mediated gene transfer by the receptor-mediated endocytosis pathway. *J. Biol. Chem.* **268,** 6866–6869.

Miller, A. D. (1992). Human gene therapy comes of age. *Nature* (*London*) **357,** 455–460.

Morgan, R. A., and French Anderson, W. (1993). Human gene therapy. *Annu. Rev. Biochem.* **62,** 191–217.

Newgard, C. B. (1992). Cellular engineering for the treatment of metabolic disorders: Prospects for therapy in diabetes. *BioTechnology* **10,** 1112–1120.

Newgard, C. B., Hwang, P. K., and Fletterick, R. J. (1989). The family of glycogen phosphorylases: Structure and function. *CRC Crit. Rev. Biochem. Mol. Biol.* **24,** 69–99.

Newgard, C. B., Hughes, S. D., Quaade, C., BeltrandelRio, H., Gómez-Foix, A. M., and Ferber, S. (1993). Molecular engineering of the pancreatic $\beta$-cell. *J. Lab. Clin. Med.* **122,** 356–363.

Nienhuis, A., Walsh, C. E., and Liu, J. (1993). Viruses as therapeutic gene transfer vectors. *Hematology* **16,** 353–414.

Paabo, S. T., Nilsson, T., and Peterson, P. A. (1986). Adenoviruses of subgenera B, C, D, and E modulate cell-surface expression of major histocompatibility complex class I antigens. *Proc. Natl. Acad. Sci. USA* **83,** 9665–9669.

Persson, H., Kvist, S., Ostberg, L., Petersson, P. A., and Phillipson, L. (1979). Adenovirus early glycoprotein E3-19K and its association with transplantation antigens. *Cold Spring Harbor Symp. Quant. Biol.* **44,** 509–517.

Ponder, K. P., Gupta, S., Leland, F., Darlington, G., Finegold, M., DeMayo, J., Ledley, F. D., Chowdhury, J. R., and Woo, S. L. C. (1991). Mouse hepatocytes migrate to liver parenchyma and function indefinitely after intrasplenic transplantation. *Proc. Natl. Acad. Sci. USA* **88,** 1217–1221.

Quantin, B., Perricaudet, L. D., Tajbakhsh, S., and Mandel, J-L. (1992). Adenovirus as an expression vector in muscle cells in vivo. *Proc. Natl. Acad. Sci. USA* **89,** 2581–2584.

Ragot, T., Vincent, N., Chafey, P., Vigne, E., Gilgenkrantz, H., Couton, D., Cartaud, J., Briand, P., Kaplan, J-C., Perricaudet, M., and Kahn, A. (1993). Efficient adenovirus-mediated transfer of a human minidystrophin gene to skeletal muscle of mdx mice. *Nature (London)* **361,** 647–650.

Ralston, R., and Bishop, J. M. (1983). The protein products of the *myc* and *myb* oncogens and adenovirus E1A are structurally related. *Nature (London)* **306,** 803–806.

Rosenfeld, M. A., Siegfried, W., Yoshimura, K., Yoneyama, K., Fukayama, M., Stier, L. E., Paakko, P. K., Filardi, P., Stratford-Perricaudet, L. D., Perricaudet, M., Jallat, S., Pavirani, A., Lecocq, J-P., and Crystal, R. G. (1991). Adenovirus-mediated transfer of a recombinant a1-antitrypsin gene to the lung epithelium in vivo. *Science* **252,** 431–434.

Rosenfeld, M. A., Yoshimura, K., Trapnell, B. C., Yoneyama, K., Rosenthal, E. R., Dalemans, W., Fukayama, M., Bargon, J., Stier, L. E., Stratford-Perricaudet, L., Perricaudet, M., Guggino, W. B., Pavirani, A., Lecocq, J-P., and Crystal, R. G. (1992). *In vivo* transfer of the human cystic fibrosis transmembrane conductance regulator gene to the airway epithelium. *Cell* **68,** 143–155.

Rowe, W. P., Huebner, R. J., Gilmore, L. K., Parrott, R. H., and Ward, T. G. (1953). Isolation of a cytopathogenic agent from human adenoids undergoing spontaneous degeneration in tissue culture. *Proc. Soc. Exp. Biol. Med.* **84,** 570–573.

Ruben, M., Bacchetti, S., and Graham, F. L. (1983). Covalently closed circles of human adenovirus type 5 DNA. *Nature (London)* **301,** 172–174.

Sambrook, J., Fritsch, E. F., and Maniatis, T. (1989). "Molecular Cloning: A Laboratory Manual." Cold Spring Harbor Laboratory Press, Cold Spring Harbor, New York.

Sen, A., Dunnmon, P., Henderson, S. A., Gerard, R. D., and Chien, K. R. (1988). Terminally differentiated neonatal rat myocardial cells proliferate and maintain specific differentiated functions following expression of SV40 large T antigen. *J. Biol. Chem.* **263,** 19132–19136.

Smith, G. L. (1991). Vaccinia virus vectors for gene expression. *Curr. Opin. Biotechnol.* **2,** 713–717.

Solnick, D. (1981). Construction of an adenovirus-SV40 recombinant producing SV40 T-antigen from adenovirus late promoter. *Cell* **24,** 135–143.

Stillman, B., Gerard, R. D., Guggenheimer, R. A., and Gluzman, Y. (1985). T antigen and template requirements for SV40 DNA replication *in vitro*. *EMBO J.* **4,** 2933–2939.

Stratford-Perricaudet, L. D., Levrero, M., Chasse, J-F., Pericaudet, M., and Briand, P. (1990). Evaluation of the transfer and expression in mice of an enzyme-encoding gene using a human adenovirus vector. *Hum. Gene Ther.* **1,** 241–256.

Stratford-Perricaudet, L. D., Makeh, I., Perricaudet, M., and Briand, P. (1992). Widespread long-term gene transfer to mouse skeletal muscles and heart. *J. Clin. Invest.* **90,** 626–630.

Thummel, C. R., Tijan, R., and Grodzicker, T. (1981). Expression of SV40 T antigen under control of adenovirus promoter. *Cell* **23,** 825–836.

Van Doren, K., and Gluzman, Y. (1984). Efficient transformation of human fibroblasts by adenovirus-simian virus 40 recombinants. *Mol. Cell. Biol.* **4,** 1653–1656.

Van Doren, K., Hanahan, D., and Gluzman, Y. (1984). Infection of eucaryotic cells by helper-independent recombinant adenoviruses: Early region 1 is not obligatory for integration of viral DNA. *J. Virol.* **50,** 606–614.

Varmus, H. (1988). Retroviruses. *Science* **240,** 1427–1435.

Wagner, E., Zenke, M., Cotten, M., Beug, H., and Birnsteil, M. L. (1990). Transferrin-polycation conjugates as carriers for DNA uptake into cells. *Proc. Natl. Acad. Sci. USA* **87,** 3410–3414.

Wagner, E., Zatloukal, K., Cotten, M., Kirlappos, H., Mechtler, K., Curiel, D. T., and Birnsteil, M. L. (1992). Coupling of adenovirus to transferrin-polylysine/DNA complexes greatly enhances receptor-mediated gene delivery and expression of transfected genes. *Proc. Natl. Acad. Sci. USA* **89,** 6099–6103.

Welsh, M., Claesson-Welsh, L., Hallberg, A., Welsh, N., Betsholtz, Ch., Arkhammar, P., Nilsson, T., Heldin, C-H., and Berggren, P-O. (1990). Coexpression of the platelet-derived growth factor (PDGF) B chain and the PDGF $\beta$ receptor in isolated pancreatic islet cells stimulates DNA synthesis. *Proc. Natl. Acad. Sci. USA* **87,** 5807–5811.

Whyte, D., Buchkovich, K. J., Horowitz, J. M., Friend, S. H., Raybuck, M., Weinberg, R. A., and Harlow, E. (1988). Association between an oncogene and an antioncogene: The adenovirus E1A proteins bind to the retinoblastoma gene product. *Nature* (*London*) **334,** 124–129.

Wilson, J. M., Chowdhury, N. R., Grossman, M., Wajsman, R., Epstein, A., Mulligan, R. C., and Chowdhury, J. R. (1990). Temporary amelioration of hyperlipidemia in low density lipoprotein receptor-deficient rabbits transplanted with genetically modified hepatocytes. *Proc. Natl. Acad. Sci. USA* **87,** 8437–8441.

Wu, G. Y., and Wu, C. H. (1987). Receptor-mediated *in vitro* gene transformation by a soluble DNA carrier system. *J. Biol. Chem.* **262,** 4429–4432.

Wu, G. Y., and Wu, C. H. (1988). Receptor-mediated gene delivery and expression *in vivo*. *J. Biol. Chem.* **263,** 14621–14624.

Yamada, M., Lewis, J. A., and Grodzicker, T. (1985). Overproduction of the protein product of a nonselected foreign gene carried by an adenovirus vector. *Proc. Natl. Acad. Sci. USA* **82,** 3567–3571.

Zhu, N., Liggitt, D., Liu, Y., and Debs, R. (1993). Systemic gene expression after intravenous DNA delivery into adult mice. *Science* **261,** 209–211.

# CHAPTER 9

# Amplicon-Based Herpes Simplex Virus Vectors

**Dora Y. Ho**

Department of Biological Sciences
Stanford University
Stanford, California 94305



## I. Introduction

The wide host range of herpes simplex virus type 1 (HSV-1), both *in vivo* and *in vitro,* and the relative ease of its genetic manipulation have made it an attractive candidate as a tool for gene transfer. In particular, the natural propensity of the virus to infect and establish life-long latent infection in postmitotic neurons has prompted much recent effort in developing HSV vectors.

Two different types of HSV vectors have been developed. For the first type, the recombinant virus vector, the gene of interest is inserted into the backbone of the viral genome by genetic recombination; this method is discussed in Chapter 10. The second type of HSV vector is amplicon-based (see subsequent discussion) and has been termed the defective virus vector (Geller and Breakefield, 1988). This chapter will focus on the components and construction of amplicon-based vectors. Recent developments, concerns, and prospects of HSV-based vectors have been reviewed extensively elsewhere (Breakefield and DeLuca, 1991; Glorioso *et al.*, 1992; Lieb and Olivo, 1993).

The development of amplicon-based vectors has capitalized on the natural occurrence of defective interfering viruses (also called DI particles) that arise during high-multiplicity propagation of viruses. In general, DI particles are subgenomic viral particles that lack an essential portion of the genome, require the complementation of homologous "helper" virus for replication, and thus "interfere" with the replication of the helpers by replicating at their expense when infecting the same cell. The natural and ubiquitous occurrence of DI particles among both RNA and DNA viruses may play a very important role in their biology and pathology; for example, both attenuation and intensification of disease processes by DI particles have been observed (for review, see Holland, 1990; Roux *et al.*, 1991).

DI particles have been identified in a number of herpesviruses, including HSV (reviewed by Frenkel, 1981), human cytomegalovirus (Stinski *et al.*, 1979), and equine herpesvirus (Henry *et al.*, 1979). The possibility of using herpes DI particles as gene transfer vectors stemmed from the work of Frenkel and colleagues, which clearly characterized the *cis* signals essential to the replication and encapsidation of the defective genome (Spaete and Frenkel, 1985). The resulting vector derived from HSV defective genomes was termed an amplicon (for cloning–amplifying vector) by Spaete and Frenkel (1982), and has been employed to express various genes in eukaryotic cells (Kwong and Frenkel, 1985; Stow *et al.*, 1986; Kaplitt *et al.*, 1991; Federoff *et al.*, 1992; Battleman *et al.*, 1993; Bergold *et al.*, 1993; Geller *et al.*, 1993; Ho *et al.*, 1993a). Essentially, the gene of interest is cloned into a bacterial-based plasmid with the origin of replication and cleavage/packaging signals from HSV. In the presence of a replication-competent helper virus, the plasmid will replicate as concatemers and will be packaged into virions. Since viral particles of DI viruses are indistinguishable from those of the parental helper virus, amplicon-derived vectors also have one of the advantages of the recombinant HSV vectors; the broad host range of the virus allows the viral vector to infect many cell types, including postmitotic neurons, which are difficult targets for most gene transfer methods. Gene transfer into neurons by DI viral vectors was first reported by Geller and Breakefield (1988) and has been successfully demonstrated by other groups both *in vivo* and *in vitro* (Kaplitt *et al.*, 1991; Federoff *et al.*, 1992; Battleman *et al.*, 1993; Bergold *et al.*, 1993; Ho *et al.*, 1993a).

The amplicon-based HSV vectors have generally been referred to as "defective vectors" (Geller and Breakefield, 1988), which is rather confusing (as also pointed out in Chapter 10). Although the vectors themselves certainly cannot replicate on their own and are thus defective, the parental viruses for constructing recombinant vectors or the helper viruses for propagating the amplicons are frequently viral mutants that are themselves defective in replication under nonpermissive conditions. Therefore, for clarity, whereas the term "amplicon" refers to the plasmid form of the vector, infectious virions containing the amplicon concatemers will be termed "defective interfering" (DI) vectors.

## II. Components of an Amplicon

The basic components of an amplicon include prokaryotic sequences that allow propagation and drug selection in bacteria, an origin of replication and a packaging signal from HSV, and a transcriptional unit for expressing the gene of interest. A schematic diagram of a basic amplicon is shown in Fig. 1.

### A. Prokaryotic Sequences

The prototype amplicons that our laboratory constructs use prokaryotic sequences from the pGEM cloning vectors (Promega, Madison, WI). The resulting amplicon plasmids are propagated in *Escherichia coli* strain DH5α. Any other cloning vectors that provide a drug selection marker and bacterial origin of replication would suffice.

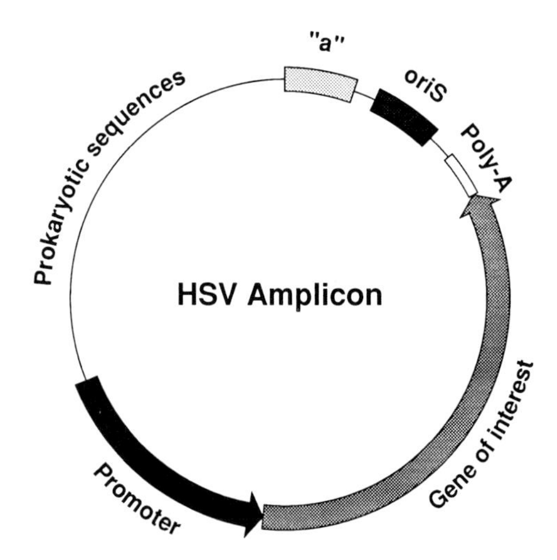


**Fig. 1** Schematic diagram of a prototype amplicon. The transcriptional unit in the amplicon is represented by the promoter, the gene of interest, and a polyadenylation signal. The two *cis*-acting sequences from HSV-1, the $ori_S$, and the *a* sequence provide necessary signals for replication and packaging into virions. The prokaryotic sequences represented by the thin line allow propagation and drug selection in bacteria. The effects or significance of the relative positions of these elements have not been investigated in detail.

## B. HSV Sequences

Before examining the functions of specific HSV components, the genome organization of HSV-1 is first briefly described (Fig. 2). The HSV-1 genome (152 kb) consists of two components, designated long (L) and short (S). Each component contains unique sequences ($U_L$ and $U_S$, respectively) flanked by inverted repeats. The repeat elements of the L component are *ab* at the terminus and *b'a'*, inverted at the internal junction; those of the S component are *a'c'* at the internal junction and *ca* at the terminus. A standard HSV-1 genome can be represented by $a_nb$-$U_L$-$b'a'_mc'$-$U_S$-$ca$. Whereas the S terminus contains only a single copy of the *a* sequence, the L–S junction and the L terminus contain a tandem array of *a* sequences, with *n* and *m* varying from 1 to greater than 10 (Roizman and Sears, 1990).

### 1. Origin of Replication

The HSV genome contains three origins of replication: one copy of $ori_L$, located in the middle of the L component, and two copies of $ori_S$, located in the repeat elements of the S component (Fig. 2). The $ori_L$ consists of an A + T-rich 144-bp sequence forming a perfect palindrome. Because of its extensive dyad symmetry, DNA fragments containing $ori_L$ tend to be unstable when propagated in *E. coli* (Weller *et al.*, 1985). The $ori_S$ has been mapped to a 90-bp sequence that contains a much shorter A + T-rich imperfect palindrome of 45 bp (Stow and McMonagle, 1983). DNA sequences containing $ori_S$ can be propagated in *E. coli* without instability. Some of the original amplicons constructed by Frenkel and colleagues consisted of $ori_L$ derived from natural

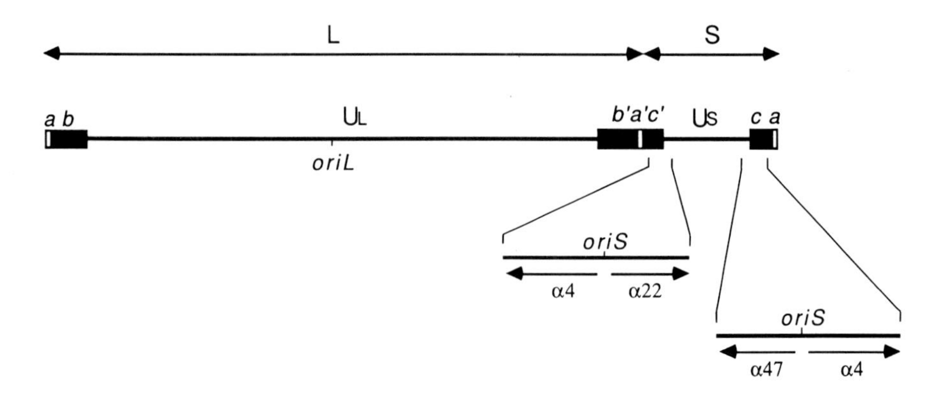


**Fig. 2** Schematic diagram of the HSV-1 genome. The HSV-1 genome consists of the long (L) and short (S) components. It is further depicted with *b* and *c* inverted sequences indicated by solid black boxes flanking the unique long ($U_L$) and unique short ($U_S$) regions, and with the terminal-repeat *a* sequences indicated by open boxes. The $ori_L$ is located in the middle of the $U_L$ region. Parts of the two short repeat sequences are expanded to show the relative positions of the α4, α22, and α47 genes and the two copies of $ori_S$.

DI particles. Interestingly, although these amplicons contained deletions in the $ori_L$ sequences, the defective genome derived from them invariably had the deletion restored (Spaete and Frenkel, 1985). Most subsequent work by other groups of researchers has employed $ori_S$ from HSV-1 (Stow *et al.*, 1986; Geller and Breakefield, 1988; Ho *et al.*, 1993a) or HSV-2 (Kaplitt *et al.*, 1991).

The $ori_S$ is located between the promoters of $\alpha4$ and $\alpha22/\alpha47$ (Fig. 2). Wong and Schaffer (1991) showed that the transcriptional regulatory elements of these promoters and the *trans*-acting factors that bind to them play a critical role in the efficiency of $ori_S$. When compared with the minimal origin of 90 bp, these elements stimulate replication greater than 80-fold. Since propagation and amplification of DI vectors rely on successful competition with the helper virus for replication machineries, the efficiency of $ori_S$ used in amplicons is crucial. Using the 90-bp minimal $ori_S$ may not be sufficient. If either the $\alpha4$ or the $\alpha22/\alpha47$ promoter is used to express the gene of interest, the $ori_S$ and the promoter can be isolated together to construct the amplicon. Such an approach can guarantee a fully functional $ori_S$ and was first employed by Geller and Breakefield (1988). When other promoters are used, a 295-bp fragment of the HSV-1 $ori_S$ (Ho *et al.*, 1993a) or a 270-bp fragment of the HSV-2 $ori_S$ have been shown to confer efficient replication (Kaplitt *et al.*, 1991).

## 2. Cleavage/Packaging Signal

HSV genomes replicate by a rolling circle mechanism (Jacob *et al.*, 1979; Rabkin and Hanlon, 1990). Newly synthesized viral DNA (from amplicons or from helper virus) is in concatemeric form and must be cleaved and packaged into preformed empty capsids. Considerable evidence suggests that these two processes are linked; the signal on the viral genome that is responsible for the cleavage/packaging event has been mapped to the *a* sequence (Deiss and Frenkel, 1986; Deiss *et al.*, 1986). The cleavage process is also accompanied by amplification of the *a* sequence (Deiss *et al.*, 1986). Studies of amplicon-based defective genomes show that cleavage of the genome can occur between any pair of *a* sequences, and amplicon units of different sizes can be packaged into nuclear capsids. However, only full-length defective genomes are found in the cytoplasmic virions (Vlazny *et al.*, 1982).

The *a* sequence consists of two unique sequences, $U_b$ and $U_c$, flanked by a 20-nucleotide direct repeat DR1. Separating $U_b$ and $U_c$ are internal repeated arrays that vary in copy number and sequence among different HSV isolates. Because of the variabilities in the number of repeat sequences, the size of the *a* sequence varies from about 250 to 500 bp from strain to strain. For example, in HSV-1 strain F, the structure of the *a* sequence can be represented as [DR1-$U_b$-DR2$_n$-DR4$_m$-$U_c$-DR1]. DR1 is a 20-bp direct repeat; DR2 is a 12-bp direct repeat ($n$ = 19–23); DR4 is a 37-bp direct repeat ($m$ = 2–3); and $U_b$ and $U_c$ are unique sequences of 65 and 58 bp, respectively. In the amplicon system, *a* sequences from HSV strain F (Kaplitt *et al.*, 1991) and strain KOS (Ho *et al.*, 1993b) have been used to confer cleavage/packaging function.

In addition to serving the purpose of cleavage and packaging, the *a* sequence is also responsible for the inversion of L and S components for the viral genome (Mocarski *et al.*, 1980; Chou and Roizman, 1985) and the circularization of the linear genome upon entry into infected cells (Mocarski and Roizman, 1982; Poffenberger and Roizman, 1985). The *a* sequence also has promoter activity (Chou and Roizman, 1986), expressing a viral gene designated $\gamma_1 34.5$, which plays a role in neurovirulence (Chou *et al.*, 1990; Chou and Roizman, 1992). How the promoter activity of the *a* sequence may affect gene expression in an amplicon (or in a DI vector) has not been investigated.

## C. Transcriptional Units

### 1. Monocistronic

Conventional amplicons contain a single monocistronic transcriptional unit. Various kinds of gene products have been expressed using such systems. Some of the specific examples include $\beta$-galactosidase ($\beta$-gal) (Geller and Breakefield, 1988; Kaplitt *et al.*, 1991), nerve growth factor (Federoff *et al.*, 1992) and its receptor (Battleman *et al.*, 1993), glucose transporter (Ho *et al.*, 1993a), glutamate receptor (Bergold *et al.*, 1993), and adenylate cyclase (Geller *et al.*, 1993). Most of these studies examined gene expression from DI vectors in neurons *in vitro* or *in vivo*, and exclusively employed either the human cytomegalovirus immediate early gene 1 (HCMV IE1) promoter (Kaplitt *et al.*, 1991; Ho *et al.*, 1993a) or the HSV-1 $\alpha22/\alpha47$ promoter (Geller and Breakefield, 1988; Federoff *et al.*, 1992; Battleman *et al.*, 1993; Bergold *et al.*, 1993; Geller *et al.*, 1993). Although the HCMV IE1 is a very strong promoter in many cell types, the $\alpha22/\alpha47$ promoter (as well as the other HSV immediate early promoters) is induced by VP16, a tegument protein of HSV-1 brought along by the virion (for review, see Roizman and Sears, 1990). Although there were some concerns that induction of the HSV immediate early promoters by VP16 may not occur in neurons (Kemp *et al.*, 1990; Lillycrop *et al.*, 1991), both the $\alpha22/\alpha47$ and the $\alpha4$ promoter (Lawrence *et al.*, 1994) have been shown to provide strong and reliable gene expression in neurons, both *in vitro* and *in vivo*. Other promoters that have been used in the amplicon system include the human metallothionein promoter (D. Y. Ho, unpublished results), the human neurofilament L promoter (Federoff *et al.*, 1990), the rat preproenkephalin promoter (Kaplitt *et al.*, 1992), and the tyrosine hydroxylase promoter (Oh *et al.*, 1992; Song *et al.*, 1992).

### 2. Bicistronic

A few laboratories have also devised amplicons for bicistronic expression. Verhaagen *et al.* (1992) employed two separate transcriptional units arranged in tandem to express the *lacZ* gene from *E. coli* and the B-50 (GAP43) gene. Both of the transcriptional units were driven by the HCMV IE1 promoter. Our laboratory has used two other approaches for bicistronic expression (Ho *et al.*, 1993b; Lawrence *et al.*, 1994). The first approach employed a single transcriptional unit with two cistrons. This was made possible by inserting the internal ribosomal entry site (IRES) from the encephalomyocarditis virus between the first and second cistrons (Jang *et al.*, 1988,1989; Duke *et al.*, 1992). IRES provides sites for internal binding of ribosomes and thus directs cap-independent translation of the second cistron. With the glucose transporter gene as the first intron and the *lacZ* gene as the second intron, we have consistently observed coexpression of both genes in the same cells by double immunofluorescence. Furthermore, although the level of gene expression varies from cell to cell, the level of glucose transporter expression in each individual cell closely correlates with that of $\beta$-gal, as judged by the intensities of respective immunofluorescence signals. In the second approach, we took advantage of the fact that the HSV-1 $\alpha 4$ and $\alpha 22/\alpha 47$ promoters are separated by $ori_S$ and are in opposite orientation (see Fig. 2). Thus, this piece of DNA can conveniently provide the promoters for two different transcriptional units in an amplicon. We are currently testing the relative strength and kinetics of the two promoters.

### D. Size Limit

Since each virion packages approximately 152 kb of DNA, each DI virion will carry multiple copies of the transcriptional unit. For example, if the amplicon is 10 kb in size, a total of 15 copies will be packaged into the virion. Theoretically, there should not be a limitation on the size of the amplicon, as long as the multiple copies add up to the packaging size of the virion (and provided that the amplicon can be stably propagated in bacteria). However, Kwong and Frenkel (1984) studied the size limit of amplicons and found that those larger than 15 kb tend to be unstable when serially propagated with the helper virus. Various forms of deletion were observed. Furthermore, it was more difficult to achieve a high ratio of DI to helper virus with larger amplicons. Conversely, amplicons less than 15 kb could be stably propagated in serially passaged viral stocks. Amplicons reported by various laboratories have not exceeded this size limit, and the best helper to DI virus ratio achieved is about 1 : 1. In their early studies, Spaete and Frenkel (1982) reported that up to 90% of virions carried amplicon sequences. However, their amplicons did not carry any transcriptional units and thus had a greater replication advantage over those with transcriptional units.

## III. Choice of Helper Virus

The helper virus provides functions in *trans* for the propagation of the amplicon. However, in most cases, the presence of replication-competent virus would result in the death of the host cells. Therefore, to limit the cytopathic effects of the virus, conditionally defective viral mutants have been employed as helpers. In most studies, the viral mutants employed are defective in $\alpha 4$, the major immediate early protein of HSV. $\alpha 4$ is an essential function, a transcriptional regulator that is responsible for the induction of the early and late genes of the virus (see Roizman and Sears, 1990, for review). Therefore, without the expression of $\alpha 4$, infection by the virus would become abortive. However, such mutants can still express the other four immediate early proteins and the large subunit of ribonucleotide reductase (DeLuca *et al.*, 1985; Paterson and Everett, 1990). Some of these gene products have been linked to cytotoxicity (Johnson *et al.*, 1992a,b). Both temperature-sensitive (Geller and Breakefield, 1988; Kaplitt *et al.*, 1991; Ho *et al.*, 1993a) and deletion (Federoff *et al.*, 1992; Bergold *et al.*, 1993; Geller *et al.*, 1993) mutants of $\alpha 4$ have been employed as helpers. Temperature-sensitive mutation tends to be leaky and has a greater chance of reversion. Furthermore, the use of such mutants may limit the experimental conditions to a higher-than-physiological temperature. Therefore, the use of deletion mutants as helpers is more favorable. The deletion mutants (and the amplicons) must be propagated in cell lines that are stably transformed with the essential viral function (i.e., $\alpha 4$ is this case). Several groups have constructed $\alpha 4$ deletion mutants and complementing cell lines that have been used successfully for the generation of DI viral vectors (Davidson and Stow, 1985; DeLuca *et al.*, 1985; Paterson and Everett, 1990).

## IV. Generation of DI Vectors

The original protocol of DI virus generation involved co-transfection of amplicon DNA with helper virus DNA (Spaete and Frenkel, 1982; Ho *et al.*, 1993a). However, the production of infectious viral DNA is rather laborious and it takes longer for the virus to develop 100% cytopathic effect (CPE; manifested as rounding up or "ballooning" of infected cells) after transfection. The current protocol that most laboratories adopt involves transfection of amplicons followed by infection by the virus.

### A. Transfection with Amplicons

Essentially any protocol that gives good transfection efficiency would suffice. We propagate the DI vectors in E5, a VERO (African Green monkey kidney cell)-based cell line stably transformed with the HSV-1 $\alpha 4$ gene (DeLuca *et al.*, 1985). Like its parental cell line, E5 is relatively resistant to transfection. Nevertheless, our laboratory has been able to generate workable stocks of DI viral vectors using calcium phosphate transfection, electroporation, or liposome-mediated transfection.

1. Calcium Phosphate Transfection

1. Seed cells in 25-cm$^2$ flask. The cells should have a confluence of 70–80% and be used within 24 hr of plating.
2. Dilute amplicon DNA (8–20 $\mu$g) to 250 $\mu$l with water; then add 250 $\mu$l 2× HEPES buffer (1× HEPES buffer: 21 m$M$ HEPES/137 m$M$ NaCl/5 m$M$ KCl/0.7 m$M$ $Na_2HPO_4$/5.6 m$M$ D-glucose, pH 7.12). If concentration of DNA is very high, the DNA can be added directly to 1× HEPES.
3. Add 30 $\mu$l 2.2 $M$ $CaCl_2$. Mix gently and incubate at room temperature for 20–30 min.
4. During incubation, remove medium from cells. Add 3 ml 200 $\mu$g/ml DEAE–dextran (500,000 MW, No. 17-0350-1; Pharmacia) in phosphate-buffered saline (PBS) and incubate for about 3 min.
5. Aspirate the DEAE–dextran solution and wash cells once with 1× HEPES.
6. Aspirate the HEPES buffer and add the DNA precipitates to the cells.
7. Add 4 ml Dulbecco's modified Eagle's medium (DMEM) + 10% Nu-Serum (Collaborative Research) to each flask and return cells to incubator.
8. After 4–6 hr of incubation, remove medium and add 3 ml 15% glycerol/ 1× HEPES to each flask. Incubate at room temperature for 1–3 min.
9. Aspirate the glycerol solution, rinse once with DMEM + 10% NuSerum, and add 4 ml DMEM + 10% NuSerum to each flask.

***Comments***

E5 and VERO cells were originally grown in DMEM + 10% fetal calf serum (FCS). We have substituted FCS with NuSerum, a less expensive alternative, and have not observed any noticeable effects on the cells.

This transfection protocol was obtained from the laboratory of Edward S. Mocarski (Stanford University). Among the three methods we have tried, this is certainly the least efficient or reproducible (transfection efficiency is approximately 1–5%). Normally, one must perform several transfections and passagings to obtain the optimal titer and helper : DI virus ratio. However, this method is also the least expensive and does not require any special reagents or equipment compared with the following two protocols.

2. Electroporation

1. Seed cells in 175-cm$^2$ flask. The cells should have a confluence of 70–80% and be used within 24 hr of plating.
2. Trypsinize cells and transfer the contents of 2–3 175-cm$^2$ flasks to sterile 50-ml conical tubes.
3. Spin down and resuspend cells in 0.5 ml Opti-MEM I medium (GIBCO/ BRL; Grand Island, NY).

4. Count cells at 1 : 40 dilution and distribute $5 \times 10^6$ cells per cuvette. We use 0.4-cm electroporation cuvettes from Bio-Rad (Richmond, CA).
5. Complete volume to 200–400 $\mu$l with Opti-MEM I reduced-serum medium.
6. Add 10–20 $\mu$g amplicon DNA. The DNA should be in less than 20 $\mu$l TE (10 m*M* Tris/1 m*M* EDTA, pH 7.5) and should be completely free of salt.
7. Shake, electroporate at 220 V, 950 mF, and shake again. (The electroporation apparatus we use is the Gene Pulser with Capacitance Extender from Bio-Rad.)
8. Add 600 $\mu$l Opti-MEM I medium within 10 min and resuspend cells thoroughly.
9. Plate the contents of each cuvette into a 25-cm$^2$ flask, which should already have 3 ml DMEM + 10% NuSerum added and equilibrated in the incubator.
10. The efficiency of transfection can be assessed 24–48 hr later.

***Comments***

This protocol was a modification of that originally perfected for VERO cells (Masse *et al.*, 1992). Although this protocol can result in considerable cell loss (up to 80%), and thus demands an initial large quantity of cells, it is rapid and transfection efficiencies of 15–30% can be achieved. Varying DNA concentration and volume may further optimize results.

### 3. Liposome-Mediated Transfection

We have tried both lipofectin and lipofectamine from GIBCO/BRL and have found that lipofectamine gives much higher transfection efficiency than lipofectin with E5 cells. Essentially, the transfection is performed according to the manufacturer's protocol. We have tried a variety of parameters and have found the optimal condition for E5 to be as follows.

1. Seed E5 cells at a density of $8 \times 10^5$ cells per 25-cm$^2$ flask.
2. On the following day (approximately 18 hr later), for each 25-cm$^2$ flask dilute 14 $\mu$l lipofectamine and 4 $\mu$g amplicon DNA each into 280 $\mu$l Opti-MEM I medium. Combine the two solutions, mix gently, and incubate at room temperature for 30 min.
3. At the end of incubation, rinse cells once with Opti-MEM I medium.
4. Add 2.2 ml Opti-MEM I medium to the DNA–lipofectamine complexes, and add the complex solution to the cells.
5. Incubate the cells for 6 hr; then add 2.7 ml DMEM with 20% NuSerum.
6. 18–24 hr after the start of transfection, one can proceed with viral infection or estimate efficiency of transfection by X-gal staining (if detecting $\beta$-gal activity is the desired end point) or by immunocytochemistry.

*Comments*

We have achieved transfection efficiency of 30–40% with this method.

## B. Superinfection with Helper Virus and Serial Passaging

The helper virus we use is *d*120 (DeLuca *et al.*, 1985), an $\alpha$4 deletion mutant kindly provided by Neal DeLuca (University of Pittsburgh). Repeated freezing and thawing of HSV stocks would significantly reduce viral titers. Therefore, the viral stocks should be kept in small aliquots and frozen at $-80°C$ or in liquid nitrogen. Addition of 10% sterile nonfat milk to the viral stocks (1 : 1) can help stabilize the virus and does not interfere with subsequent experiments for some applications.

1. Between 24 and 48 hr post-transfection, remove medium from cells and infect cells with the helper virus (e.g., *d*120) at a multiplicity of infection (moi) of 10–20. Infection should be done in a small volume (e.g., 1.5 ml for a 25-$cm^2$ flask) to ensure efficient viral adsorption. If an incubator with a shaking platform is unavailable, the cultures should be agitated periodically.
2. Allow viral adsorption to proceed for 90 min, then remove inoculum and replace with 4 ml fresh medium.
3. When 100% CPE develops (24–30 hr postinfection), freeze cells at $-80°C$.
4. To harvest cells, thaw them slightly at room temperature, and use the "ice" to scrape cells off the wall by shaking the flask.
5. When the cells are almost completely thawed, transfer the cell lysate into a 5-ml vial and sonicate on ice. (We use a sonicator model XL2020 from Heat Systems; Farmingdale, NY; power output level at 3.2, 3× 5-sec pulses, 5-sec rest in between.) This is a viral stock of passage 1 (P1).
6. To propagate the virus stock, infect fresh E5 cells at 1 : 4 dilution (i.e., use 1 ml P1 viral stock with 0.5 ml fresh medium as the next inoculum). Freeze and harvest cells when 100% CPE is observed.
7. Continue serial passaging 3 or 4 times.

*Comments*

Initial infection with high moi would inevitably result in very high helper virus titer in the first passage and thus requires serial passaging of the viral stock to obtain a good DI : helper ratio. Using an moi of 10–20, we usually obtain the highest DI : helper ratio at passage 3 or 4. We have recently obtained a protocol from Laura Greenlund (Washington University) that involves transfection of E5 cells with lipofectin and subsequent viral infection with an moi of 0.3. Following this suggestion, we have tried much lower mois and have found that, using an moi of 0.05–0.3 following lipofectamine transfection, we can obtain viral stocks with a DI : helper ratio up to 1 : 1 in the very first passage.

This method reduces serial passaging and minimizes the occurrence of natural DI particles, which is encouraged by infection with high moi.

## C. Purifying and Concentrating Viral Stocks

Viral stocks prepared directly from cell lysate as just described contain cellular debris and have titers ranging from $10^5$ to $10^7$ plaque-forming units (pfu) per ml. For some applications, such titers (see subsequent discussion) are sufficiently high and there is no need to purify the virus from the cellular debris. Under other circumstances, further purification and/or concentration of the viral stock may be desirable. We have observed acute cytotoxicity caused by viral stocks in primary neuronal cultures. However, the majority (but not all) of cytopathic effects can be accounted for by the metabolites of the cell lysates and can be blocked by kynurenic acid, an antagonist of glutamate receptor (Ho *et al.*, 1994). This observation is striking because excessive accumulation of glutamate and stimulation of glutamate receptors appear to be central to the neuronal death caused by a wide variety of neurological diseases (see Choi, 1990, for review). In this regard, purifying and/or concentrating the viral stocks may be a necessity for studies in neuronal cultures. As a very simple procedure, use of a concentrator device (MW cutoff 100K) has allowed us to concentrate the virus stock from 5- to 10-fold.

Alternatively, the virus stock can be concentrated and partially purified with the following protocol. (All steps performed on ice or at 4°C.)

1. After sonication of the viral stock (prepared from a 75-$cm^2$ flask), clear large cellular debris by centrifugation at about 1,800 *g* for 10 min.
2. Layer supernatant onto 2 ml 25% sucrose in PBS and spin in an SW41 rotor at 75,000 *g* overnight.
3. Remove supernatant and resuspend viral pellet in PBS (100–300 $\mu$l).
4. A 15- to 40-fold concentration of the viral stock can be achieved.

## D. Titering Defective and Helper Virus

Titering the helper virus is usually performed by standard plaque assay. For titering DI vectors, two different approaches have been reported. In the first approach, cells infected with the DI vectors are processed for immunocytochemistry (or for X-gal staining, if $\beta$-gal expression is to be detected). The titer is then quantified directly by counting the number of cells that are stained positively. In the second approach, the titer is estimated indirectly; this method involves extraction of DNA from the viral stocks, followed by serial dilution, transfer to solid support, and hybridization to a specific probe corresponding to the gene of interest. The titer is then estimated by the intensity of hybridization compared to a control (Bergold *et al.*, 1993). Unless antibodies against the gene of interest are unavailable, or endogenous expression of the gene in host cells renders it impossible to detect expression from DI vectors, the first approach is certainly more accurate and less labor intensive than the second. We have used the first approach exclusively to titer the DI vectors.

1. Prepare 48-well plates of VERO cells and 12-well or 24-well plates of E5 cells. The monolayers should be 95–100% confluent on the day of use.
2. Serially dilute viral stocks (from $10^{-1}$ to $10^{-6}$ dilutions) in DMEM + 10% NuSerum. Plate out 0.2 ml dilutions onto VERO cells and 0.4 ml or 1 ml dilutions onto E5, depending on the size of the plate that is used.
4. Allow viral adsorption to proceed for 90 min, with occasional swirling if possible; then remove the inoculum. For VERO cells, replace with regular medium. For E5 cells, add medium containing 0.1% immune human serum (gamma globulin) [we use Immune Globulin (human)-GAMMER® (Armour Pharmaceutical Co., Kankakee, IL)] to inhibit the secondary spread of infectious virus.
5. For titering helper virus on E5 cells, count plaques 2–3 days postinfection. The plaques should be big enough to be seen by the naked eye but not fused to one another. For titering DI virus, the VERO cells are usually processed for X-gal staining (if detecting $\beta$-gal expression) or for immunocytochemistry 8–24 hr postinfection, depending on the strength and the kinetics of the promoter used. The titer of the DI virus is then estimated by the number of positively stained cells.

To ensure that there are no revertants of the helper virus present, viral stocks should be routinely checked by plating onto VERO cells.

## V. Gene Expression in Neurons

One of the factors that may affect gene expression from HSV vectors is the virion host shut-off (*vhs*) function (Read and Frenkel, 1983), a component of the HSV virion. It has been shown to cause translational shut-off and mRNA degradation of host cells and has been mapped to the UL41 open reading frame of the HSV-1 genome (Smibert *et al.*, 1992). Although the effect of *vhs* on transgene expression is certainly a concern, preliminary evidence suggests that *vhs* does not function in sympathetic and sensory neurons (Nichol *et al.*, 1994). Since inhibition of protein synthesis is certainly very deleterious to the host cells, the propensity of the virus to establish latent infection in neurons argues for the possible silence of *vhs* in this particular cell type.

Although the virus can infect both neurons and astrocytes, we have consistently observed that in a mixed primary culture of both cell types, the majority of infected cells are neurons, based on their cellular morphology (D. Y. Ho, unpublished data). Geller and colleagues have reported long-term $\beta$-gal expression from DI vectors driven by the HSV-1 $\alpha22/\alpha47$ promoter. $\beta$-gal expression could be observed in 60–70% of the neurons in mixed cultures for at least 2 wk after initial infection (Geller and Breakefield, 1988).

To study the kinetics and pattern of gene expression from DI vectors *in vivo,* we have looked at $\beta$-gal expression from the HCMV IE1 or the HSV-1 $\alpha4$ promoters in the rat striatum and hippocampus after stereotaxic injection of the virus (D. Y. Ho, unpublished data). Regardless of the promoter used or the brain region studied, the majority of $\beta$-gal-expressing cells are neurons, accounting for 70–90% of total $\beta$-gal-positive cells. Furthermore, although the numbers of $\beta$-gal-expressing cells vary from animal to animal, the efficiency of infection approaches 60% at the peak of expression. For the HCMV IE1 promoter, expression peaked at day 2 postinfection, whereas that of the HSV1 $\alpha4$ promoter peaked at day 1 postinfection. However, by day 7 postinfection (the longest we have examined), although $\beta$-gal-expressing cells could still be observed with either of the promoters, the number of positive cells had declined to less than 5% of the peak. Using a similar HCMV IE1-*lacZ* construct, Kaplitt *et al.* (1991) also studied the duration of gene expression from DI vectors *in vivo*. $\beta$-gal expression from the vector was readily detectable in the hypothalamus and hippocampus of rats 18 hr or 3 days postinfection. However, 2 wk postinfection $\beta$-gal expression could be observed only in the hypothalamus but not in the hippocampus. Although long-term gene expression from DI vectors has been reported, it is by no means effective. Use of promoters from housekeeping genes or the promoter of the HSV latency-associated transcripts may be a good alternative and awaits further testing.

Several groups (Chiocca *et al.*, 1990; Kaplitt *et al.*, 1991; Le Gal La Salle *et al.*, 1993) reported that no adverse pathological effects were observed after stereotaxic infusion of HSV vectors (or other viral vectors) into rat brains. Such conclusions were drawn based on the absence of behavioral or neurological abnormalities and lack of enlargement of the ventricles or other gross anatomical changes in infected animals. To look further at the cytotoxicity of HSV vectors *in vivo,* we have performed stringent counting of surviving neurons around the injection site in the rat striatum and hippocampus. In the striatum, stereotaxic injection of up to $5 \times 10^7$ pfu of *d*120 or *ts*756 (an HSV-1 temperature-sensitive mutant deficient in $\alpha4$; Hughes and Munyon, 1975) did not cause any significant damage compared with injection of PBS or E5 cell lysates. However, similar injection of *ts*756, but not *d*120, resulted in a significant amount of cell loss in CA4 and the dentate gyrus of the hippocampus (Ho *et al.*, 1994). Since injection with either virus did not result in any observable behavioral, physiological, or neuroanatomical changes in the animals, these results strongly suggest that more stringent criteria must be used to assess the cytotoxicity of a particular viral vector.

## VI. Final Remark: Choosing between DI and Recombinant HSV Vectors

The construction and design of HSV recombinant vectors and amplicon-based DI vectors are detailed in Chapter 10 of this volume and in this chapter, respectively. Both types of vectors share the advantages and disadvantages of HSV vectors that have been discussed in other reviews (Breakefield and DeLuca, 1991; Glorioso *et al.,* 1992; Lieb and Olivo, 1993). For example, both types of vectors have the ability to infect a vast number of cell types, including postmitotic neurons, and both can manifest cytopathic effects in infected cultures. In choosing the type of vectors to use, several aspects should be considered.

Gene expression from recombinant vectors depends not only on the choice of the promoter, but also on the insertion site in the viral genome. For example, although the HCMV IE1 promoter works well in the *Bam*HI *z* fragment of the $U_s$ region, as does the SV40 early promoter in the thymidine kinase gene locus, the latter promoter does not give rise to any detectable expression when inserted into the *Bam*HI *z* fragment (Roemer *et al.,* 1991; Johnson *et al.,* 1992a). In another example, Goodart *et al.* (1992) studied expression of the HSV-1 UL38 promoter in recombinant viruses and found that genome position strongly influences the level of expression but not the kinetics of the promoter. These studies demonstrate that designing a recombinant vector requires more careful consideration of the influence of the viral genome than designing an amplicon does. Unfortunately, the behavior of a particular promoter at a certain locus has not been easily predictable.

Despite the difficulty of choosing the "perfect" genomic locus for gene insertion, the recombinant vector may still be a better candidate for long-term gene expression *in vivo*. It is known that when HSV establishes latency in neurons, the genome probably persists as an episome (Rock and Fraser, 1985; Efstathiou *et al.,* 1986; Mellerick and Fraser, 1987); long-term gene expression from the virus has been observed in neurons of latently infected animals (Ho and Mocarski, 1989; Dobson *et al.,* 1990). In contrast, although long-term gene expression with DI vectors has been claimed, most of the reports studied neurons in culture, and the physical status of the DI genome in infected cells has not been examined closely.

The cytotoxicity induced by HSV in infected cells is certainly a major drawback to its usefulness as a gene transfer vector. If cytotoxicity is caused by immediate early gene expression of the virus, as suggested by Johnson *et al.* (1992a), this disadvantage would pertain to both the recombinant vectors and the helper virus that co-exists with the DI vectors. However, if a low moi is used, concomitant infection by both helper and DI virus in the same cell would be rare and cells that receive the DI virus alone would not be subjected to the deleterious effects of the helper virus genome. This is certainly an advantage of DI vectors over recombinant vectors. Nevertheless, it should still be cautioned that HSV virions contain a minimum of some 33 different proteins (Roizman and Sears, 1990), some of which may still exert unknown effects on the physiology of cells.

Finally, from a technical standpoint, DI vectors are easier to construct. Once an amplicon plasmid is made, a DI virus stock can be prepared within a few days, whereas screening and purifying a recombinant virus can take weeks. This is particularly important if various combinations of genes and promoters are to be tested. However, once a recombinant virus is purified, the virus can be propagated and an unlimited supply of a homogeneous stock will be available. In contrast, because of the properties of DI vectors, the titers and helper:DI virus ratios of the viral stocks vary from passage to passage. Thus, it is necessary to regenerate stocks of desirable titers and ratios constantly and to deal with the unpredictability of every passage.

Obviously both types of vectors have their own advantages and disadvantages. The final choice rests on the application of the vectors and the expertise of the laboratory. HSV vectors remain among the few tools for gene transfer into postmitotic neurons. The number of laboratories using them has increased dramatically in recent years. However, the vectors are by no means perfect; in particular, to be useful for human gene therapy many hurdles still must be overcome. Whether the virus is studied as a pathogen causing unsightly blisters and occasional encephalitis or as a potential tool to cure neurodegenerative diseases, research efforts to further understand its basic biology will be crucial to the development of HSV vectors.

## Acknowledgments

I am deeply in debt to Dr. Robert M. Sapolsky for his unfailing support and encouragement. Much of the observations and methodology described in this chapter are summaries of experience and work of his laboratory. I would like to thank its members Sheri Fink, Matthew Lawrence, Timothy Meier, Tippi Saydam, and Jeremy Tompkins for their contribution and useful discussion; Dr. M. J. Masse and L. Greenlund for their protocols; and Dr. Gregory M. Duke of Stanford University for providing the EMC-*lacZ* construct.

D. Y. Ho is supported by a Huntington's Disease Society of America fellowship.

## References

Battleman, D. S., Geller, A. I., and Chao, M. V. (1993). HSV-1 vector-mediated gene transfer of the human nerve growth factor receptor p75hNGFR defines high-affinity NGF binding. *J. Neurosci.* **13,** 941–951.

Bergold, P. J., Casaccia-Bonnefil, P., Xiu-Liu, Z., and Federoff, H. J. (1993). Transsynaptic neuronal loss induced in hippocampal slice cultures by a herpes simplex virus vector expressing the GluR6 subunit of the kainate receptor. *Proc. Natl. Acad. Sci. USA* **90,** 6165–6169.

Breakefield, X. O., and Deluca, N. A. (1991). Herpes simplex virus for gene delivery to neurons. *New Biol.* **3,** 203–218.

Chiocca, E. A., Choi, B. B., Weizhong, N. A., DeLuca, N. A., Schaffer, P. A., DeFiglia, M., Breakefield, X. A., and Martuza, R. L. (1990). Transfer and expression of the *lacZ* gene in rat brain neurons mediated by herpes simplex virus mutants. *New Biol.* **2,** 739–746.

Choi, D. W. (1990). Cerebral hypoxia: Some new approaches and unanswered questions. *J. Neurosci.* **10,** 2493–2501.

Chou, J., and Roizman, B. (1985). The isomerization of the herpes simplex type 1 genome: Identification of the *cis*-acting and recombination sites within the domain of the *a* sequence. *Cell* **41,** 803–811.

Chou, J., and Roizman, B. (1986). The terminal *a* sequence of the herpes simplex virus genome contains the promoter of a gene located in the repeat sequences of the L component. *J. Virol.* **57,** 629–637.

Chou, J., and Roizman, B. (1992). The $\gamma_1$ 34.5 gene of herpes simplex virus precludes neuroblastoma cells from triggering total shutoff of protein synthesis characteristic of programmed cell death in neuronal cells. *Proc. Natl. Acad. Sci. USA* **89,** 3266–3270.

Chou, J., Kern, E. R., Whitley, R. J., and Roizman, B. (1990). Mapping of HSV-1 neurovirulence to $gamma_1$ 34.5, a gene nonessential for growth in culture. *Science* **250,** 1262–1266.

Davidson, L., and Stow, N. D. (1985). Expression of an immediate early polypeptide and activation of a viral origin of replication in cells containing a fragment of herpes simplex virus DNA. *Virology* **141,** 77–88.

Deiss, L. P. and Frenkel, N. (1986). Herpes simplex virus amplicon: Cleavage of concatermeric DNA is linked to packaging and involves amplification of the terminally reiterated *a* sequence. *J. Virol.* **57,** 933–941.

Deiss, L. P., Chou, J., and Frenkel, N. (1986). Functional domains within the *a* sequence involved in the cleavage-packaging of herpes simplex virus DNA. *J. Virol.* **59,** 605–618.

DeLuca, N. A., McCarthy, A. M., and Schaffer, P. A. (1985). Isolation and characterization of deletion mutants of herpes simplex virus type 1 in the gene encoding immediate-early regulatory protein ICP4. *J. Virol.* **56,** 558–570.

Dobson, A. T., Margolis, T. P., Sedarati, F., Stevens, J. G., and Feldman, L. T. (1990). A latent, nonpathogenic HSV-1-derived vector stably expresses $\beta$-galactactosidase in mouse neurons. *Neuron* **5,** 353–360.

Duke, G. M., Hoffman, M. A., and Palmenberg, A. C. (1992). Sequence and structural elements that contribute to efficient encephalomyocarditis virus RNA translation. *J. Virol.* **66,** 1602–1609.

Efstathiou, S., Minton, A. C., Field, H. J., Anderson, J. R., and Wildy, P. (1986). Detection of herpes simplex virus-specific DNA sequences in latently infected mice and in humans. *J. Virol.* **57,** 446–455.

Federoff, H., Geschwind, M., Geller, A., and Lessler, J. (1992). Expression of NGF *in vivo* from a defective HSV-1 vector, prevents effects of axotomy on sympathetic ganglia. *Proc. Natl. Acad. Sci. USA* **89,** 1636–1640.

Federoff, N. J., Geller, A., and Lu, B. (1990). Neuronal specific expression of human neurofilament L promoter in a HSV-1 vector. *20th Annu. Mtg. Soc. Neurosci. Abstr.* 154.2.

Frenkel, N. (1981). Defective interfering herpesviruses. *In* "The Human Herpesviruses—An Interdisciplinary Prospective" (A. J. Nahmias, W. R. Dowdle, and R. F. Schinazi, eds.), pp. 91–120. Elsevier, New York.

Geller, A. I., and Breakefield, X. O. (1988). A defective HSV-1 vector expresses *Escherichia coli* $\beta$-galactosidase in cultured peripheral neurons. *Science* **241,** 1667–1669.

Geller, A. I., During, M. J., Haycock, J. W., Freese, A., and Neve, R. (1993). Long-term increases in neurotransmitter release from neuronal cells expressing a constitutively active adenylate cyclase from a herpes simplex virus type 1 vector. *Proc. Natl. Acad. Sci. USA* **90,** 7603–7607.

Glorioso, J. G., Goins, W. F., and Fink, D. J. (1992). Herpes simplex virus-based vectors. *Sem. Virol.* **3,** 265–276.

Goodart, S. A., Guzowski, J. F., Rice, M. K., and Wagner, E. K. (1992). Effect of genomic location on expression on $\beta$-galactosidase mRNA controlled by the herpes simplex virus type 1 UL38 promoter. *J. Virol.* **66,** 2973–2981.

Henry, B. E., Newcomb, W. W., and O'Callaghan, D. J. (1979). Biological and biochemical properties of defective interfering particles of equine herpesvirus type 1. *Virology* **92,** 495–506.

Ho, D. Y., and Mocarski, E. S. (1989). Herpes simplex virus latent RNA (LAT) is not required for latent infection in the mouse. *Proc. Natl. Acad. Sci. USA* **86,** 7596–7600.

Ho, D. Y., Mocarski, E. S., and Sapolsky, R. M. (1993a). Altering central nervous system physiology with a defective herpes simplex virus vector expressing the glucose transporter gene. *Proc. Natl. Acad. Sci. USA* **90,** 3655–3659.

Ho, D. Y., Lawrence, M. S., and Sapolsky, R. M. (1993b). A defective herpes simplex virus vector expressing the glucose transporter gene protects neurons from metabolic decline and cell death during glucose deprivation. *23rd Annu. Mtg. Soc. Neurosci. Abstr.* 420.2.

Ho, D. Y., Fink, S. L., Lawrence, M. S., Meier, T. J., Saydam, T. C., Dash, R., and Sapolsky, R. M. (1994). Herpes simplex virus vector system: Analysis of its in vivo and in vitro cytopathic effects. (*submitted*).

Holland, J. J. (1990). Defective viral genome. *In* "Virology" (B. N. Fields and D. M. Knipe, eds.) pp. 151–165. Raven Press, New York.

Hughes, R. G., Jr., and Munyon, W. H. (1975). Temperature-sensitive mutants of herpes simplex virus type 1 defective in lysis but not in transformation. *J. Virol.* **16,** 275–283.

Jacob, R. J., Morse, L. S., and Roizman, B. (1979). Anatomy of herpes simplex virus DNA. XIII. Accumulation of head to tail concatemers in nuclei of infected cells and their role in the generation of the four isomeric arrangements of viral DNA. *J. Virol.* **29,** 448–457.

Jang, S. K., Kräusslich, H., Nicklin, M. J. H., Duke, G. M., Palmenberg, A. C., and Wimmer, E. (1988). A segment of the 5′ nontranslated region of encephalomyocarditis virus RNA directs internal entry of ribosomes during *in vitro* translation. *J. Virol.* **62,** 2636–2643.

Jang, S. K., Davies, M. V., Kaufman, R. J., and Wimmer, E. (1989). Initiation of protein synthesis by internal entry of ribosomes into the 5′ nontranslated region of encephalomyocarditis virus RNA *in vivo*. *J. Virol.* **63,** 1651–1660.

Johnson, P. A., Miyanohara, A., Levine, F., Cahill, T., and Friedmann, T. (1992a). Cytotoxicity of a replication-defective mutant of herpes simplex virus type 1. *J. Virol.* **66,** 2952–2965.

Johnson, P. A., Yoshida, K., Gage, F. H., and Friedmann, T. (1992b). Effects of gene transfer into cultured CNS neurons with a replication-defective herpes simplex virus type 1 vector. *Mol. Brain Res.* **12,** 95–102.

Kaplitt, M. G., Pfaus, J. G., Kleopoulos, S. P., Hanlon, B. A., Rabkin, S. D., and Pfaff, D. W. (1991). Expression of a functional foreign gene in adult mammalian brain following *in vivo* transfer via a herpes simplex virus type 1 defective viral vector. *Mol. Cell. Neurosci.* **2,** 320–330.

Kaplitt, M. G., Lipworth, L., Rabsin, S. D., and Pfaff, D. W. (1992). *In vivo* analysis of the rat preproenkephalin promoter using an HSV defective viral vector. *22nd Annu. Mtg. Soc. Neurosci. Abstr.* 254.5.

Kemp, L. M., Dent, C. L., and Latchman, D. S. (1990). Octamer motif mediates transcriptional repression of HSV immediate-early genes and octamer-containing cellular promoters in neuronal cells. *Neuron* **4,** 215–222.

Kwong, A. D., and Frenkel, N. (1984). Herpes simplex virus amplicon: Effect of size on replication of constructed defective genomes containing eucaryotic DNA sequences. *J. Virol.* **51,** 595–603.

Kwong, A. D., and Frenkel, N. (1985). The herpes simplex virus amplicon: IV. Efficient expression of a chimeric chicken ovalbumin gene amplified within defective virus genome. *Virology* **142,** 421–425.

Lawrence, M. S., Ho, D. Y., Dash, R., and Sapolsky, R. (1994). Herpes simplex virus vectors overexpressing the glucose transporter gene protects against excitotoxin seizures. (*submitted*).

Le Gal Le Salle, G., Robert, J. J., Berrard, S., Ridoux, V., Stratford-Perricaudet, L. D., Perricaudet, M., and Mallet, J. (1993). An adenovirus vector for gene transfer into neurons and glia in the brain. *Science* **259,** 988–990.

Lieb, D. A., and Olivo, P. D. (1993). Gene delivery to neurons: Is herpes simplex virus the right tool for the job? *BioEssays* **15,** 547–554.

Lillycrop, K. A., Dent, C. L., Wheatley, S. C., Beech, M. N., Ninkina, N. N., Wood, J. N., and Latchman, D. S. (1991). The octamer-binding protein oct-2 represses HSV immediate-early genes in cell lines derived from latently infectable sensory neurons. *Neuron* **7,** 381–390.

Masse, M. J., Karlin, S., Schachtel, G. A., and Mocarski, E. S. (1992). Human cytomegalovirus origin of DNA replication (*oriLyt*) resides within a highly complex repetitive region. *Proc. Natl. Acad. Sci. USA* **89,** 5246–5250.

Mellerick, D. M., and Fraser, N. W. (1987). Physical state of the latent herpes simplex virus genome in a mouse model system: Evidence suggesting an episome state. *Virology* **158,** 265–275.

Mocarski, E. S., and Roizman, B. (1982). Structure and role of the herpes simplex virus DNA termini in inversion, circularization and generation of virion DNA. *Cell* **31,** 89–97.

Mocarski, E. S., Post, L. E., and Roizman, B. (1980). Molecular engineering of the herpes simplex virus genome: Insertion of a second L-S junction into the genome causes additional genome inversion. *Cell* **22,** 243–255.

Nichol, P. F., Chang, J. Y., Johnson, E. M., and Olivo, P. D. (1994). Infection of sympathetic and sensory neurons with herpes simplex virus does not elicit a shutoff of cellular protein synthesis: Implications for viral latency and herpes vectors. (submitted).

Oh, Y. J., Wong, S. C., Moffat, M., Ullrey, D., Geller, A. I., and O'Malley, K. L. (1992). Delineation of CNS and PNS DNA response elements responsible for cell-type specific expression of tyrosine hydroxylase. *22nd Annu. Mtg. Soc. Neurosci. Abstr.* 578.14.

Paterson, T., and Everett, R. D. (1990). A prominent serine-rich region in Vmw175, the major regulatory protein of herpes simplex virus type 1 is not essential for virus growth in tissue culture. *J. Gen. Virol.* **71,** 1775–1783.

Poffenberger, K. L., and Roizman, B. (1985). Studies on non-inverting genome of a viable herpes simplex virus 1. Presence of head-to-tail linkages in packaged genomes and requirements for circularization after infection. *J. Virol.* **53,** 589–595.

Rabkin, S. D., and Hanlon, B. (1990). Herpes simplex virus DNA synthesis at a preformed replication fork *in vitro*. *J. Virol.* **64,** 4957–4967.

Read, G. S., and Frenkel, N. (1983). Herpes simplex virus mutants defective in the virion-associated shutoff of host polypeptide synthesis and exhibiting abnormal synthesis of alpha (immediate early) viral polypeptides. *J. Virol.* **46,** 498–512.

Rock, D. L., and Fraser, N. W. (1985). Latent herpes simplex virus type 1 DNA contains two copies of the virion DNA joint region. *J. Virol.* **55,** 8449–8452.

Roemer, K., Johnson, P. A., and Friedmann, T. (1991). Activity of the simian virus 40 early promoter-enhancer in herpes simplex virus type 1 vectors is dependent on its position, the infected cell type, and the presence of Vmw174. *J. Virol.* **65,** 6900–6912.

Roizman, B., and Sears, A. E. (1990). Herpesviruses and their replication. *In* "Virology" (B. N. Fields and D. M. Knipe, eds.), pp. 1795–1841. Raven Press, New York.

Roux, L., Simon, A. E., and Holland, J. J. (1991). Effects of defective interfering viruses on virus replication and pathogenesis *in vitro* and *in vivo*. *Adv. Virus Res.* **40,** 181–211.

Smibert, C. A., Johnson, D. C., and Smiley, J. R. (1992). Identification and characterization of the virion-induced host shutoff product of herpes simplex virus gene UL41. *J. Gen. Virol.* **73,** 467–470.

Song, S., Hartley, D., Bryan, J., Ullrey, D., Ashe, O., O'Malley, K., Neve, R., Geller, A., and During, M. A. (1992). HSV-1 vector expressing an unregulated protein kinase C from the tyrosine hydroxylase promoter causes rotational behavior following stereotaxic injection into the substantia nigra compacta of unlesion rats. *22nd Annu. Mtg. Soc. Neurosci. Abstr.* 363.17.

Spaete, R. R., and Frenkel, N. (1982). The herpes simplex virus amplicon: A new eucaryotic defective-virus cloning-amplifying vector. *Cell* **30,** 295–304.

Spaete, R. R., and Frenkel, N. (1985). The herpes simplex virus amplicon: Analyses of *cis*-acting replication functions. *Proc. Natl. Acad. Sci. USA* **82,** 694–698.

Stinski, M. F., Mocarski, E. S., and Thomsen, D. R. (1979). DNA of human cytomegalovirus; Size heterogeneity and defectiveness resulting from serial undiluted passage. *J. Virol.* **31,** 231–239.

Stow, N. D., and McMonagle, E. C. (1983). Characterization of the TRs/IRs origin of DNA replication of herpes simplex virus type 1. *Virology* **130,** 427–438.

Stow, N. D., Murray, M. D., and Stow, E. C. (1986). Cis-acting signals involved in the replication and packaging of herpes simplex virus type-1 DNA. *In* "Cancer Cells 4/DNA Tumor Viruses" (M. Botchan, T. Grodzicker, and P. Sharp, eds.), pp. 497–507. Cold Spring Harbor Laboratory Press, Cold Spring Harbor, New York.

Verhaagen, J., Gispen, W. H., Hermens, W., Rabkin, S. D., Pfaff, D. W., and Kaplitt, M. G. (1992). Expression of B-50 (GAP43) via a defective herpes simplex virus vector in cultured non-neuronal cells. *22nd Annu. Mtg. Soc. Neurosci. Abstr.* 262.10.

Vlazny, D. A., Kwong, A., and Frenkel, N. (1982). Site-specific cleavage/packaging of herpes simplex virus DNA and the selective maturation of nucleocapsids containing full-length viral DNA. *Proc. Natl. Acad. Sci. USA* **79,** 1423–1427.

Weller, S. K., Spadoro, A., Schaffer, J. E., Murray, A. W., Maxam, A. M., and Schaffer, P. A. (1985). Cloning, sequencing, and functional analysis of $ori_L$, a herpes simplex virus type 1 origin of DNA synthesis. *Mol. Cell. Biol.* **5,** 930–942.

Wong, S. W., and Schaffer, P. A. (1991). Elements in the transcriptional regulatory region flanking herpes simplex virus type 1 $ori_S$ stimulate origin function. *J. Virol.* **65,** 2601–2611.

# CHAPTER 10

# Replication-Defective Recombinant Herpes Simplex Virus Vectors

**Paul A. Johnson and Theodore Friedmann**

Department of Pediatrics
Center for Molecular Genetics
University of California, San Diego
La Jolla, California 92093



## I. Introduction

This chapter discusses the design and construction of replication-defective herpes simplex virus type 1 (HSV-1) vectors for a variety of purposes. Two reviews address in detail the development of HSV-1 as a vector for gene delivery to neurons (Breakefield and DeLuca, 1991; Leib and Olivo, 1993). These reviews also contain useful descriptions of the virus life cycle and latency, knowledge of which is essential for understanding the uses and limitations of HSV-1 vectors.

## A. What Is the Need for HSV-1 Vectors?

HSV-1 can efficiently infect a wide variety of cells. The initial attachment of the virus appears to be via heparin sulfate, a ubiquitous cell surface molecule. This attachment is followed by stable binding to an unknown cellular receptor(s), fusion with the plasma membrane, and release of the nucleocapsid into the cytoplasm. The encapsidated viral genome is transported and released into the cell nucleus where it can be expressed, even if the cell is fully differentiated and postmitotic. For the purpose of introducing new genetic material into cells, this is one of the most useful properties of HSV-1, since nondividing cells are refractory to the more common methods of gene transfer (Roemer and Friedmann, 1992). Since HSV-1 can remain latent in sensory neurons for the lifetime of an individual, it has been the expectation that vectors derived from HSV-1 can also be stably maintained in infected cells (for a review of latency, see Ho, 1992). During latency, a portion of the viral genome remains transcriptionally active and produces the so-called latency-associated transcripts (LATs). The aim of many researchers in the HSV-1 vector field has been to utilize and extend this property of latency, so different types of neurons and even nonneuronal cells may be infected by a vector that causes no serious physiological changes in the infected cell and stably expresses the introduced foreign genetic material.

Most of the published work to date concerning vectors derived from HSV-1 has focused on the parameters governing infection and transgene expression, usually involving the use of reporter genes that encode proteins with easily detectable functions. However, the broader applications for HSV-1 vectors, either current or anticipated, include

1. studies of gene function. HSV-1 vectors can be used to transfer genes into postmitotic cells such as neurons to facilitate investigations of cellular functions in cells that are otherwise refractory to genetic manipulation.
2. somatic cell gene therapy. Some investigators have suggested that it may be possible to use HSV-1 vectors to introduce "therapeutic genes" directly into differentiated cell types without the need for *ex vivo* manipulations.
3. "live" vaccination, for the production of proteins inside infected cells that will result in the presentation of antigens to both the cellular and the humoral arm of the immune system.

Undoubtedly there will be more uses for HSV-1 vectors than those just listed; conversely, there may be other viral vector systems that can out-perform HSV-1 for a specific application. Some of the advantages and disadvantages for using HSV-1 as a vector are discussed in this chapter, but it is important to realize that the HSV-1 vector field is still relatively new. Currently there are important limitations that diminish the usefulness of HSV-1 for certain applications, but the existing technological problems at the time of writing are likely to be overcome to enable HSV-1 vectors to become useful for basic and applied (therapeutic) purposes.

## B. Differences between Amplicon Vectors and Recombinant Viral Vectors

HSV-1 amplicon vectors, discussed in Chapter 9, are essentially plasmid molecules that contain appropriate HSV-1 signals to allow their replication and packaging by an HSV-1 helper virus (Kwong and Frenkel, 1984). Amplicon vectors are sometimes referred to as "defective" (Geller and Breakefield, 1988; Leib and Olivo, 1993), but this term can be ambiguous since the virus for an amplicon or a recombinant virus vector is also frequently "defective." Amplicon vectors have two distinct advantages over recombinant virus vectors. Firstly, it is much easier to clone a gene into an amplicon than to recombine it into a virus. Secondly, in a cell infected with an amplicon vector (without concomitant helper virus infection) there should not be expression of HSV-1 genes that could have deleterious effects on cell physiology. However, these advantages can only be realized if it is possible to obtain an amplicon vector stock with sufficiently high titer and a favorable ratio of vector to helper virus, a task not easily achievable at present. In contrast, recombinant HSV-1 vectors do not contain a mixed genetic population and are therefore much more uniform in their effects on infected cells.

## C. Advantages and Disadvantages of HSV-1 as a Gene Transfer Vehicle

HSV-1 has a wide target cell range and, most importantly, can infect nondividing cells (see Fig. 1). The HSV-1 genome can stably persist in neurons during a latent infection and produce the LAT RNAs. The virus can accept fairly large inserts of foreign DNA (15 kb or more), depending on the amount of nonessential viral DNA that has been deleted (Longnecker and Roizman, 1987). It is also possible to insert multiple foreign DNA sequences into the HSV-1 genome. Manipulation of the HSV-1 genome is facilitated because its 150-kb double-stranded DNA sequence has been completely determined and the locations of most of its 72-plus genes are known (McGeoch *et al.,* 1988). The virus can be grown to high titers; $10^8$ plaque-forming units (pfu)/ml or more are readily achievable for many mutant strains. Finally, HSV-1 is unlikely to cause transformation since it does not usually integrate into the cellular genome and no association has been demonstrated with virally encoded transforming genes.

Three inherent disadvantages of replication-defective HSV-1 vectors follow. (1) The large size of the HSV-1 genome requires more complex procedures for its manipulation than simple cloning; (2) Although HSV-1 can efficiently infect dividing cells, it makes little sense to use a replication-defective HSV-1 vector

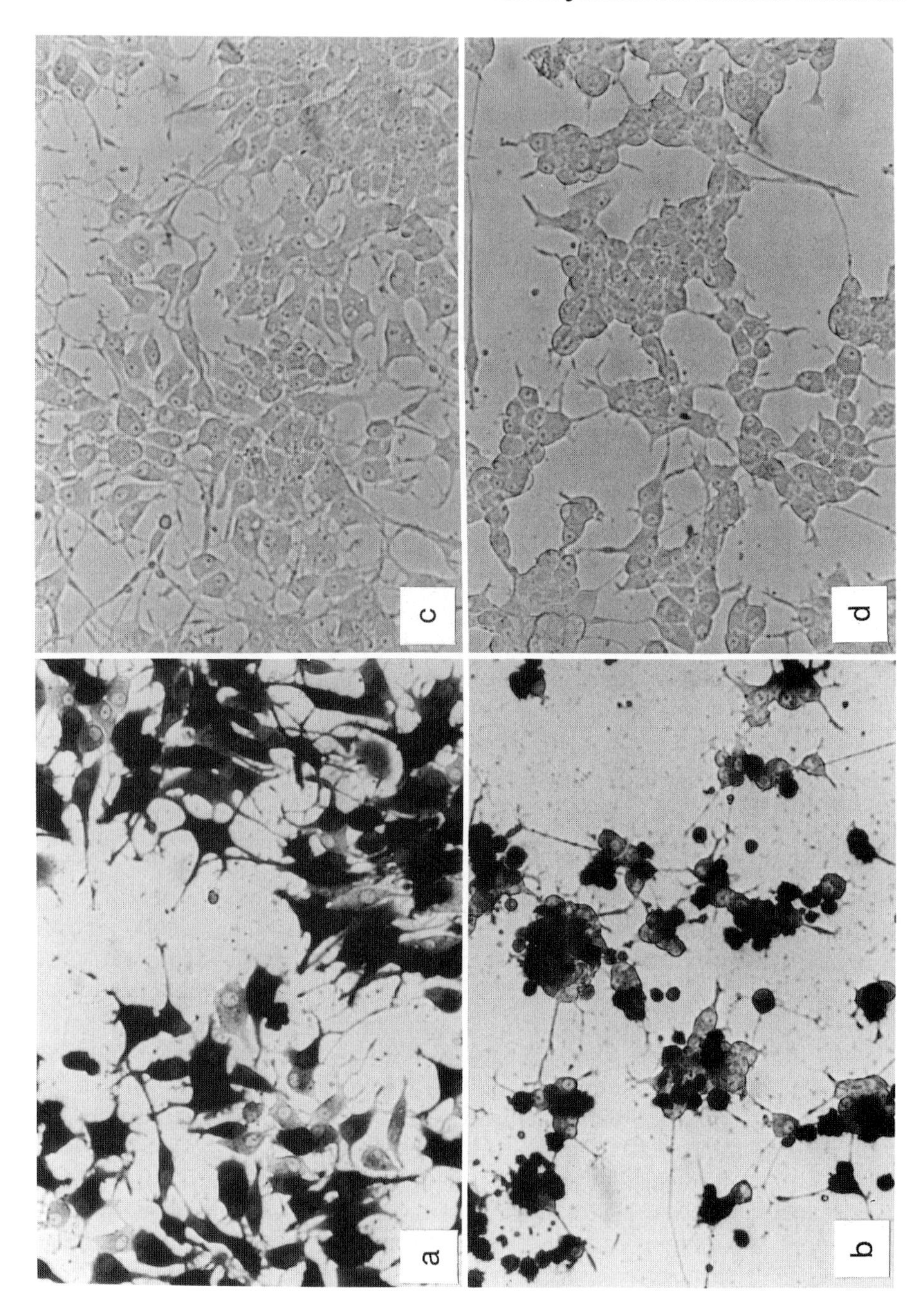
c
d
a
b

to infect such cells for long-term experiments, since the introduced DNA cannot be transmitted to all the daughter cells except in rare cases following recombination with the cellular genome (Roemer *et al.*, 1992). (3) HSV-1 vectors may have the potential to cause reactivation of latent virus *in vivo*.

Additionally, there are two important problems with the current recombinant HSV-1 vectors that severely limit the anticipated potential of this virus as a gene transfer tool: (1) cytotoxicity, which is discussed subsequently (see also Fig. 1b), and (2) the difficulty of obtaining high levels of long-term gene expression. Clearly, to overcome these problems, a better understanding of latency is critical. How is HSV-1 able to remain latent in the natural situation without killing the host cell? What determines the shut-down of most viral gene expression? What enables the LAT region to remain transcriptionally active, and what is its function?

Some clues to the answers to these questions may lie in the fact that latency *in vivo* occurs specifically in sensory neurons; for instance, some neuron-specific isoforms of the cellular transcription factor Oct-2 have been shown to repress immediate early (IE) gene transcription (Wood *et al.*, 1992). The establishment of latency appears to require little or no viral gene expression, since even an IE 3 (ICP4) mutant can establish latency (Sedarati *et al.*, 1993), and LAT expression itself is not necessary for latency. The problems in realizing the full potential of HSV-1 vectors are how to achieve shut-off of IE gene expression, yet retain expression of the transgene in a broad range of target cell types.

## D. Harmful Properties of HSV-1

### 1. Virulence

Virus that can replicate has the potential to be virulent. A productive infection with wild-type HSV-1 invariably leads to destruction of the host cell within 24 hr and to uncontrolled spreading as adjacent and connecting cells become infected and die. Although uncontrolled replication is unlikely to occur with most HSV-1 vectors, poor choice of mutant or inadequate quality control of stocks may inadvertently lead to the generation of virulent recombinants. Replication competent strains may differ in their neurovirulence (virus replication in the brain) and neuroinvasiveness (the ability of the virus to reach the brain and replicate after inoculation at a peripheral site). Lesions in a number of

---

**Fig. 1** Infection of PC12 cells with CgalΔ3. PC12 cells differentiated with nerve growth factor were infected with the IE 3-deficient recombinant HSV-1 vector CgalΔ3 (Johnson *et al.*, 1992a), at an moi of 5 pfu per cell (a,b) or mock infected (c,d). Cells were stained with X-gal to detect expression of the virally transduced *E. coli lacZ* gene at 1 day (a,c) and 3 days (b,d) postinfection. Nearly all the cells in the infected samples show strong staining due to *lacZ* expression, but cytopathic effects are also evident at 3 days postinfection.

HSV-1 genes can cause attenuation either by restricting the host cell range in which the virus can replicate or by reducing its ability to replicate at a low multiplicity of infection (moi). For instance, many of the HSV-1-encoded enzymes involved in DNA metabolism are dispensable for virus growth in tissue culture, but are needed for virus replication in nondividing cells such as neurons. Lesions in HSV-1 gene regulatory functions, IE 1 (ICP0) (Stow and Stow, 1986) and VP16 (Ace *et al.*, 1989), impair the ability of the virus to grow at a low moi, and severely reduce virulence *in vivo*. Attenuated mutants have been considered as a way to increase the number of cells infected relative to replication-defective vectors. However, attenuated vectors have not yet been shown to demonstrate a significant advantage over replication-defective mutants in terms of efficiency of gene delivery or duration of transgene expression (Chiocca *et al.*, 1990; Chang *et al.*, 1991; Fink *et al.*, 1992; Huang *et al.*, 1992), and necrosis to various degrees was associated with those mutants that were stereotactically injected into rodent brains. Although IE 1 gene mutants are attractive since they are unable to reactivate endogenous latent virus, mutation of the IE 1 gene alone in a vector is insufficient to prevent a ''smoldering'' infection in the brain (Huang *et al.*, 1992).

On the other hand, attenuated HSV-1 strains have been proposed as vaccine vehicles (Longnecker *et al.*, 1988) and for antitumor therapy (Martuza *et al.*, 1991). Martuza *et al.* exploited the restricted host range of a thymidine-kinase deficient ($TK^-$) mutant of HSV-1 to destroy glioma tissue selectively in the brain while preventing progression of infection with the virus to an encephalitis (Martuza *et al.*, 1991). In this case, the destructive properties of HSV-1 were being used to an advantage. How can we make the virus safe? Roizman and colleagues have identified a gene lying in the inverted repeats flanking the long unique sequence of the HSV-1 genome that encodes a very potent neurovirulence factor termed ICP34.5 (Chou *et al.*, 1990). Mutation of this gene can decrease the $LD_{50}$ of HSV-1 (following intracerebral inoculation of mouse brain) by a very significant amount, in the case of HSV-1 strain $17^+$ from less than 10 pfu to more than $10^6$ (MacLean *et al.*, 1991). Thus, for vectors that have the potential to replicate *in vivo*, it may be desirable to incorporate a mutation in the ICP34.5 gene.

## 2. Cytotoxicity

Although it is evident that HSV-1 replication destroys infected cells, it is also true that if the replication cycle is blocked, the host cell can still suffer cytopathic effects (CPE). Although infected neurons survive during latency, it has been observed that replication-defective mutants of HSV-1 are capable of inducing CPE in a variety of cells that are nonpermissive for replication; viral gene expression is a major factor for this effect (Johnson *et al.*, 1992a,b). Since even an IE 3 deletion mutant is cytotoxic to most cell types, yet is unable to express early and late classes of viral genes, it is likely that one or more of the products of the remaining four IE genes are responsible for inducing CPE. Known IE gene functions that could adversely affect cell survival include the IE 1 gene product, which is a general transactivator of transcription (Everett, 1987), and the IE 2 gene product, which appears to cause disruption of RNA splicing by the host cell (Sandri-Goldin and Mendoza, 1992).

Two components of the HSV-1 virion might also contribute to the cytotoxic potential of HSV-1. The first is the virion component VP16, which functions to stimulate expression of the HSV-1 IE genes. A virus that has an insertion mutation in the gene encoding VP16 (called *in*1814) has an increased propensity to enter latency rather than undergo lytic replication, even though it is replication competent at high moi (Ace *et al.*, 1989; Harris and Preston, 1991). A second candidate is the virion host shut-off function (*vhs*) (Read and Frenkel, 1983). The *vhs* function is not essential for virus growth *in vitro* and its role is to increase the turnover of both viral and cellular mRNAs (Oroskar and Read, 1989). The strength of the *vhs* function is variable according to virus strain, being stronger in HSV-2 strain G and HSV-1 KOS than in HSV-1 strain $17^+$ (Read and Frenkel, 1983; Fenwick and Everett, 1990). We have found that in the background of a strain $17^+$ IE 3 deletion mutant, creation of an insertion mutation in the coding region of *vhs* does not appear to reduce virus-induced CPE further (P. A. Johnson, unpublished observations). We have also combined the VP16 mutation of *in*1814 with a deletion of IE 3; the resulting double-mutant virus is completely replication defective and significantly less cytotoxic than a single mutant for IE 3 at an moi of less than 5 pfu/cell. However, this mutation has unexpectedly reduced the level of transgene expression in the double-mutant constructs we have examined to date (P. A. Johnson, unpublished observations).

Leib and Olivo (1993) suggest that it may not be possible to generate an infectious HSV-1-based vector that completely lacks any detrimental effects. The route to a more benign vector will probably involve deletion of multiple IE genes, but this will also require the concomitant generation of a complementary cell line in which to grow such a crippled mutant. Whether such a virus can be generated, and whether it will grow sufficiently well to be amenable to further routine genetic manipulation as a vector, remains to be seen.

## II. Designing an HSV-1 Vector

### A. Choice of Vector Backbone

Foreign genes can be inserted into essentially any type of HSV-1 strain or mutant. Some HSV-1 strains have been better characterized than others. For example, strains F (Chicago), KOS (Boston), and $17^+$ (Glasgow) have been extensively studied in laboratories at the indicated geographical locations and a wide range of mutants has been isolated and characterized in these background strains. Strain differences can affect such properties as neurovirulence, neuroinvasiveness, syncytium formation, or degree of host shut-off. Additional strains with reduced virulence [e.g., HSV-1 ANG (Izumi and Stevens, 1990) and HSV-1 HFEM (Spivack and Fraser, 1988)] have also been reported. Since some of the genes responsible for these properties are known, it may be possible to engineer a vector that has an improved genetic make-up or is designed so that deleterious gene functions are not expressed.

Replication-defective mutants HSV-1 may have conditional-lethal mutations in any one of a number of essential gene functions. Temperature-sensitive mutants are probably not as useful as engineered mutants that, by virtue of having either a large deletion or an insertion, are much less likely to revert and therefore will provide a tighter "null" phenotype. Such mutants are grown on specially constructed cell lines that provide the missing function in *trans*. Deletion mutants have been engineered in all different classes of HSV-1 genes, including IE gene 3 (DeLuca *et al.*, 1985; Paterson and Everett, 1990), IE gene 2 (McCarthy *et al.*, 1989; Rice and Knipe, 1990), early genes encoding the DNA polymerase accessory protein UL42 (Johnson *et al.*, 1991), the major DNA binding protein ICP8 gene (de Bruyn Kops and Knipe, 1988), the glycoprotein B gene (Cai *et al.*, 1987), and a late gene encoding glycoprotein H (Forrester *et al.*, 1992). Potentially any of these mutants could serve as a vector, but only mutants that are defective in IE gene 3 express relatively few other viral genes, since the product of IE 3 is essential for early and late gene expression.

## B. Characteristics of IE 3 (ICP4) Mutants

Mutants of IE 3 do not represent the most ideal backbone for HSV-1 vectors since they are capable of expressing the four remaining IE genes and the large subunit of ribonucleotide reductase (DeLuca *et al.*, 1985; Paterson and Everett, 1990), but for most purposes they are probably the most useful type of mutant currently available. Vectors derived from IE 3 deletion mutants have been shown to achieve high but transient levels of transgene expression in central nervous system (CNS) neurons in culture (Johnson *et al.*, 1992b) and *in vivo* (Chiocca *et al.*, 1990) (see also Fig. 1). Long-term transgene expression was reported following the introduction of an IE 3 mutant vector into the peripheral nervous system of mice, but is was detectable only in a limited number of cells (Dobson *et al.*, 1990). The potential for gene therapy has also been investigated using an IE 3 mutant vector to deliver the canine blood clotting factor IX gene to mouse liver by direct injection (Miyanohara *et al.*, 1992). Although expression was transient, it was possible to demonstrate the production of physiologically relevant levels of canine factor IX in the circulation of recipient mice.

Cell lines capable of supporting growth in IE 3 mutants include E5 (DeLuca and Schaffer, 1987), M64A (Davidson and Stow, 1985), and RR1 (Johnson *et al.*, 1992a). M64A cells were generated with a larger HSV-1 IE 3-containing DNA fragment than RR1 and E5 cells, and consequently give rise to "wild-type" revertants at a higher frequency (1 in $10^4$ versus $<1$ in $10^6$). The presence of replication-competent virus at a frequency between 1 in $10^6$ and 1 in $10^7$, although detectable, is not high enough to cause concern in most experiments. However, depending on the background strain of the mutant virus (see previous discussion), a few pfu of "wild-type" revertants could pose a serious problem following an intracerebral inoculation. The properties of RR1 and E5 cells are slightly different. In our hands, we find that the BHK-derived RR1 cells transfect better (a necessary step in genetic manipulation of the virus), but the VERO-derived E5 cells produce better defined plaques (useful for screening and purifying recombinants). Therefore it is useful to have both cell types on hand for generating new vectors.

## C. Choice of Promoter

Numerous different promoters have been used to drive transgene expression in replication-defective vectors, but to date an ideal candidate for general use has not been identified. Reports of long-term expression using the *Escherichia coli lacZ* gene as reporter must be viewed with some caution because of the relatively stable nature of the *lacZ* gene product. High levels of gene expression have been obtained from the strong promoter of the human cytomegalovirus (HCMV) IE control region (Johnson *et al.*, 1992b; Miyanohara *et al.*, 1992), the HSV-1 promoter for the large subunit of ribonucleotide reductase (ICP6) (Chiocca *et al.*, 1990) which is expressed as an IE gene (DeLuca *et al.*, 1985), and the SV40 early control region (Roemer *et al.*, 1991). However, in all cases expression was very transient (not more than a few days). Long-term expression has been reported using the promoter of the Moloney murine leukemia virus long terminal repeat (MoMLV LTR) in sensory neurons, but not in other cell types (Dobson *et al.*, 1990). Expression of the rabbit $\beta$-globin gene and the human $\alpha$2-globin gene under control of their own promoters in an IE 3 deletion mutant background only occurred when the IE 3 product was supplied in *trans* (Smiley and Duncan, 1992). Activity of the rat neuron-specific enolase (NSE) promoter in an IE 3 deletion mutant was found to be low and not neuron specific (K. Roemer and P. A. Johnson, unpublished observations), although some degree of prolonged and tissue-specific expression has been reported when using this promoter in a $TK^-$ attenuated vector (Anderson *et al.*, 1992).

The LAT region of the HSV-1 genome has been the subject of intense interest since it is the only region of the viral genome to be actively transcribed during latency. The sequences that control LAT expression and permit the LATs to escape the general shut-down experienced by other transcription units in the HSV-1 genome have not yet been defined. Two groups showed that insertion of foreign sequences downstream from either the two potential LAT promoters resulted in stable transgene expression during a latent infection in sensory neurons *in vivo* (Dobson *et al.*, 1989; Ho and Mocarski, 1989). However, LAT promoter activity appears to function best in sensory neurons. Attempts to use the LAT promoter to drive transgene expression in other types of neuron (Wolfe *et al.*, 1992) and in nonneuronal cells (Miyanohara *et al.*, 1992) have resulted in weak or transient expression. Currently, attempts are being made in a number of laboratories to broaden the tissue-specific range of the LAT promoter by inserting heterologous transcription regulatory sequences into the LAT region.

Unfortunately, one must conclude by saying that is not currently known how to achieve useful levels of long-term gene expression in cells other than sensory neurons. Until recently, the identification of suitable regulatory sequences was hampered by the problem of vector-induced cytotoxicity. In cells dying from vector infection, long-term gene expression was irrelevant. However, with the generation of less cytotoxic vectors, it is possible to screen various promoters for long-term activity.

## D. Choice of Insertion Site

There are numerous sites in the HSV-1 genome into which foreign DNA may be inserted. The site of insertion can affect transgene expression although the parameters governing this action are far from understood (Roemer *et al.*, 1991). The TK gene has been a popular location since insertional inactivation of TK permits the selection of recombinants in the presence of nucleoside analogs that are converted by TK into products that inhibit HSV-1 DNA replication (Roizman and Jenkins, 1985). A well-tested alternative location is in the *Bam*HI *z* fragment of the short unique portion of the viral genome. The HCMV promoter appears to work well at the *Bam*HI *z* location in the background of an IE 3 deletion mutant (Johnson *et al.*, 1992a; Miyanohara *et al.*, 1992). However, Roemer *et al.* (1991) found than an SV40 early promoter–enhancer-driven construct was not expressed at this location (except after a transient inhibition of protein synthesis), but was expressed when inserted into the TK gene. Other locations for inserting transgenes have been described, often involving replacement of a viral gene (Breakefield and DeLuca, 1991). Insertion of genes into the LAT locus is slightly more complicated than insertion into unique locations such as TK or *Bam*HI *z* since the LAT locus is diploid; it is therefore necessary to purify a virus having an insertion at both sites.

The most commonly used method of inserting foreign DNA into the HSV-1 genome is by homologous recombination. For this to take place it is necessary to flank the foreign sequences with 500 bp or more of HSV-1 DNA. For example, a foreign gene can be inserted into the *Bam*HI *z* fragment of the HSV-1 genome using the plasmid pGX40Xba (Rixon and McLauchlan, 1991). As shown in Fig. 2, this fragment has a unique *Xba*I site lying between the 3′ co-termini of genes US 8 and 9 and US 10, 11, and 12 into which the gene of interest, linked to appropriate signals to control its expression and 3′-end processing, can be inserted. A detailed procedure for generating an HSV-1 vector by homologous recombination, using plaque hybridization as a method to detect recombinants, is described in the following section.

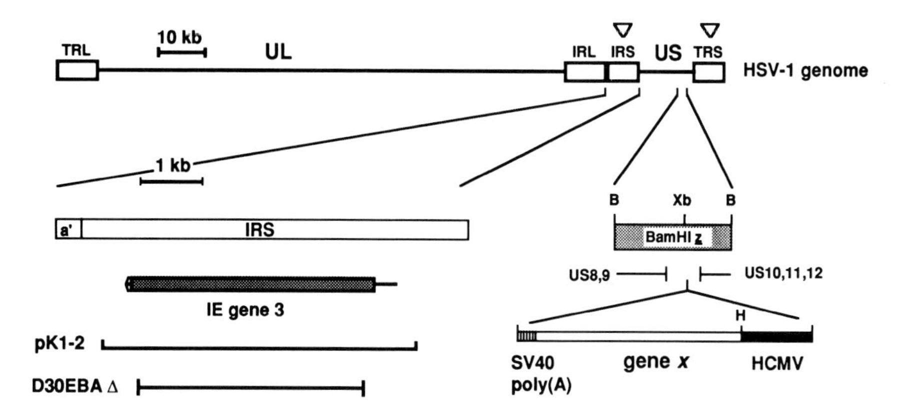


**Fig. 2** Genomic structure of a recombinant HSV-1 vector. (*Top*) The general structure of the HSV-1 genome consisting of a long unique (UL) and a short unique (US) sequence, each of which is flanked by terminal (T) and internal (I) repeated sequences (TRL, IRL and TRS, IRS, respectively). The locations of the two IE 3 genes in the short repeats, which are deleted in the mutant D30EBA (Paterson and Everett, 1990) and its derivatives, are indicated by inverted triangles. (*Lower left*) An expansion of one copy of the IE 3 gene (in IRS), the extent of the deletion in D30EBA, and the region encompassing IE 3 in the plasmid pK1-2 which was used to derive E5 cells (DeLuca and Schaffer, 1987) and RR1 cells (Johnson *et al.*, 1992a). Note that identical sequences are present in TRS. The box at the left-hand side of IRS represents the "a" sequence or packaging signal; it is also present at both termini of the genome. (*Lower right*) The *BamHI z* fragment derived from the plasmid pGX40Xba (Rixon and McLauchlan, 1991), which has a unique *XbaI* site located between the 3′ co-termini of genes US8,9 and US10,11,12. A foreign gene ("*x*"), shown here driven by the HCMV promoter and terminated with the SV40 polyadenylation signal, can be inserted into this *XbaI* site for recombination into the viral genome. Restriction sites: *BamHI* (B), *XbaI* (Xb).

# III. Methods

This section describes how to generate a recombinant HSV-1 vector using, for example, *BamHI z* as the region for insertion and an IE 3 deletion mutant as the parental virus (See Fig. 2).

## A. Generation of Recombinant HSV-1 Vectors

### 1. Cells, Virus, and Reagents

RR1 (Johnson *et al.*, 1992a), E5 (DeLuca and Schaffer, 1987), and African Green monkey kidney (VERO) cells, grown in Dulbecco's modified Eagle's medium (DMEM) containing 10% fetal calf serum (FCS). RR1 and E5 cells are passaged in medium containing 400 μg/ml G418 (GIBCO/BRL, Grand Island, NY) but G418 is not used during virus preparations. E5 cells were a gift from N. A. DeLuca (University of Pittsburgh).

D30EBA, an IE 3 deletion mutant of HSV-1 strain $17^+$ (Paterson and Everett, 1990) and a gift from R. D. Everett (Medical Research Council Virology Unit, Glasgow, Scotland).

Transcription unit containing a promoter, polyadenlyation signal, and gene "*x*", cloned into the HSV-1 *Bam*HI *z* fragment cloned in pGX40Xba (Rixon and McLauchlan, 1991) or pGXA (Johnson *et al.,* 1992a) (Fig. 2).

Hexamethylene bisacetamide (HMBA) stock solution (500 m*M*), prepared in water and sterilized by filtering through a 0.2-μm filter; can be stored at room temperature for at least 1 mo.

D30EBA genomic DNA, prepared from infected E5 or RR1 cells following standard procedures for the isolation of cellular DNA; remaining cellular DNA is minimal when DNA is prepared from cells after extensive CPE has appeared.

## 2. Procedures

1. Day 1. Approximately $1 \times 10^5$ RR1 cells are seeded onto a 35-mm petri dish in 2 ml medium.
2. Day 2. The RR1 cells are transfected with 2 μg plasmid DNA (e.g., pGXA + gene) and 2 μg D30EBA genomic DNA, by standard calcium phosphate co-precipitation (Graham and Van der Eb, 1973).
3. Day 3. The transfection medium is replaced with fresh medium containing 5 m*M* HMBA (optional) to stimulate IE gene expression from the transfected viral DNA.
4. Days 4–10. Replace the medium (no HMBA) on day 4 or 5, and examine the dish each day for the appearance of plaques.
5. Harvest the virus once plaques are evident by pipetting the loose cells off into the medium and transferring into a sterile freezing tube. Freeze/thaw the cells and store this primary stock (1°) at −70°C.

## 3. Comments

1. The integrity of the viral DNA can be checked on a 0.7% agarose gel prior to transfection. Digestion of HSV-1 genomic DNA with *Bam*HI should yield a distinct characteristic banding pattern visible by staining with ethidium bromide (see Davison and Wilkie, 1981). A high molecular-weight background smear indicates the degree of cellular DNA present, but minor contamination with cellular DNA does not affect the procedure.

2. It may be helpful to try different quantities and ratios of plasmid and viral DNA, but keep the total amount of DNA between 2 and 5 μg per 35-mm well, and use one-fifth of the volume of the solutions that would be used for transfection of a 100-mm dish. Linearization of the plasmid DNA at the junction

between the HSV-1 and the vector sequences may also facilitate recombination events.

3. HMBA stimulates HSV-1 IE gene transcription in the absence of VP16 (McFarlane *et al.*, 1992). It may therefore help to increase the infectivity of transfected HSV-1 DNA although it is not essential for the generation of plaques. Because it is slightly toxic, HMBA should not be left on cells for more than 1-2 days.

4. The chances of isolating a recombinant are better when a large number of plaques arises from the transfection. If few or no plaques appear, consider repeating the transfection with a fresh preparation of viral DNA.

## B. Detection and Isolation of Recombinants

### 1. Materials

Agarose overlay: Make up 100-ml stocks of 2.5% electrophoresis-grade agarose in distilled water, autoclave, and store at room temperature.

Nylon filter discs: 87-mm, 0.45-$\mu$m pore size (Schleicher and Schuell, Keene, NH).

Whatmann 3-mm paper.

[$\alpha$-$^{32}$P]dATP-labeled DNA probe corresponding to gene "*x*".

### 2. Procedures

1. Day 1. Seed six 100-mm dishes with $2 \times 10^6$ E5 cells per dish.

2. Day 2. The cells should be 80–90% confluent. Infect 5 plates with 0.1, 1, 5, 10, and 50 $\mu$l volumes of the 1° vector stock. Infect the sixth plate with 500 pfu of the parental vector (D30EBA). The infections are performed in a volume of 5 ml cell culture medium for 1 hr, after which an additional 5 ml is added.

3. Day 3. To prepare the 0.8% agarose overlay, melt the stock agarose and allow to cool to approximately 60°C. Meanwhile, decant 70 ml cell growth medium into a 100-ml bottle and allow that to reach 37°C. When ready, dilute the prewarmed medium with 30 ml melted agarose, and perform the next step rapidly.

4. Remove the medium from the cells and overlay with the agarose–medium mixture. Continue to incubate plates for 2–4 days, until plaques are well formed.

5. Rim the plates with a spatula to loosen the agarose. Then transfer the agarose plugs, upside down (monolayer facing up), to the lids of the same plates. After the blotting is complete (Step 6), place the bottom half of the plate over the agarose, seal with parafilm and store at 4°C.

6. Label the nylon filter discs with a pencil, and place one filter over each agarose plug with the pencilled side facing down. Make alignment marks through

the nylon filter and into the agarose with India ink using a syringe and a narrow gauge needle. Leave the filter in contact with the agarose for 2–4 min.

7. Place the filter, monolayer side facing up, on Whatman filter paper (3-mm) saturated with 0.5 *M* NaOH, for 1 min. Remove the filter and allow to air dry on a paper towel. Repeat this step one time.

8. Neutralize the filters by blotting on 3-mm paper saturated with 1 *M* Tris-HCl (pH 7.5) for 1 min.

9. Immerse the filters in 0.1 *M* Tris-HCl/0.15 *M* NaCl (pH 7.5); then air dry. Fix the DNA to the membrane by baking or by exposure to UV light.

10. Probe the filters with a radiolabeled fragment from "gene *x*". It is convenient to radiolabel probes with [$\alpha$-$^{32}$P]dATP by the method of Feinberg and Vogelstein (1983), and to utilize the hybridization and washing conditions suggested by the manufacturers of Nytran (Schleicher and Schuell). Expose the filters to X-ray film.

11. After developing the autoradiograph, the film should be examined for evidence of signals corresponding to plaques from the 1° stock that are stronger than those of the negative control. Align the film with the stored plates and, using a needle, mark a number of plaques that show strong hybridization. The purification will be more efficient by picking plaques on less densely infected plates.

12. Pick up to 12 different plaques using a sterile pasteur pipet and bulb. Eject the agarose plug into an eppendorf tube containing 0.5 ml medium. Vortex and store at −70°C.

### 3. Comments

1. It is very important to include D30EBA as a negative control for the plaque hybridization procedure. The titer of the 1° vector stock could be determined first to allow screening of a more appropriate range of pfu, but this will lengthen the overall time period for isolating the new vector for minor benefit. Some plates will need to be discarded because of total cell lysis or because there are too few plaques. It is desirable to screen plates with a range of several hundred to several thousand plaques.

2. It is critical that the agarose–medium mixture is not too hot or the cells will die (and not too cold or it well set while decanting!).

## C. Checking Purity and Genomic Struture

### 1. Materials

[$\alpha$-$^{32}$P]dATP-labeled probe corresponding to *Bam*HI *z*

Agarose gel electrophoresis and Southern blotting apparatus

## 2. Procedures

1. Infect RR1 or E5 cells on a 24-well plate with 20 $\mu$l of each plaque isolate. Continue to incubate the cells until large plaques have formed.

2. Harvest the cells and prepare viral DNA as for cellular DNA. Digest the DNA with an enzyme that will give rise to bands that allow differentiation between the 1.85-kb wild-type *Bam*HI *z* fragment and the presence of gene "*x*" in *Bam*HI *z;* usually *Bam*HI will work well. Include D30EBA DNA as a negative control, and ~50 ng starting plasmid DNA digested with appropriate enzymes as a positive control.

3. The digested DNAs are separated by agarose gel electrophoreisis and then transferred to a nylon membrane. Probe the blot with the radiolabeled *Bam*HI *z* fragment and autoradiograph after hybridization and washing.

4. Examine the plaque isolates for the presence of the appropriate new band detected by the *Bam*HI *z* probe. Choose those plaques that have the best ratio of new band to wild-type *Bam*HI *z* band for further purification.

5. Infect monolayers of E5 cells on 100-mm dishes with the plaque isolates chosen for further purification, and overlay with agarose-containing medium the next day.

6. If the presence of the new *Bam*HI *z* band was detectable but weak, repeat the plaque hybridization procedure. If the novel band to original band ratio was 1 : 4 or better, it is probably faster to simply pick well-isolated plaques and repeat the check on the genomic structure without performing the plaque-hybridization procedure.

7. Continue the plaque-purification procedure until no wild-type *Bam*HI *z* band is detectable. Then perform an additional plaque-purification step, making sure there are very few extremely well-isolated plaques on the plate. Use this plaque to make a seed stock of vector and a final DNA preparation.

## 3. Comments

1. In the long run, it is faster to check for the presence of the correct genomic structure in a mixed plaque at this stage before performing multiple plaque purification steps.

2. If there are too few plaques or if the cells grow too fast, the whole monolayer may not lyse. However, there is usually sufficient viral DNA from a few large plaques to determine the genomic structure of recombinants by Southern blotting.

3. The plaque hybridization procedure can cause some diffusion of plaques so, once the frequency of recombinants is high enough, it is probably better to pick well-isolated plaques "blindly" and then check the DNA by Southern blotting.

## D. Preparation of Stocks

### 1. Procedures

1. Day 1. Seed six 100-mm dishes with $1.5 \times 10^6$ RR1 or E5 cells per dish. To make the seed vector stock, use only one 100-mm dish.

2. Day 2. Infect the monolayers, having reached 70–80% confluence, with the seed vector stock at an moi of 0.003 pfu per cell. To make the seed vector stock, use 50–100 $\mu$l final plaque isolate. Infect the cells in a volume of 5 ml, and distribute the virus well. Initially, the cells can be left in this volume of medium.

3. Days 3–7. Examine the plates daily for evidence of virus growth and the state of the cells. Replace or add medium if necessary.

4. Harvest the virus when the CPE is completely generalized. Pipet the cells into a 50-ml conical tube. Pellet the cells by low-speed centrifugation at room temperature. Transfer the supernatant to a new tube and store on ice. Vortex the pellet and freeze it on dry ice. Thaw the frozen pellet at 37°C and vortex. Repeat the freeze/thaw procedure one more time. Add back the supernatant to the thawed pellet, mix well, and repellet cell debris by low-speed centrifugation at room temperature. Decant the supernatant into a sterile tube suitable for use in an SS34 rotor (Sorvall; Dupont, Delaware). Spin down the virus on the SS34 rotor at 12,000 rpm (17,000 *g*) for 2 hr at 4°C. Discard the supernatant and resuspend virus in 3–6 ml medium by gentle pipetting. Pellet remaining cell debris by an additional low-speed centrifugation step; then aliquot the virus into freezing vials and store at −70°C.

5. The titer of the vector is determined by infecting 90% confluent monolayers of E5 cells on 35-mm dishes with a 10-fold serial dilutions of the vector stock. The vector is allowed to adsorb to the cells in 0.5 ml medium for 1 hr with occasional swirling. The inoculum is then replaced with 2 ml fresh medium containing 10% human serum to neutralize virus released into the medium. The monolayers are incubated for an additional 3–5 days. Plaques can be counted directly or after staining cells with Giemsa.

6. The presence of any IE $3^+$ recombinants or revertants should be tested for by infecting a subconfluent 100-mm plate of VERO cells with $10^6$ pfu of vector.

### 2. Comments

1. When preparing vector stocks it is important to infect the cells at a low moi; otherwise the generation of unintentional variants is encouraged.

2. Virus can be released from cells using a sonicating water bath instead of the freeze/thaw procedure if preferred.

3. Because of the toxicity of the IE 3 mutant, it is necessary to infect a large number of VERO cells to detect rare IE $3^+$ recombinants without killing the

whole cell monolayer. If the frequency of IE $3^+$ recombinants is greater than 1 in $10^6$, the vector should be plaque purified again.

## IV. Prospects

The use of HSV-1 vectors for long-term gene transfer in cells other than sensory neurons will require specific adaptations to the virus to avoid cytotoxicity and specific promoter modifications to avoid shut-down of foreign gene expression. Since some of the parameters governing cytotoxicity are beginning to be understood, the generation of less cytotoxic vectors is underway. However, elucidation of the parameters governing gene expression in the latent virus genome will be essential to the development of vectors suitable for long-term gene expression. In the meantime, replication-defective recombinant HSV-1 vectors still offer one of the few efficient means available to achieve high levels of even transient expression of foreign proteins in nondividing cells.

### Acknowledgments

We thank Drs. Ming Jing Wang and Atsushi Miyanohara for critical reading of the manuscript. This work was supported by NIH Grant HD #20034 and by a grant from the Charles H. and Anna S. Stern Foundation.

### References

Ace, C. I., McKee, T. A., Ryan, J. M., Cameron, J. M., and Preston, C. M. (1989). Construction and characterization of a herpes simplex virus type 1 mutant unable to transinduce immediate-early gene expression. *J. Virol.* **63,** 2260–2269.

Andersen, J. K., Garber, D. A., Meaney, C. A., and Breakefield, X. O. (1992). Gene transfer into mammalian central nervous system using herpes virus vectors: Extended expression of bacterial lacZ in neurons using the neuron-specific enolase promoter. *Hum. Gene Ther.* **3,** 487–499.

Breakefield, X. O., and DeLuca, N. A. (1991). Herpes simplex virus for gene delivery to neurons. *New Biol.* **3,** 203–218.

Cai, W., Person, S., Warner, S. C., Zhou, J., and Deluca, N. A. (1987). Linker-insertion nonsense and restriction-site deletion mutations of the gB glycoprotein gene of herpes simplex virus type 1. *J. Virol.* **61,** 714–721.

Chang, J. Y., Johnson, E. M., Jr., and Olivo, P. D. (1991). A gene delivery/recall system for neurons which utilizes ribonucleotide reductase-negative herpes simplex viruses. *Virology* **185,** 437–440.

Chiocca, E. A., Choi, B. B., Cai, W., DeLuca, N. A., Schaffer, P. A., DiFiglia, M., Breakefield, X. O., and Martuza, R. L. (1990). Transfer and expression of the *lacZ* gene in rat brain neurons mediated by herpes simplex virus mutants. *New Biol.* **2,** 739–746.

Chou, J., Kern, E. R., Whitley, R. J., and Roizman, B. (1990). Mapping of herpex simplex virus-1 neurovirulence to γ1 34.5, a gene nonessential for growth in culture. *Science* **250,** 1262–1266.

Davidson, I., and Stow, N. D. (1985). Expression of an immediate early polypeptide and activation of a viral origin of replication in cells containing a fragment of herpes simplex virus DNA. *Virology* **141,** 77–88.

Davison, A. J., and Wilkie, N. M. (1981). Nucleotide sequences of the joint between the L and S segments of herpes simplex virus types 1 and 2. *J. Gen. Virol.* **55,** 315–331.

de Bruyn Kops, A., and Knipe, D. M. (1988). Formation of DNA replication structures in herpes simplex virus-infected cells requires a viral DNA binding protein. *Cell* **55,** 857–868.

DeLuca, N. A., and Schaffer, P. A. (1987). Activities of herpes simplex virus type 1 (HSV-1) ICP4 gene specifying nonsense peptides. *Nucleic Acids Res.* **15,** 4491–4511.

DeLuca, N. A., McCarthy, A. M., and Schaffer, P. A. (1985). Isolation and characterization of deletion mutants of herpes simplex virus type 1 in the gene encoding immediate-early regulatory protein ICP4. *J. Virol.* **56,** 558–570.

Dobson, A. T., Sederati, F., Devi-Rao, G., Flanagan, W. M., Farrell, M. J., Stevens, J. G., Wagner, E. K., and Feldman, L. T. (1989). Identification of the latency-associated transcript promoter by expression of rabbit beta-globin mRNA in mouse sensory nerve ganglia latently infected with a recombinant herpes simplex virus. *J. Virol.* **63,** 3844–3851.

Dobson, A. T., Margolis, T. P., Sedarati, F., Stevens, J. G., and Feldman, L. T. (1990). A latent, nonpathogenic HSV-1 derived vector stably expresses $\beta$-galactosidase in mouse neurons. *Neuron* **5,** 353–360.

Everett, R. D. (1987). The regulation of transcription of viral and cellular genes by herpesvirus immediate-early gene products. *Anticancer Res.* **7,** 589–604.

Feinberg, A. P., and Vogelstein, B. (1983). A technique for radiolabeling DNA restriction endonuclease fragments to high specific activity. *Anal. Biochem.* **132,** 6–13.

Fenwick, M. L., and Everett, R. D. (1990). Transfer of UL41, the gene controlling virion-associated host cell shutoff between different strains of herpes simplex virus. *J. Gen. Virol.* **71,** 411–418.

Fink, D. J., Sternberg, L. R., Weber, P. C., Mata, M., Goins, W. G., and Glorioso, J. C. (1992). In vivo expression of $\beta$-galactosidase in hippocampal neurons by HSV-1 mediated gene transfer. *Hum. Gene Ther.* **3,** 11–19.

Forrester, A., Farrell, H., Wilkinson, G., Kaye, J., Davis-Poynter, N., and Minson, T. (1992). Construction and properties of a mutant of herpes simplex virus type 1 with glycoprotein H coding sequences deleted. *J. Virol.* **66,** 341–348.

Geller, A. I., and Breakefield, X. O. (1988). A defective HSV-1 vector expresses *Escherichia coli* $\beta$-galactosidase in culture peripheral neurons. *Science* **241,** 1667–1669.

Grahman, F. L., and Van der Eb, A. J. (1973). A new technique for the assay of infectivity of human adenovirus 5 DNA. *Virology* **52,** 456–467.

Harris, R. A., and Preston, C. M. (1991). Establishment of latency *in vitro* by the herpes simplex virus type 1 mutant *in* 1814. *J. Gen. Virol.* **72,** 907–913.

Ho, D. Y. (1992). Herpes simplex virus latency: Molecular aspects. *Prog. Med. Virol.* **39,** 76–115.

Ho, D. Y., and Mocarski, E. S. (1989). Herpes simplex virus latent RNA (LAT) is not required for latent infection in the mouse. *Proc. Natl. Acad. Sci. USA* **86,** 7596–7600.

Huang, Q., Vonsattel, J-P., Schaffer, P. A., Martuza, R. L., Breakefield, X. O., and DiFiglia, M. (1992). Introduction of a foreign gene (*Escherichia coli lacZ*) into rat neostriatal neurons using herpes simplex virus mutants: A light and electron microscopic study. *Exp. Neurobiol.* **115,** 303–316.

Izumi, K. M., and Stevens, J. G. (1990). Molecular and biological characterization of a herpes simplex virus type 1 (HSV-1) neuroinvasiveness gene. *J. Exp. Med.* **172,** 487–496.

Johnson, P. A., Best, M. G., Friedmann, T., and Parris, D. S. (1991). Isolation of a herpes simplex virus type 1 mutant deleted for the essential UL42 gene and characterization of its null phenotype. *J. Virol.* **65,** 700–710.

Johnson, P. A., Miyanohara, A., Levine, F., Cahill, T., and Friedmann, T. (1992a). Cytotoxicity of a replication defective mutant of herpes simplex virus type 1. *J. Virol.* **66,** 2952–2965.

Johnson, P. A., Yoshida, K., Gage, F. H., and Friedmann, T. (1992b). Effects of gene transfer into cultured CNS neurons with a replication-defective herpes simplex virus type 1 vector. *Mol. Brain Res.* **12,** 95–102.

Kwong, A. D., and Frenkel, N. (1984). Herpes simplex virus amplicon: Effect of size on replication of constructed defective genomes containing eucaryotic DNA sequences. *J. Virol.* **51,** 595–603.

Leib, D. A., and Olivo, P. D. (1993). Gene delivery to neurons: Is herpes simplex virus the right tool for the job? *BioEssays* **15,** 547–554.

Longnecker, R., and Roizman, B. (1987). Clustering of genes dispensable for growth in culture in the small component of the herpes simplex virus 1 genome. *Science* **236,** 573–576.

Longnecker, R., Roizman, B., and Meignier, B. (1988). Herpes simplex viruses as vectors: Properties of a prototype vaccine strain suitable for use as a vector. *In* "Viral Vectors" (Y. Gluzman and S. H. Hughes, eds.), pp. 68–72. Cold Spring Harbor Laboratory Press, Cold Spring Harbor, New York.

MacLean, A. R., Ul-Fareed, M., Robertson, L., Harland, J., and Brown, S. M. (1991). Herpes simplex virus type 1 deletion variants 1714 and 1716 pinpoint neurovirulence-related sequences in Glasgow strain $17^+$ between immediate early gene 1 and the 'a' sequence. *J. Gen. Virol.* **72,** 631–639.

Martuza, R. L., Malick, A., Markert, J. M., Ruffner, K. L., and Coen, D. M. (1991). Experimental therapy of human glioma by means of a genetically engineered virus mutant. *Science* **252,** 854–856.

McCarthy, A. M., McMahan, L., and Schaffer, P. A. (1989). Herpes simplex virus type 1 ICP27 deletion mutants exhibit altered patterns of transcription and are DNA deficient. *J. Virol.* **63,** 18–27.

McFarlane, M., Daksis, J. I., and Preston, C. M. (1992). Hexamethylene bisacetamide stimulates herpes simplex virus immediate early gene expression in the absence of trans-induction by Vmw65. *J. Gen. Viol.* **73,** 285–292.

McGeoch, D. J., Dalrymple, M. A., Davison, A. J., Dolan, A., Frame, M. C., McNab, D., Perry, L. J., Scott, J. E., and Taylor, P. (1988). The complete DNA sequence of the long unique region in the genome of herpes simplex virus type 1. *J. Gen. Virol.* **69,** 1531–1574.

Miyanohara, A., Johnson, P. A., Elam, R. L., Dai, Y., Witztum, J. L., Verma, I. M., and Friedmann, T. (1992). Direct gene transfer to the liver with herpes simplex virus type 1 vectors: Transient production of physiologically relevant levels of circulating factor IX. *New Biol.* **4,** 238–246.

Oroskar, A. A., and Read, G. S. (1989). Control of mRNA stability by the virion host shutoff function of herpes simplex virus. *J. Virol.* **63,** 1897–1906.

Paterson, T., and Everett, R. D. (1990). A prominent serine-rich region in Vmw175, the major regulatory protein of herpes simplex virus type 1 is not essential for virus growth in tissue culture. *J. Gen. Virol.* **71,** 1775–1783.

Read, G. S., and Frenkel, N. (1983). Herpes simplex virus mutants defective in the virion-associated shutoff of host polypeptide synthesis and exhibiting abnormal synthesis of alpha (immediate early) viral polypeptides. *J. Virol.* **46,** 498–512.

Rice, S. A., and Knipe, D. M. (1990). Genetic evidence for two distinct transactivation functions of the herpes simplex virus a protein ICP27. *J. Virol.* **64,** 1704–1715.

Rixon, F. J., and McLauchlan, J. (1991). Insertion of DNA sequences at a unique restriction enzyme site engineered for vector purposes into the genome of herpes simplex virus type 1. *J. Gen. Virol.* **71,** 2931–2939.

Roemer, K., and Friedmann, T. (1992). Concepts and strategies for human gene therapy. *Eur. J. Biochem.* **208,** 211–225.

Roemer, K., Johnson, P. A., and Friedmann, T. (1991). Activity of the Simian Virus 40 early promoter–enhancer in herpes simplex virus type 1 vectors in dependent on its position, the infected cell type and the presence of Vmw175. *J. Virol.* **65,** 6900–6912.

Roemer, K., Johnson, P. A., and Friedmann, T. (1992). Recombination between a herpes simplex virus type 1 vector deleted for immediate early gene 3 and the infected cell genome. *J. Gen. Virol.* **73,** 1553–1558.

Roizman, B., and Jenkins, F. J. (1985). Genetic engineering of novel genomes of large DNA viruses. *Science* **229,** 1208–1214.

Sandri-Goldin, R. M., and Mendoza, G. E. (1992). A herpesvirus regulatory protein appears to act post-transcriptionally by affecting mRNA processing. *Genes Dev.* **6,** 848–863.

Sedarati, F., Margolis, T. P., and Stevens, J. G. (1993). Latent infection can be established with drastically restricted transcription and replication of the HSV-1 genome. *Virology* 192, 687–691.

Smiley, J. R., and Duncan, J. (1992). The herpes simplex virus type 1 immediate-early polypeptide ICP4 is required for expression of globin genes located in the viral genome. *Virology* 190, 538–541.

Spivack, J. G., and Fraser, N. W. (1988). Expression of herpes simplex virus type 1 (HSV-1) latency-associated transcripts and transcripts affected by the deletion in avirulent mutant HFEM: Evidence for a new class of HSV-1 gene. *J. Virol.* **62,** 3281–3287.

Stow, N. D., and Stow, E. C. (1986). Isolation and characterization of a herpes simplex virus type 1 mutant containing a deletion within the gene encoding the immediate early polypeptide Vmw110. *J. Gen. Virol.* **67,** 2571–2585.

Wolfe, J. H., Deshmane, S. L., and Fraser, N. W. (1992). Herpesvirus vector gene transfer and expression of beta-glucoronidase in the central nervous system of MPS VII mice. *Nature Gen.* **1,** 379–384.

Wood, J. N., Lillycrop, K. A., Dent, C. L., Ninkina, N. N., Beech, M. M., Willloughby, J. J., Winter, J., and Latchman, D. S. (1992). Regulation of expression of the neuronal POU protein Oct-2 by nerve growth factor. *J. Biol. Chem.* **267,** 17787–17791.

# PART II

## The Use of Expression Plasmids in Continuous Cell Lines

CHAPTER 11

# Cytomegalovirus Plasmid Vectors for Permanent Lines of Polarized Epithelial Cells

**Colleen B. Brewer**
Department of Biochemistry
University of Texas Southwestern Medical Center
Dallas, Texas 75235



## I. Introduction

Early attempts to produce stably transfected Madin–Darby canine kidney (MDCK) cell lines for studies of epithelial polarity were made difficult by the relative inefficiency of transfection of MDCK cells. This problem can be overcome, at least partially, by using plasmid expression vectors that have a strong promoter for transcription of the gene of interest and that also direct expression of a selectable marker. Such expression vectors can yield the maximum benefit from each DNA uptake event. A good promoter for this purpose

METHODS IN CELL BIOLOGY, VOL. 43

is the major immediate early promoter of the human cytomegalovirus (CMV) (Towne strain), which was cloned and identified by Stinski and colleagues (Thomsen and Stinski, 1981; Thomsen *et al.*, 1984). This strong promoter is active in many cell types and is not dependent on the presence of viral proteins (Stinski, 1983, and references therein; Stinski *et al.*, 1983). This chapter will describe some vectors that incorporate this CMV promoter into a primary expression cassette and that express a selectable marker from a secondary expression cassette. The use of these vectors to make permanently transfected polarized epithelial cell lines will be discussed in detail.

These CMV-promoter vectors with selectable markers are being widely used. Some groups that formerly used retroviral vectors to efficiently introduce DNA into MDCK cells (Hunziker and Mellman, 1989; Breitfeld *et al.*, 1990; Casanova *et al.*, 1991) are now using the CMV vectors, which require less time and effort but produce equally useful cell lines (Hunziker *et al.*, 1991; Matter *et al.*, 1992; Aroeti *et al.*, 1994). The same advantages that make the CMV vectors desirable for making polarized cell lines—transfection efficiency, high expression level, and convenience—have also led to their widespread use in other cell types.

## II. Features of Some CMV Plasmid Vectors

### A. pCMV1, pCB6, and pCB7

In the laboratory of David Russell (University of Texas Southwestern), an expression vector called pCMV1 was constructed using a large portion of the multifunctional plasmid pTZ18R (Pharmacia, Piscataway, NJ) and incorporating nucleotides −760 to +3 of the CMV-promoter regulatory region provided by Stinski (Thomsen *et al.*, 1984), followed by a polylinker for cDNA insertion and a segment of DNA containing transcription termination and polyadenylation signals from the human growth hormone (hGH) gene. This portion of the hGH gene, base pairs 1533–2157 (Seeburg, 1982), was provided by Richard Palmiter (University of Washington, Seattle). The pCMV1 plasmid also contains the SV40 origin of replication and the early region promoter–enhancer, obtained from plasmid pcD-X (Okayama and Berg, 1983). The SV40 sequence has two functions in pCMV1. First, it elicits efficient amplification of the plasmid in cells that produce SV40 large T antigen, making pCMV1 suitable for transient expression in COS cells (Gluzman, 1981). Second, it can promote transcription of a selectable marker DNA that may be inserted downstream.

Two plasmids were made from pCMV1 in the laboratory of Michael Roth (University of Texas Southwestern) by the insertion of two different selectable markers. pCB6 was made by inserting the Tn 5 neomycin resistance gene, followed by the SV40 poly(A) addition sequence, after the SV40 promoter to form a second expression cassette. pCB7 was made from pCB6 by replacing the neomycin resistance (*neo*) gene with a bacterial plasmid-encoded hygromycin B resistance (*hygro*) gene. In both pCB6 and pCB7, some unnecessary sequences were removed to make additional restriction sites unique. pCB6 and pCB7 are shown in Fig. 1. The sequence of the polylinker region (common to pCB6 and pCB7) is shown in Fig. 2, with enough flanking sequence to use in designing oligonucleotide primers. A partial set of restriction site information is presented for pCB6 in Table I and for pCB7 in Table II. Although some of these sites have been confirmed by actual digestions, the majority are theoretical sites located by computer analysis of sequence data that may be flawed. Hence, the tables should be regarded as incomplete and possibly somewhat inaccurate.

A potential source of problems with the use of pCMV1 and its derivatives as expression vectors is the occurrence of the start-of-translation sequence ATG between the *Hin*d III and *Sse*8387 I sites in the polylinker. However, if this segment of the polylinker is removed during subcloning or if cDNA is inserted ahead of this sequence, potential difficulties are avoided. Another possible pitfall is the presence of a human middle repetitive sequence of the Alu family between nucleotides 1170 and 1460 in the hGH-derived portion of the vector. This should not be a problem unless the plasmid is used to probe genomic DNA.

## B. pCB6+ and pJB20

Two other useful derivatives of pCB6 are pCB6+ and pJB20, each of which incorporates a unique *Eco* RI site into the polylinker region. pCB6+ was

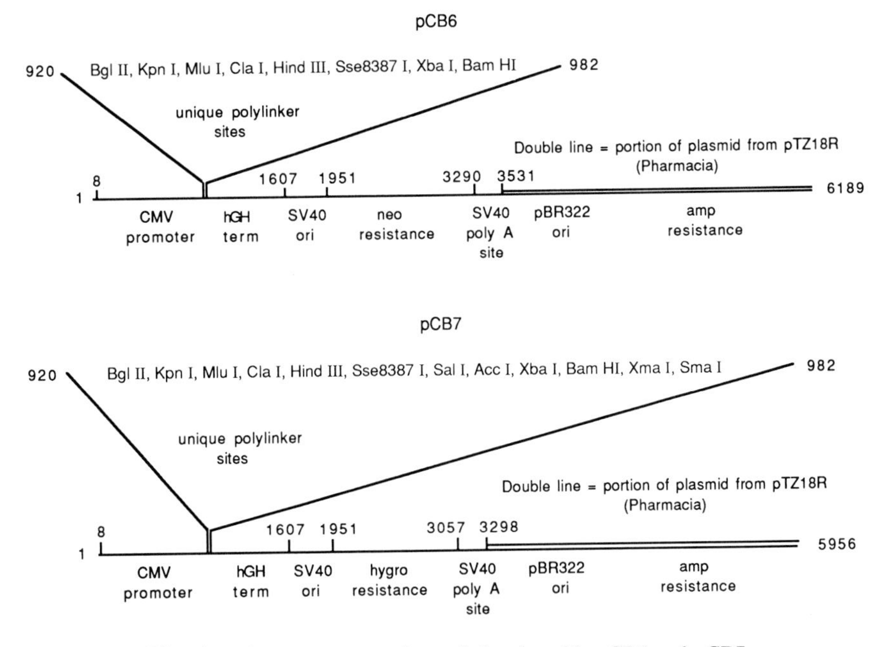


**Fig. 1** Linear representations of the plasmids pCB6 and pCB7.

Sequence of polylinker from pCB6/pCB7 and surrounding region

820

GTCGTAATAACCCCGCCCCGTTGACGCAAATGGGCGGTAGGCGTGTACGGTGGGAGGTCTAT

ATAGCAGAGCTCGTTTAGTGAACCGTCAGAATTAATTCAGATCTGGTACCACGCGTATCGA

TAAGCTTGCATGCCTGCAGGTCGACTCTAGAGGATCCCCGGGTGGCATCCCTGTGACCCCTCC

CCAGTGCCTCTCCTGGCCTTGGAAGTTGCCACTCCAGTGCCCACCAGCCTTGTCCTAATAAAA

TTAAGTTGCATC

1080

Sequence of polylinker in pCB6+

920

AGATCTGAATTCGGTACCGCGGCCGCATCGATGATATCAAGCTTGCATGCCTGCAGGTCGAC
982

**Fig. 2** Selected sequence data.

constructed by Andrea Gashler in the laboratory of Vikas Sukhatme (current affiliation: Beth Israel Hospital, Boston). This plasmid was made by the insertion of a synthetic polylinker into the *Bgl* II–*Hin*d III sites of pCB6 (Gashler *et al.*, 1993). The resulting polylinker no longer has a site for *Mlu* I but has, in order, unique sites for *Bgl* II, *Eco* RI, *Kpn* I, *Not* I, *Cla* I, *Eco* RV, and *Hin*d III in addition to those that follow the *Hin*d III site in pCB6. The polylinker sequence of pCB6+ is shown in Fig. 2. pJB20 was made in the laboratory of J. F. Sambrook (University of Texas Southwestern) by Pamela Beck (Beck *et al.*, 1990) and can be obtained from Steven Lacey (University of Texas Southwestern) (Matsue *et al.*, 1992) or from Beck at Texas Biotechnology Corporation in Houston. Construction of pJB20 was in two steps. First, *Eco* RI linkers were inserted into the *Bgl* II site of the pCB6 polylinker. This destroyed the *Bgl* II site and introduced a unique *Eco* RI site at position 920. Second, the resulting plasmid was cut with *Hin*d III and *Bam* HI and a *Hin*d III–*Ba*m HI fragment containing the SV40 small T intron and SV40 poly(A) addition site from pSVT7 (Sambrook *et al.*, 1989) was inserted. The unique polylinker sites that could be useful for subcloning in pJB20 are, in order, *Eco* RI, *Kpn* I, *Mlu* I, *Cla* I, and *Hin*d III.

## C. Commercially Available Vectors

Some similar vectors are available from commercial sources at reasonable prices, for example, pRc/CMV from Invitrogen (San Diego, CA) and pBK-CMV from Stratagene (La Jolla, CA). These should also be suitable for making permanent cell lines from polarized cells.

**Table I**
**Selected Restriction Enzyme Sites in pCB6**[a]

| Enzyme | Number of sites | Location(s) | Enzyme | Number of sites | Location(s) |
|---|---|---|---|---|---|
| *Afl* II | 0 | | **Bam* HI | 1 | 974 |
| *Asc* I | 0 | | **Bgl* II | 1 | 920 |
| *Avr* II | 0 | | *Bss* HII | 1 | 2523 |
| *Bsp* MII | 0 | | *Bst* BI | 1 | 2808 |
| *Bst* EII | 0 | | **Cla* I | 1 | 939 |
| *Eco* NI | 0 | | *Eag* I | 1 | 2032 |
| *Eco* RI | 0 | | **Hind* III | 1 | 944 |
| *Eco* RV | 0 | | *Hpa* I | 1 | 3396 |
| *Nhe* I | 0 | | **Kpn* I | 1 | 930 |
| *Not* I | 0 | | **Mlu* I | 1 | 932 |
| *Pac* I | 0 | | *Nde* I | 1 | 559 |
| *Pma* CI | 0 | | *Nru* I | 1 | 2981 |
| *Pme* I | 0 | | *Rsr* II | 1 | 2642 |
| *Ppu* MI | 0 | | *Sac* I | 1 | 892 |
| *Sac* II | 0 | | *Sca* I | 1 | 5284 |
| *Sci* I | 0 | | *Sfi* I | 1 | 1901 |
| *Sgr* AI | 0 | | *Sna* BI | 1 | 665 |
| *Spl* I | 0 | | *Spe* I | 1 | 325 |
| *Stu*I | 0 | | **Sse*8387 I | 1 | 960 |
| *Swa* I | 0 | | **Xba* I | 1 | 968 |
| *Xca* I | 0 | | | | |
| *Xho* I | 0 | | | | |
| **Acc* I | 2 | 963, 3132 | **Ava* I | 3 | 979, 1275, 2963 |
| **Afl* III | 2 | 932, 3911 | *Bcl* I | 3 | 1951, 1967, 3290 |
| *Apa* LI | 2 | 4225, 5471 | *Bsp* HI | 3 | 3188, 4631, 5639 |
| *Bsu*36 I | 2 | 3002, 3215 | *Drd* I | 3 | 2153, 4019, 6000 |
| *Dra* III | 2 | 3227, 5956 | *Fsp* I | 3 | 21, 2228, 5026 |
| *Eco*78 I | 2 | 2125, 3242 | *Nae* I | 3 | 2628, 2911, 5850 |
| *Nar* I | 2 | 2126, 3243 | *Nco* I | 3 | 685, 1855, 2558 |
| *Pvu* I | 2 | 42, 5174 | **Pst* I | 3 | 960, 2179, 3102 |
| **Sal* I | 2 | 962, 3131 | | | |
| **Sma* I | 2 | 981, 2965 | | | |
| **Xma* I | 2 | 979, 2963 | *Eco*57 I | 4 | 2272, 2704, 4459, 5471 |
| *Xmn* I | 2 | 914, 5401 | *Pvu* II | 4 | 71, 2232, 2994, 3735 |

[a] Enzymes marked * have a site in the polylinker.

## III. Madin–Darby Canine Kidney Cell Lines

Inducing cells to express exogenous proteins by transient transfection has major advantages over the alternative procedure—making permanent cell lines. Transient transfection gives results in days instead of weeks and does not require the large amount of work and expense of producing and maintaining a

**Table II**
**Selected Restriction Enzyme Sites in pCB7**[a]

| Enzyme | Number of sites | Location(s) | Enzyme | Number of sites | Location(s) |
|---|---|---|---|---|---|
| *Afl* II | 0 | | **Acc* I | 1 | 963 |
| *Aoc* I | 0 | | **Bam* HI | 1 | 974 |
| *Asc* I | 0 | | **Bgl* II | 1 | 920 |
| *Avr* II | 0 | | **Cla* I | 1 | 939 |
| *Bcl* I | 0 | | *Eco* RI | 1 | 2229 |
| *Bss* HII | 0 | | **Hind* III | 1 | 944 |
| *Bst* BI | 0 | | *Hpa* I | 1 | 3163 |
| *Bst* EII | 0 | | **Kpn* I | 1 | 930 |
| *Eco*47 III | 0 | | **Mlu* I | 1 | 932 |
| *Eco* NI | 0 | | *Rsr* II | 1 | 2394 |
| *Eco* RV | 0 | | *Sac* I | 1 | 892 |
| *Nar* I | 0 | | *Sac* II | 1 | 2766 |
| *Nhe* I | 0 | | **Sal* I | 1 | 962 |
| *Not* I | 0 | | *Sap* I | 1 | 3562 |
| *Nru* I | 0 | | *Sfi* I | 1 | 1901 |
| *Pac* I | 0 | | **Sma* I | 1 | 981 |
| *Pma* CI | 0 | | *Sna* BI | 1 | 665 |
| *Pme* I | 0 | | *Spe* I | 1 | 325 |
| *Ppu* MI | 0 | | **Sse*8387 I | 1 | 960 |
| *Sgr* AI | 0 | | **Xba* I | 1 | 968 |
| *Spl* I | 0 | | **Xma* I | 1 | 979 |
| *Stu* I | 0 | | | | |
| *Swa* I | 0 | | | | |
| *Xca* I | 0 | | | | |
| *Xho* I | 0 | | | | |
| **Afl* III | 2 | 932, 3678 | *Dra* III | 3 | 2274, 2567, 5723 |
| **Ava* I | 2 | 979, 1275 | *Ear* I | 3 | 59, 3562, 5366 |
| *Fsp* I | 2 | 21, 4793 | *Nco* I | 3 | 685, 1855, 2338 |
| *Nde* I | 2 | 559, 2436 | *Pvu* I | 3 | 42, 2350, 4941 |
| **Pst* I | 2 | 960, 2323 | | | |
| *Pvu* II | 2 | 71, 3502 | *Apa* LI | 4 | 2266, 2568, 3992, 5238 |
| *Sca* I | 2 | 2956, 5051 | *Drd* I | 4 | 2490, 2871, 3786, 5767 |
| | | | *Msc* I | 4 | 183, 237, 1359, 1519 |

[a] Enzymes marked * have a site in the polylinker.

separate cell line for each protein. It would therefore be desirable, if possible, to use transient transfection to determine the polarity of exogenous proteins in MDCK cells. However, attempts to transfect confluent monolayers of MDCK cells on filters have been unsuccessful (C. Brewer, unpublished data). If the cells are transfected and immediately cultured on filters for an experiment, exogenous protein expression has already decreased to below usable levels within 48 hr (D. Thomas, unpublished data), long before the cells have become fully polarized (Balcarova-Ständer *et al.,* 1984). In one experiment, 36 hr after transfection expression levels were similar to those in permanent cell lines, but both apical and basolateral surface markers were sorted with less fidelity than when they were expressed in cell lines (D. Thomas, unpublished data). Although one group used transient transfection for a study of polarity in MDCK cells and reported as a control the polarized secretion of an endogenous protein after only 24–36 hr of culture on filters (Gottlieb *et al.*, 1986), subsequent work in the same laboratory employed permanently transfected cell lines (Gonzalez *et al.*, 1987; Compton *et al.*, 1989), as have the polarity studies in numerous other laboratories. The remainder of this chapter describes the making of such permanent cell lines in polarized epithelial cells, with emphasis on MDCK cells, the most commonly used polarized cultured cell model.

## A. Methods

### 1. DNA

CMV plasmid vectors have been amplified in several strains of *Escherichia coli,* including (but not limited to) DH1, DH5$\alpha$, and XL1 blue. Plasmids for transfection are commonly prepared by the alkaline lysis method and purified by polyethylene glycol (PEG) precipitation, as described by Sambrook *et al.* (1989). Commercial kits such as those from QIAGEN (Chatsworth, CA) have also been used to produce DNA for transfection. Purification by CsCl gradient centrifugation is not necessary.

### 2. Transfection

For MDCK cells, transfection using the polycationic compound polybrene is easy and inexpensive, and gives relatively good results. Selection survival frequencies of 1 cell per 2000–5000 are common. A comparison of the polybrene, electroporation, and liposome-mediated methods of introducing DNA, evaluated by chloramphenicol acetyltransferase (CAT) assays (Gorman *et al.*, 1982a,b) of transiently transfected MDCK cells, showed that polybrene was somewhat more efficient than these other methods, at least in terms of total expression in an unselected population of cells (H. Arizpe, C. Brewer, and M. Roth, unpublished results). Calcium phosphate methods have also been used to transfect MDCK cells, but the transfection frequencies obtained with these methods are no better than the frequency with the polybrene method (Roman and Garoff, 1986; Rodriguez-Boulan *et al.*, 1989).

The following protocol for transfection using polybrene, adapted from Chaney *et al.* (1986), is for one 10-cm dish of MDCK cells.

1. Plate cells 1 day before transfection, to be about 25% confluent at transfection.
2. In a tube, mix 10 $\mu$g supercoiled DNA into 3 ml complete growth medium.

3. Add 30 $\mu$g polybrene to the DNA mixture and mix again. Polybrene stock can be made as concentrated as 10 mg/ml in water and kept at 4°C.
4. Replace medium with DNA mixture.
5. Incubate 8 hr in incubator, rocking gently every 2 hr.
6. Make 5 ml 30% dimethylsulfoxide (DMSO) in complete medium.
7. Carefully replace polybrene mixture with DMSO mixture, allowing liquid to flow slowly and evenly down the side of the dish.
8. Leave DMSO mixture on cells exactly 5 min at room temperature.
9. Remove DMSO mixture, carefully wash once with 10 ml complete medium, then replace with fresh 10 ml medium for growth overnight.
10. Cells should recover enough to begin selection the next day.

### 3. Selection and Cloning

Once cells have recovered from transfection, that is, they look normal and healthy, they can be selected for expression of the *neo* gene using the neomycin analog G418 sulfate (GIBCO BRL, Grand Island, NY) or for expression of the *hygro* gene using hygromycin. For MDCK cells, 250–300 $\mu$g/ml "active" G418 or 200–300 $\mu$g/ml hygromycin B (Sigma, St. Louis, MO) in regular growth medium has been used to obtain positive colonies in 10–14 days. (The fraction of G418 that is active is given on each bottle.) It is necessary to determine experimentally for each cell line the concentration of drug that will kill all untransfected cells before a transfection is begun. In addition, each transfection and selection should be paralleled by a plate of mock-transfected control cells.

For subsequent cloning of colonies using cloning rings, the transfected cells are removed from the plate with trypsin and EDTA and are replated at about three different densities for selection, to insure a sufficient number of colonies that are spaced far enough apart. Based on the expected survival frequency, it is reasonable to plate cells for selection at $1 \times 10^4$, $4 \times 10^4$, and $20 \times 10^4$ per 10-cm dish. The medium should be replaced with fresh drug-containing medium about every 3 days during selection. Stock solutions of drugs are kept at −20°C and should not be refrozen after thawing.

Colonies are removed from dishes with cloning rings, cultured in 24-well dishes, and then expanded so they can be screened for expression level and polarity (see subsequent section). Selective drugs are removed one or two passages after cloning and before screening.

## B. Expression Levels

Considerable heterogeneity in expression level is usual among the clones obtained from one transfection. On average, about 10% of clones express usable amounts of protein and an additional 25% produce faintly detectable amounts

(D. Thomas and C. Brewer, unpublished). Rarely a clone is found that expresses a protein with an anomalous electrophoretic migration, possibly due to damage to the transfected DNA. Generally the stability of expression level after removal of clones from drug selection is good; usable levels often persist for at least 25 passages. However, cell lines expressing certain proteins seem to lose expression much more rapidly; this effect has been attributed to possible deleterious effects of these proteins.

Because of the amount of work necessitated by cloning the selected cells, it would be desirable to avoid this step if possible. However, the average expression level of the uncloned survivors of selection is rather low. One attempt to circumvent this problem was to elevate the amount of G418 in the medium at the end of the normal selection period to final concentrations of 1–8 mg/ml, in the hope that the transfected DNA might be amplified by the cells. This method appeared to work temporarily, since rather high average expression levels could be seen. However, the cells lost all detectable expression immediately after the level of G418 was dropped (C. Brewer and M. Roth, unpublished). Another strategy that has been successful in some instances is the use of sodium butyrate induction (5–10 m*M* for 12–16 hr) immediately before each experiment. However, this treatment causes a partial loss of polarity of expression for some, but not all, proteins (D. Thomas, unpublished results). Another alternative to cloning and screening numerous transfectants is to pool the selected cells and to obtain high expressors by fluorescence-activated cell sorting (FACS). Use of this strategy depends on having a reasonably high surface expression of the transfected protein and an antibody that binds very specifically to it. This method has been used successfully for both MDCK and LLC-PK1 cells, using a monoclonal antibody (C. Gottardi, personal communication).

A more efficient method of cloning colonies that survive drug selection is to perform a preliminary screen for expression as part of the cloning procedure, and thus to eliminate the cloning of many negative colonies. Again, the protein must be expressed on the cell surface, and the antibody preparation must not bind to other cell surface components. As described by Le Bivic *et al.* (1991), G418-resistant colonies were treated with monoclonal primary antibodies, then with erythrocytes coated with rabbit anti-mouse antibodies. Positive colonies, identified by erythrocyte rosettes, were cloned using cloning cylinders.

## C. Polarity of Transfected Cell Lines

MDCK cells, as supplied by the American Type Culture Collection (ATCC, Rockville, MD), consist of an uncloned mixture of epithelial cells from dog kidney. When these cells are cloned, some clones are found that do not properly polarize proteins (Roman and Garoff, 1986), particularly after the parent cells have been maintained in culture for a long period of time (B. Aroeti, personal communication). In one case, the fraction of unpolarized clones of MDCK cells was observed to increase from less than 10% to about 40% during 1 yr in passage, regardless of whether the cells were transfected prior to cloning (C. Brewer, unpublished results). A good strategy is to clone untransfected MDCK cells and screen the clones for proper polarity of endogenous markers (see subsequent discussion), and then to use one or two satisfactory clones for transfections. This strategy can yield uniformly polar transfected cell lines (C. Brewer and D. Thomas, unpublished data). It is also useful to thaw cells of relatively low passage number to use for transfection.

Tests of MDCK cell polarity are generally performed on monolayers grown on permeable filters and require that transmonolayer passage of reagents and proteins be slow on the time scale of the experiment. Therefore transmonolayer tightness is usually assayed before testing for polarity. Tightness is commonly measured as resistance to passage of electrical current (Balcarova-Ständer *et al.*, 1984) or radioactive inulin (Caplan *et al.*, 1986; Lisanti *et al.*, 1989). A simple check that is informative but not quantitative is to fill up the chamber above the cells with medium, leaving the bottom chamber empty, and then to check visually for leakage after 10–30 min.

Of the several tests of polarity that have been described, two are recommended here because they are easy to perform and they do not require antibodies or other special reagents. A measure of proper apical sorting is the ratio of apical secretion to basal secretion of gp80, the major secreted protein from MDCK cells (Urban *et al.*, 1987). This protein can easily be quantified from apical or basal medium, after a pulse–chase, by trichloroacetic acid (TCA) precipitation, gel electrophoresis, and fluorography. Alternatively, TCA precipitation may be avoided by chasing with a small volume of medium (e.g., 50 $\mu$l on each side) and by loading 10% of each medium onto the gel (Breitfeld *et al.*, 1990). A good assay for proper sorting of basolateral proteins in MDCK cells is to test the polarity of uptake of [$^{35}$S]methionine/cysteine (Balcarova-Ständer *et al.*, 1984). A simplified protocol is given here.

1. Wash monolayers 4 times with PBS + calcium + magnesium.
2. Prestarve 30 min in medium lacking methionine and cysteine.
3. Apply prestarve medium containing about 15 $\mu$Ci/ml Tran$^{35}$Slabel (ICN Biomedicals, Irvine, CA) to the apical side of one monolayer and to the basal side of another for 10 min at 37°C. Keep the opposite side wet with prestarve medium alone.
4. Wash 4 times with regular medium at 4°C.
5. Lyse and scrape cells in 0.5 ml lysis buffer on ice, and microfuge lysate at maximum speed for 20 min.
6. Spot aliquots of 10% of supernatants onto filter paper strips that have been marked off into equal rectangles with pencil.
7. Let strips air dry (0.5–1 hr); then soak strips in ice-cold 10% TCA for 30 min. (Include 1–2 filter segments in the strip with nothing spotted on them as a control for TCA-soluble counts sticking to the filters.)
8. Wash 5 min in 50% ethanol:50% ether, then 5 min in 100% ether. Then air dry 5–10 min.

9. Cut apart filter segments and count in scintillation counter. (Apical − blank)/(basal − blank) should be 8% or less.

The choice of a maximum of 8% apical uptake as the criterion for acceptable cell lines is based on repeated measurements in polarized and unpolarized cell lines. The apical methionine uptake in unpolarized cell lines was quite variable and sometimes was as little as 12% of the basal uptake. In well-polarized cell lines, apical uptake was much more repeatable and seldom exceeded 8% (C. Brewer, unpublished data).

## IV. LLC-PK1 Cell Lines

pCB6 has been used to express several constructs in LLC-PK1 cells by transfection with calcium phosphate, selection in G418, and isolation of clones with cloning rings (Gottardi and Caplan, 1993). Transfection efficiency is similar to that in MDCK cells, and the incidence of positives is generally 10–20%. The average expression level seems to be significantly higher in LLC-PK1 than in MDCK cells for the same construct. Screening of positives for polarity of the endogenous $Na^+,K^+$ ATPase has shown that all these clones sort this basolateral marker properly (C. J. Gottardi, personal communication).

## V. Summary

Several versions of plasmid vectors that incorporate CMV immediate early promoters are now in use. Of particular utility and convenience for making permanently transfected polarized cell lines are those that also direct expression of a selectable marker. Several methods of transfecting cells are available, but the polybrene method is recommended for MDCK cells because it is effective, easy, and inexpensive. After transfection, cells are replated in a selective drug for 10–14 days to kill untransfected cells; then surviving colonies are cloned with cloning rings. Screening of these colonies for expression of the desired protein ordinarily yields 10–15% cell lines with sufficiently high expression to be useful. It should not be assumed that every clone of a polarized cell line will be properly polarized, particularly in the case of MDCK cells. However, assays for correct sorting of endogenous markers can be used to verify proper polarity of transfectants or to identify well-polarized untransfected clones to be transfected. Using these methods and CMV vectors, one can easily establish one or more permanently transfected polarized cell lines within about 1 mo.

### Acknowledgments

I am grateful to Dr. Scott Rose for sequencing the polylinker region of pCB6 and to Dr. Michael Roth for helpful suggestions. I thank all those who shared unpublished information, especially Dr.

David Russell and D'Nette Thomas who is in the laboratory of Dr. Roth. This work was supported by Grant GM37547 from the National Institutes of Health.

## References

Aroeti, B., Kosen, P. A., Kuntz, I. D., Cohen, F. E., and Mostov, K. E. (1993). Mutational and secondary structural analysis of the basolateral sorting signal of the polymeric immunoglobulin receptor. *J. Cell Biol.* **123,** 1149–1160.

Balcarova-Ständer, J., Pfeiffer, S. E., Fuller, S. D., and Simons, K. (1984). Development of cell surface polarity in the epithelial Madin–Darby canine kidney (MDCK) cell line. *EMBO J.* **3,** 2687–2694.

Beck, P. J., Orlean, P., Albright, C., Robbins, P. W., Gething, M.-J., and Sambrook, J. F. (1990). The *Saccharomyces cerevisiae* DPM1 gene encoding dolichol-phosphate-mannose synthase is able to complement a glycosylation-defective mammalian cell line. *Mol. Cell. Biol.* **10,** 4612–4622.

Breitfeld, P. P., Casanova, J. E., McKinnon, W. C., and Mostov, K. E. (1990). Deletions in the cytoplasmic domain of the polymeric immunoglobulin receptor differentially affect endocytotic rate and postendocytotic traffic. *J. Biol. Chem.* **265,** 13750–13757.

Caplan, M. J., Anderson, H. C., Palade, G. E., and Jamieson, J. D. (1986). Intracellular sorting and polarized cell surface delivery of ($NA^+$,$K^+$) ATPase, an endogenous component of MDCK cell basolateral plasma membranes. *Cell* **46,** 623–631.

Casanova, J. E., Apodaca, G., and Mostov, K. E. (1991). An autonomous signal for basolateral sorting in the cytoplasmic domain of the polymeric immunoglobulin receptor. *Cell* **66,** 65–75.

Chaney, W. G., Howard, D. R., Pollard, J. W., Sallustio, S., and Stanley, P. (1986). High-frequency transfection of CHO cells using Polybrene. *Somatic Cell Mol. Genet.* **12,** 237–244.

Compton, T., Ivanov, I. E., Gottlieb, T., Rindler, M., Adesnik, M., and Sabatini, D. D. (1989). A sorting signal for the basolateral delivery of the vesicular stomatitis virus (VSV) G protein lies in its luminal domain: Analysis of the targeting of VSV G-influenza hemagglutinin chimeras. *Proc. Natl. Acad. Sci. USA* **86,** 4112–4116.

Gashler, A. L., Swaminathan, S., and Sukhatme, V. P. (1993). A novel repression module, an extensive activation domain, and a bipartite nuclear localization signal defined in the immediate-early transcription factor Egr-1. *Mol. Cell. Biol.* **13,** 4556–4571.

Gluzman, Y. (1981). SV40-transformed simian cells support the replication of early SV40 mutants. *Cell* **23,** 175–182.

Gonzalez, A., Rizzolo, L., Rindler, M., Adesnik, M., Sabatini, D. D., and Gottlieb, T. (1987). Nonpolarized secretion of truncated forms of the influenza hemagglutinin and the vesicular stomatitis virus G protein from MDCK cells. *Proc. Natl. Acad. Sci. USA* **84,** 3738–3742.

Gorman, C. M., Merlino, G. T., Willingham, M. C., Pastan, I., and Howard, B. H. (1982a). The Rous sarcoma virus long terminal repeat is a strong promoter when introduced into a variety of eukaryotic cells by DNA-mediated transfection. *Proc. Natl. Acad. Sci. USA* **79,** 6777–6781.

Gorman, C. M., Moffat, L. F., and Howard, B. H. (1982b). Recombinant genomes which express chloramphenicol acetyltransferase in mammalian cells. *Mol. Cell. Biol.* **2,** 1044–1051.

Gottardi, C. J., and Caplan, M. J. (1993). An ion-transporting ATPase encodes multiple apical localization signals. *J. Cell Biol.* **121,** 283–293.

Gottlieb, T. A., Beaudry, G., Rizzolo, L., Colman, A., Rindler, M., Adesnik, M., and Sabatini, D. D. (1986). Secretion of endogenous and exogenous proteins from polarized MDCK cell monolayers. *Proc. Natl. Acad. Sci. USA* **83,** 2100–2104.

Hunziker, W., and Mellman, I. (1989). Expression of macrophage–lymphocyte Fc receptors in MDCK cells: Polarity and transcytosis differ for isoforms with or without coated pit localization domains. *J. Cell Biol.* **109,** 3291–3302.

Hunziker, W., Harter, C., Matter, K., and Mellman, I. (1991). Basolateral sorting in MDCK cells requires a distinct cytoplasmic domain determinant. *Cell* **66,** 907–920.

Le Bivic, A., Sambuy, Y., Patzak, A., Patil, N., Chao, M., and Rodriguez-Boulan, E. (1991). An internal deletion in the cytoplasmic tail reverses the apical localization of human NGF receptor in transfected MDCK cells. *J. Cell Biol.* **115,** 607–618.

Lisanti, M. P., Caras, I. W., Davitz, M. A., and Rodriguez-Boulan, E. (1989). A glycophospholipid membrane anchor acts as an apical targeting signal in polarized epithelial cells. *J. Cell Biol.* **109,** 2145–2156.

Matsue, H., Rothberg, K. G., Takashima, A., Kamen, B., Anderson, R. G. W., and Lacey, S. W. (1992). Folate receptor allows cells to grow in low concentrations of 5-methyltetrahydrofolate. *Proc. Natl. Acad. Sci. USA* **89,** 6006–6009.

Matter, K., Hunziker, W., and Mellman, I. (1992). Basolateral sorting of LDL receptor in MDCK cells: The cytoplasmic domain contains two tyrosine-dependent targeting determinants. *Cell* **71,** 741–753.

Okayama, H., and Berg, P. (1983). A cDNA cloning vector that permits expression of cDNA inserts in mammalian cells. *Mol. Cell. Biol.* **3,** 280–289.

Rodriguez-Boulan, E., Salas, P. J., Sargiacomo, M., Lisanti, M., Le Bivic, A., Sambuy, Y., Vega-Salas, D., and Graeve, L. (1989). Methods to estimate the polarized distribution of surface antigens in cultured epithelial cells. *Meth. Cell Biol.* **32,** 37–56.

Roman, L. M., and Garoff, H. (1986). Alteration of the cytoplasmic domain of the membrane-spanning glycoprotein p62 of Semliki Forest Virus does not affect its polar distribution in established lines of Madin–Darby canine kidney cells. *J. Cell Biol.* **103,** 2607–2618.

Sambrook, J., Fritsch, E. F., and Maniatis, T. (1989). "Molecular Cloning: A Laboratory Manual," 2d Ed. Cold Spring Harbor Laboratory Press, Cold Spring Harbor, New York.

Seeburg, P. H. (1982). The human growth hormone gene family: Nucleotide sequences show recent divergence and predict a new polypeptide hormone. *DNA* **1,** 239–249.

Stinski, M. F. (1983). Molecular biology of cytomegaloviruses. *In* "The Herpesviruses" (B. Roizman, ed.), Vol. 2, pp. 67–113. Plenum Press, New York.

Stinski, M. F., Thomsen, D. R., Stenberg, R. M., and Goldstein, L. C. (1983). Organization and expression of the immediate early genes of human cytomegalovirus. *J. Virol.* 46, 1–14.

Thomsen, D. R., and Stinski, M. F. (1981). Cloning of the human cytomegalovirus genome as endonuclease *Xba* I fragments. *Gene* **16,** 207–216.

Thomsen, D. R., Stenberg, R. M., Goins, W. F., and Stinski, M. F. (1984). Promoter-regulatory region of the major immediate early gene of human cytomegalovirus. *Proc. Natl. Acad. Sci. USA* **81,** 659–663.

Urban, J., Parczyk, K., Leutz, A., Kayne, M., and Kondor-Koch, C. (1987). Constitutive apical secretion of an 80-kD sulfated glycoprotein complex in the polarized epithelial Madin–Darby canine kidney cell line. *J. Cell Biol.* **105,** 2735–2743.

## CHAPTER 12

# Inducible Protein Expression Using a Glucocorticoid-Sensitive Vector

**Robert P. Hirt,[*,‡] Nicolas Fasel,[*] and Jean-Pierre Kraehenbuhl[*,†]**

[*] Institute of Biochemistry
University of Lausanne
CH-1066 Epalinges, Switzerland

[†] Swiss Institute for Experimental Cancer Research
CH-1066 Epalinges, Switzerland



## I. Introduction

The use of recombinant DNA techniques has significantly contributed to our understanding of membrane trafficking in eukaryotic cells (Garoff, 1985; see also other chapters in this volume). This approach relies on the ability to express heterologous proteins in a given cell line from the corresponding gene or cDNA using various eukaryotic expression vectors. Previously, optimal expression was often obtained by compromising the choice of cells with the expression

[‡] Present address: Microbiology Group, Department of Zoology, The Natural History Museum, London SW7 5BD, United Kingdom.

system (Kaufman, 1990). However, to analyze membrane trafficking in a given cell type such as epithelial cells, the expression system must fit the cell of interest. Various expression vectors have been developed to express membrane proteins in epithelial cells (for review, see Roth, 1989, and other chapters in this volume). In this chapter, we shall briefly review the properties of inducible protein expression vectors and describe in detail a recombinant plasmid vector that was developed to induce expression of foreign proteins in polarized epithelial cells (Hirt *et al.,* 1992).

At least four types of inducible expression systems controlled by external stimuli have been used successfully to express foreign proteins in various mammalian cell lines (Kaufman, 1987,1990). In these four systems, protein expression from transfected DNA is directed by the interferon $\beta$ (Samanta *et al.,* 1986), the heat-shock (Wurm *et al.,* 1986), the methallothionein (Roman and Garoff, 1986), or the mouse mammary tumor virus (MMTV) long terminal repeat (LTR) promoter (Lee *et al.,* 1981). Such inducible vectors are particularly well suited to transient or stable expression of proteins that are potentially cytotoxic to the cell or that impair specific cellular functions (Garoff, 1985; Kaufman, 1990). The ability to induce foreign protein expression is also desirable in investigating the role of specific components of complex cellular functions such as cell proliferation and the cell cycle, metabolic pathways, membrane trafficking, and organelle biogenesis. Inducible systems are not restricted to the up-regulation of foreign protein expression, but can also down-regulate the expression of distinct gene products by controlled destabilization of the corresponding mRNA species using antisense RNA. To date, only a limited number of inducible expression systems have been adapted to epithelial cells. They include two vectors using the mouse or the human methallothionein promoter (Romand and Garoff, 1986; Roth, 1989) and a third that uses a kidney-derived MMTV LTR promoter (Wellinger *et al.,* 1986) controlled by glucocorticoids (Hirt *et al.,* 1992). Although the need for inducible vectors to study membrane trafficking in epithelial cells has been recognized for a long time (Roth, 1989), the degree of induction of the available expression systems was usually low and, in addition, the vectors were leaky in the absence of the external stimulus. For instance, a 3- to 5-fold induction was reported in stably transfected Madin–Darby canine kidney (MDCK) cells using a metallothionein promoter (Romand and Garoff, 1986).

Glucocorticoid hormones modulate physiological responses and developmental functions in target cells via specific receptors that act as transcription factors (Burnstein and Cidlowski, 1989). After binding to their cognate cytoplasmic receptors, glucocorticoids can stimulate or repress the transcription of glucocorticoid-regulated genes by binding to specific DNA sequences, that is, the glucocorticoid-responsive elements in their promoter region (Wahli and Martinez, 1991). The best-studied glucocorticoid-inducible promoter is the MMTV LTR that contains steroid-responsive elements acting as enhancers (Buetti and Kühnel, 1986; Meulia and Diggelmann, 1990). Several cell lines have been transfected with vectors containing the MMTV LTR promoter with a several-fold induction of protein expression when stimulated with dexamethasone, a synthetic glucocorticoid (Lee *et al.*, 1981; Klessig *et al.*, 1984; Cai *et al.*, 1990). There are, however, some problems and limitations inherent in inducible expression systems, especially in glucocorticoid-inducible expression vectors. For instance, glucocorticoids can regulate the transcription of genes whose gene products modulate other genes at transcriptional, post-transcriptional, translational, and post-translational levels, thus altering the phenotype of the transfected cell line. In certain cell types such as lymphocytes, glucocorticoids are known to induce drastic changes including programmed cell death (Migliorati *et al.*, 1992; Cohen, 1993). Glucocorticoids also affect the intracellular trafficking of membrane proteins by regulating their transport and processing (Haffar *et al.*, 1988; Goodman and Firestone, 1993). Formation of tight junctions in epithelial cell layers has been shown to be controlled by glucocorticoids (Zettl *et al.*, 1992). Therefore it should be kept in mind that the effects of glucocorticoids in the cells to be transfected must be carefully analyzed and shown not to interfere with the gene product of the foreign DNA.

## II. pLK-neo Inducible Expression Vector

MMTV, a murine retrovirus, is expressed in many different cell types including B and T lymphocytes (Tsubura *et al.*, 1988; Rollini *et al.*, 1992), mammary gland (Dickson, 1987; Mok *et al.*, 1992), salivary gland (Rollini *et al.*, 1992), and kidney epithelial cells (Wellinger *et al.*, 1986). Tissue-specific expression of MMTV has been correlated with unique sequences in the viral LTR (Wellinger *et al.*, 1986). One can take advantage of these unique features of the promoters from MMTV variants to confer tissue specificity on different inducible expression vectors. The inducible expression system described in this chapter has been primarly designed for the kidney-derived MDCK epithelial cell line, which is one of the most frequently studied lines (Simons and Wandinger-Ness, 1990). Our strategy was based on the assumption that a hormone-responsive *cis*-acting element combined with a putative tissue-specific enhancer should confer a high level of inducible gene expression on transfected MDCK cells (Hirt *et al.*, 1992). Low protein expression under noninduced conditions should allow the transfected epithelial cells to proliferate, differentiate, and establish membrane polarity before the foreign protein accumulates in the cell. The various elements that are required to construct an efficient inducible vector have been reviewed elsewhere (Kriegler, 1990). The 7.1-kb pLK-neo plasmid that we have constructed contains most of these elements, and contains a unique glucocorticoid-inducible LTR promoter (K-LTR) derived from a variant MMTV proviral DNA that was isolated from a kidney adenocarcinoma (Wellinger *et al.*, 1986). The K-LTR has a 91-bp sequence of unique DNA of unknown origin (Wellinger *et al.*, 1986). In addition to the K-

LTR, the pLK-neo vector carries a multiple cloning site for insertion of foreign DNA, a SV40 DNA segment required for efficient polyadenylation, an origin of replication, an ampicillin resistance gene for growth and selection in bacteria, and an aminoglycoside phosphotransferase gene placed in an SV40 transcription unit that allows selection of stably transfected animal cells with neomycin or its analog G418 (Fig. 1).

Both epithelial and nonepithelial cells have been transfected with the inducible pLK-neo expression system. The transfection procedures and the type of proteins expressed in epithelial and nonepithelial cells are listed in Table I.

## III. Inducible Protein Expression in Epithelial Cells

To analyze the inducibility of protein expression in epithelial cells, two reporter cDNAs were chosen encoding the polymeric immunoglobulin (poly-Ig) receptor (Schaerer *et al.*, 1990) or the glycophosphatidylinositol (GPI)-anchored membrane protein Thy-1. These genes were selected because they encode membrane proteins that require a polarized cell for correct targeting. The kinetics and dose dependence of membrane protein accumulation following dexamethasone stimulation, as well as the intracellular routing of the two membrane proteins, were analyzed in transfected MDCK cells (Hirt *et al.*, 1992,1993). Figure 2 illustrates the time course of intracellular poly-Ig receptor accumulation. In the absence of dexamethasone, no receptor was found in the transfected cells, indicating that the conditional vector is tight and that expression requires external stimulation (Fig. 2). Following addition of dexamethasone, the receptor

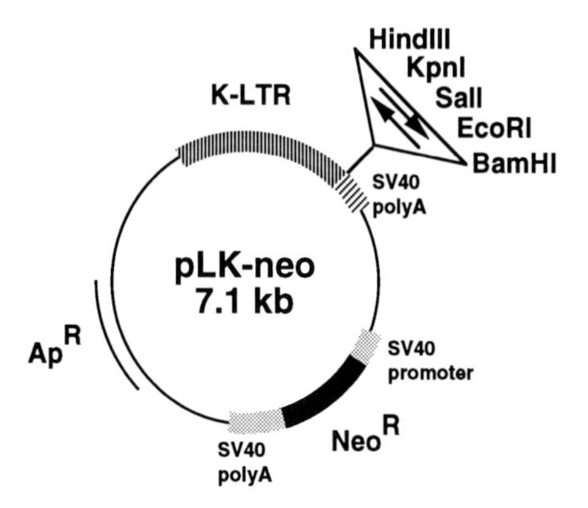


**Fig. 1** The pLK-neo expression vector contains a hormone-inducible LTR promoter derived from an MMTV of kidney origin (K-LTR), an SV40 polyadenylation sequence downstream of a polylinker with *Hin*dIII, *Kpn*I, *Sal*I, *Eco*RI, and *Bam*HI restriction sites for foreign DNA insertion and two markers for selection in bacteria (ampicillin resistance gene, $Ap^R$) and mammalian cells (neomycin resistance gene, $neo^R$).

**Table I**
**Cell Lines Studied Using the pLK-neo Expression Vector[a]**

| Cell line | Method of transfection[b] | Expressed protein[c] | References |
|---|---|---|---|
| Epithelial cells | | | |
| MDCK I | Polybrene | poly-Ig-R, Thy-1 | Hirt *et al.* (1992,1993) |
| | Lipofection | NaPi-1 | Biber *et al.* (unpublished data) |
| | Electroporation | annexin II–p11 chimera | Harder and Gerke (1993) |
| TBM | Lipofection | 11β-HSD | Duperrex *et al.* (1993) |
| LLC-PKI/4 | Lipofection | PH-R | Biber *et al.* (unpublished data) |
| Nonepithelial cells | | | |
| Ltk$^-$ | $Ca^{2+}$-phosphate | Thy-1 | Hirt *et al.* (1992) |
| | | FGF-R | Fasel *et al.* (unpublished data) |
| P815 | $Ca^{2+}$-phosphate | Thy-1 | Fasel *et al.* (unpublished data) |
| PC12 | $Ca^{2+}$-phosphate | acidic FGF | Renaud and Laurent (unpublished data) |

[a] Three epithelial cell lines derived from dog (MDCK) or pig (LLC-PK) kidney and toad (TBM) bladder have been successfully transfected with the pLK-neo vector. Receptors (poly-Ig receptor, parathyroid hormone receptor), GPI-anchored proteins (Thy-1), ion transporters ($Na^+$-dependent-phosphate transporter), enzymes (11β-hydroxysteroid dehydrogenase (Duperrex *et al.*, 1992), and a cytoskeletal associated protein (annexin II) have been expressed following dexamethasone induction. At least three different transfection protocols—electroporation, lipofection, and polybrene transfection—yielded stable transfected colonies.

[b] For a reference describing transfection methods, see Keown *et al.* (1990).

[c] Abbreviations: poly-Ig-R, polymeric immunoglobulin receptor; NaPi-1, $Na^+$-dependent phosphate transporter system 1; 11β-HSD, 11β-hydroxysteroid dehydrogenase; PH-R, parathyroid hormone receptor; FGF-R, fibroblast growth factor receptor.

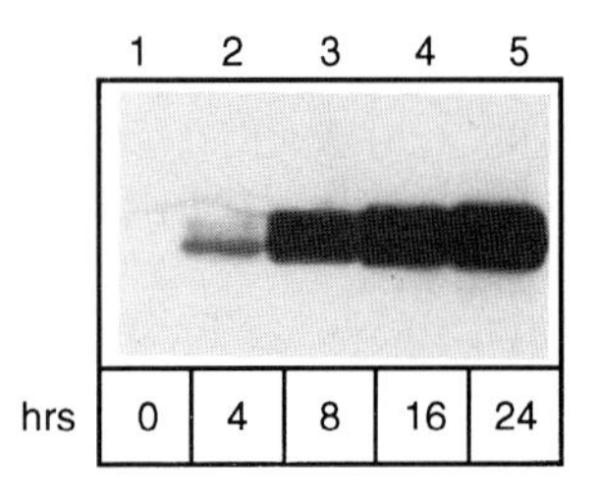


**Fig. 2** Kinetic of accumulation of the poly-Ig-receptor in MDCK cells. MDCK cells were grown to confluence on Transwell filters. Receptor expression was induced for increasing periods of time with dexamethasone (DEX). The cells were lysed and the proteins analyzed for the presence of poly-Ig receptor by Western blot using a monoclonal antibody specific for the receptor's cytoplasmic tail. At 4 hr, a $10^5$-kDa band corresponding to the core-glycosylated receptor can be seen. With longer stimulation time, there is a progressive accumulation of core and terminally glycosylated receptor. Steady-state level of expression is reached after ~20 hr.

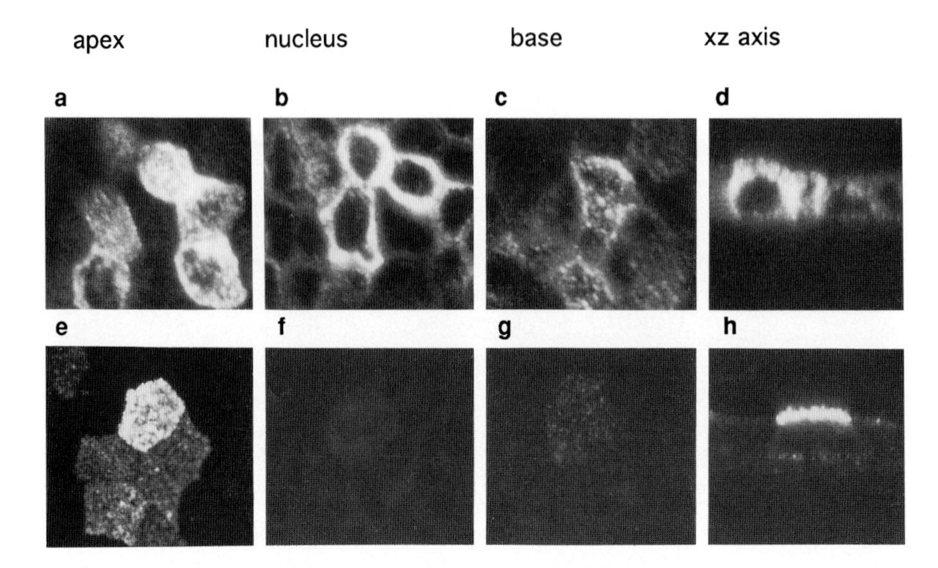


**Fig. 3** Confocal laser scanning microscopy of MDCK cells transfected with the poly-Ig receptor and Thy-1. MDCK cells transfected with poly-Ig receptor (a–d) or Thy-1 (e–h) cDNA inserted into the pLK-neo plasmid were grown to confluence on Transwell filters for 4 days and stimulated with 1 $\mu M$ dexamethasone for 16 hr. The monolayers were fixed, permeabilized with saponin, and incubated with fluoresceinated monoclonal antibodies directed against the poly-Ig receptor or Thy-1. Optical sections were made through the apex (a,h), the nucleus (b,f), and the base (c,g) of the cells. XZ reconstructions are shown in d and e. Note that Thy-1 is restricted to the apex of the cells, whereas the poly-Ig receptor is also found intracellularly and along the basolateral cell surface. Magnification, 500×.

appeared in its core-glycosylated for 4 hr after stimulation. Accumulation was then linear over the next several hours and steady-state levels were reached after 24 hr (Fig. 2). At 24 hr, the amount of receptor in the stimulated cells was >200-fold higher than in nonstimulated cells, as previously reported (Renegar and Small, 1991). These high levels were maintained for up to 1 month. The inducible system was successfully used to analyze membrane trafficking of the receptor and its cell surface delivery in the absence of the ligand, that is, the IgA antibodies. The role of phosphorylation of the receptor's cytoplasmic tail in sorting and targeting of the receptor was also analyzed using the pLK-neo vector, and has been reported elsewhere (Hirt *et al.,* 1993).

Expression of Thy-1 in MDCK cells under the control of a constitutive SV40 promoter resulted in the progressive loss of the stably transfected cells (Wilson *et al.,* 1990), probably as a result of the cytotoxicity of Thy-1, as reported for other proteins (Klessig *et al.,* 1984; Cai *et al.,* 1990). In contrast, cells transfected with cDNA for Thy-1 inserted in the pLK-neo vector could be passaged in the absence of dexamethasone. Following dexamethasone stimulation, expression of the GPI-anchored protein was high and delivery of Thy-1 was restricted to the apical plasma membrane when the transfected cells were grown on Transwell filters (Costar, Cambridge, MA), as demonstrated for other GPI-anchored proteins (Lisanti et al., 1990).

Since expression of the poly-Ig receptor was high in induced MDCK cells grown on Transwell filters, we analyzed the function mediated by the receptor and tested whether the degree of protein expression in the transfected cells paralleled the capacity to transcytose IgA antibodies. Secretory IgA (sIgA) antibodies, the effector molecules in secretions of the digestive, respiratory, and urogenital tracts, are produced by cooperation between two cell types—a plasma cell that produces dimeric IgA (dIgA) and an epithelial cell that contributed SC, the cleaved protion of the poly-Ig receptor required for transepithelial transport (Kraehenbuhl and Neutra, 1992). To replicate the normal *in vivo* system, we developed an efficient *in vitro* transepithelial transport system that produces sIgA. Hybridoma cells that provide a continuous supply of dIgA were cocultivated with MDCK cells stably transfected with cDNA for the poly-Ig receptor inserted into the pLK-neo plasmid. Hybridoma cells producing dIgA antibodies embedded in a thin layer of polymerized collagen were cultivated separately for 4–5 days in the absence of dexamethasone and then cocultivated with MDCK cells in the presence of dexamethasone (1 $\mu M$) to induce receptor expression. Poly-Ig receptor expression and dIgA transport were assessed biochemically by enzyme-linked immunosorbent assay (ELISA; Fig. 4A) and morphologically by confocal microscopy (Fig. 4B). In the absence of dexamethasone, the amount of dIgA recovered from the apcial medium was low (Fig. 4A) but it was significantly higher than in nontransfected MDCK cells (Hirt *et al.*, 1993), reflecting the low level of expression of poly-Ig receptor in noninduced cells, as mentioned earlier. The amount of transport observed in nontransfected cells represents dIgA transported by bulk transcytosis, as reported for other proteins (Brändli *et al.*, 1990). Little dIgA was detected in noninduced cells by confocal fluorescence microscopy of permeabilized monolayers stained with anti-IgA antibodies (Fig. 4B). After induction of poly-Ig receptor, the rate of dIgA transcytosis through the monolayer increased linearly to reach a level of ~25 ng per $10^6$ cells per 24 hr and remained constant for up to 15 days (Hirt *et al.*, 1993). In dexamethasone-stimulated MDCK cells, dIgA accumulated in small vesicles along the lateral plasma membrane and in the apical cytoplasm, while fewer vesicles were found in the basal portion of the cells (Fig. 4B).

We also compared the rate of dIgA transcytosis in MDCK cells transfected with the poly-Ig receptor cDNA inserted into either the inducible pLK-neo vector or a constitutive strong expression system with a cytomegalovirus (CMV) promoter (Hunziker *et al.*, 1990). When maximally stimulated, the level of poly-Ig receptor accumulation and the rate of dIgA transcytosis in cells transfected with pLK-neo were ~5-fold lower than those obtained with the CMV vector.

## IV. Inducible Protein Expression in Nonepithelial Cells

Inducible expression of heterologous proteins was also analyzed in several nonepithelial cells, as listed in Table 1.

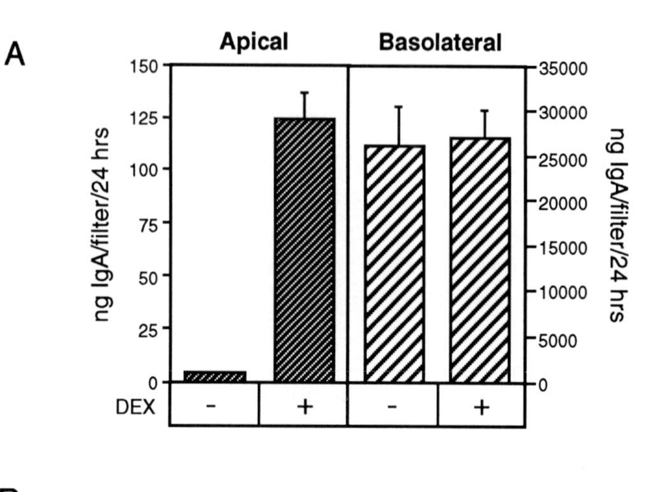

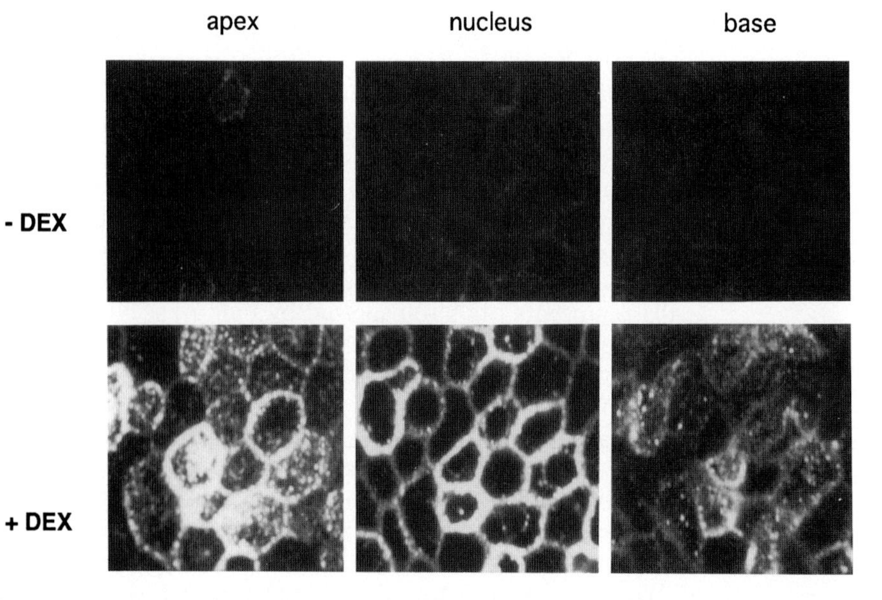


**Fig. 4** Transcytosis of dimeric IgA (dIgA) by poly-Ig receptor-expressing MDCK cells using the coculture system. (A) Inducible IgA transcytosis in the coculture system. Stably transfected MDCK cells were grown to confluence on Transwell filters and transferred onto wells containing IgA-producing hybridoma cells coated in a collagen matrix. Cocultures were incubated in absence (−) or presence (+) of dexamethasone (DEX). IgA accumulation in the apical and basolateral medium was measured by ELISA. (B) Confocal laser scanning microscopy. Cocultures were incubated in the absence (− DEX) or the presence (+ DEX) of dexamethasone for 24 hr. The cells were fixed, permeabilized with saponin, and labeled with fluoresceinated anti-IgA antibodies. Optical sections were made through the apex, the nucleus, and the base of the cells. Magnification, 500×.

### 1. Mouse Fibroblast Ltk Cells

We previously reported the expression of Thy-1 antigen at the mRNA and protein level in mouse fibroblast Ltk$^-$ cells (Hirt *et al.*, 1992). We also expressed various cDNAs overlapping the ectodomain of the fibroblast growth factor receptor (FGF-R) and produced the soluble corresponding fragments that we used to characterize the binding domain of the receptor (Fasel et al., 1991). Mouse fibroblast Ltk$^-$ cells express at their cell surface different forms of the FGF-R encoded by mRNAs ~4.5 kb in size (Fig. 5, lane 7). The cells were transfected with different cDNAs inserted in the pLK-neo vector. The analysis of steady-state levels of RNA accumulation in cells transfected with a cDNA encoding one of the truncated forms of FGF-R (MB1 FGF-R) is shown in Fig. 5). Independent G418-resistant clones were tested for the presence of MB1 transcripts in the presence or absence of dexamethasone (Fig. 5). MBI FGF-R transcripts from two different clones were detected at similar levels in the absence of dexamethasone (Fig. 5, lanes 1 and 3). When stimulated, the level of induction also varied from clone to clone when normalized to the level of endogenous FGF-R mRNA accumulation (Fig. 5, compare lanes 2 and 4). Note, however, that some G418-resistant clones do not express MB1 FGF-R in the absence or in the presence of dexamethasone (Fig. 5, lanes 5 and 6). This clearly illustrates the heterogeneity of expression using the pLK-neo vector, both under noninduced and induced conditions. Therefore, several clones should be analyzed to select those with tight control of expression under nonstimulated conditions and high levels of expression in the presence of glucocorticoids.

Note also that the hormonal effect can occur at a translational or posttranslational level. For example, accumulation Thy-1 mRNA in L cells following dexamethasone stimulation was only 2-fold, whereas Thy-1 protein synthesis

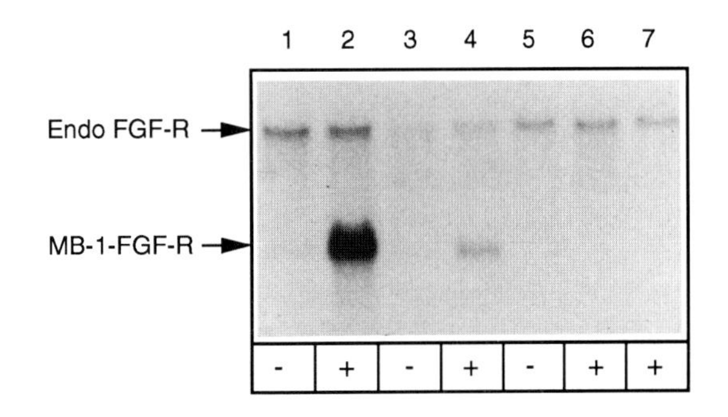


**Fig. 5.** Accumulation of FGF receptor mRNA in mouse Ltk$^-$ cells. Transfected cells (lanes 1–6) were incubated in absence (−) or presence (+) of DEX for 24 hr; nontransfected cells were used as a control (lane 7). Cytoplasmic RNA was subjected to Northern blot analysis using Riboprobes. Arrows indicate the positions of the endogenous FGF-R (Endo FGF-R) and the transfected MB-1-FGF-R mRNA.

was increased more than 60-fold. Such an effect on Thy-1 synthesis was observed in L cells but not in MDCK cells (Hirt et al., 1992). The reason for such a discrepancy is presently not known.

### 2. P815 Mastocytoma Cells

Dexamethasone-dependent synthesis of Thy-1 cell surface antigen was also documented in the mouse mastocytoma cell line P815. Quantitative analysis of Thy-1 expression was assessed both at the mRNA and the protein level. We estimated from fluorescence-activated cell sorting measurements that up to $10^6$ Thy-1 molecules were present at the cell surface (N. Déglon and M. Fasel, unpublished manuscript).

### 3. PC12 Cells

The pLK-neo vector was also used to elicit inducible expression of mouse acidic FGF in rat pheochromocytoma PC12 cells (Renaud and Laurent, personal communication). Although quantification studies have not been performed, preliminary results indicate that the level of expression of the foreign gene product is increased by hormonal stimulation. Thus, it becomes possible to investigate the role of growth factor(s) in the differentiation process of PC12 cells.

## V. Perspectives

Inducible expression systems such as the pLK-neo vector should be particularly useful tools for antisense strategies. Such strategies have been used to study the role of distinct gene products in complex cellular functions as well as in the development (Green *et al.*, 1986; Coleman, 1990). Specific mRNA species can be destroyed in a cell by expressing an RNA molecule from an antisence cDNA inserted into an expression vector (Green *et al.*, 1986). Induced expression of thymidine kinase (TK) antisense RNA under the control of an MMTV-LTR promoter resulted in more than 90% inhibition of TK expression and enzyme activity in mouse L cells (Izant and Weintraub, 1985). More recently, antisense strategies have been applied to the analysis of membrane trafficking. In such studies, oligonucleotides complementary to mRNAs encoding proteins that mediate sorting and targeting functions have been shown to disrupt trafficking when introduced into cells (Lledo *et al.*, 1993; Osen-Sand *et al.*, 1993). Based on such pioneering studies, the use of inducible expression systems in which the expression can be tightly controlled is likely to be valuable for dissecting the multiple steps involved in membrane trafficking in all cells, and in sorting and polarized delivery of membrane proteins in epithelial cells.

# VI. Protocols

## A. DNA Constructs and Transfection

### 1. Reagents and Solutions

Tris–mEDTA (TE) buffer: 10 m*M* Tris-HCl, pH 7.5, 0.1 m*M* EDTA

Complete Dulbecca's modified Eagle's medium (DMEM): DMEM containing 10 m*M* HEPES, pH 7.4 (Seromed, Berlin, Germany), complemented with 10% fetal calf serum (FCS; Inotech), 2 m*M* gluatmine (GIBCO/BRL, Grand Island, NY) penicillin/streptomycin (50 IU/50 μg/ml; GIBCO)

25% (v/v) dimethylsulfoxide (DMSO)–complete DMEM, prepared in advance to allow the solution to equilibrate at room temperature

Circular plasmid DNA, recovered from large-scale preparation, suspended in TE buffer (usually at 1 μg/μl)

100 mg/ml polybrene (hexadimetrine-bromide; FLUKA, Buchs, Switzerland) in $H_2O$, sterilized by filtration through 0.22-μm Millipore filters, and stored in 100-μl aliquots that were used only once upon thawing.

Dexamethasone, dissolved at 1 m*M* in ethanol and stored at −20°C

### 2. DNA Constructs

cDNAs for the various genes were inserted into the pLK-neo vector using the restriction sites of its polylinker (Fig. 1). All constructions were used to transform DH5a *Escherichia coli* cells using standard protocols (Sambrook *et al.*, 1989). Plasmids were purified either on CsCl gradients or QIAGEN columns (QIAGEN, Chatsworth, CA) resuspended in sterile Tris–EDTA (TE) buffer, and used directly for transfection (see next section).

### 3. Transfection

Cells were transfected following procedures listed in Table I.

For MDCK cells, an adaptation (Wessels *et al.*, 1990) of the polybrene technique (Kawai and Nishizawa, 1984) was used.

1. MDCK I cells at low passage numbers were thawed, cultured in complete DMEM, and passaged once in 25-cm$^2$ bottles (Costar). At confluence, the cells were trypsinized. About 7 × 105 cells were seeded in small (10-cm) petri dishes (Costar) and incubated at 37°C.

2. The medium was discarded and replaced with 2.5 ml complete DMEM 24 hr later. Circular plasmid DNA (30 μg) was added to the cells and the plates were shaken to spread the DNA evenly. After adding 30 μg polybrene and additional stirring, the cells were incubated for another 6 hr at 37°C and stirred every 30 min.

3. The medium was then replaced by 5 ml complete DMEM containing 25% DMSO at room temperature. The cells were incubated for exactly 3 min, gently washed once with 10 ml serum free DMEM, and grown for 24 hr in 10 ml complete DMEM. The medium was replaced and the cells were cultivated for an additional 24 hr.

4. Cells were then trypsinized, suspended in complete DMEM, split into 20 petri dishes (10-cm), and grown in selective medium containing 300 $\mu$g/ml G418 (GIBCO). Selection was maintained for 2–3 wk and the medium changed every 3–5 days. After 2–3 wk, conspicuous colonies appeared and were isolated using sterilized cloning rings. The efficiency of the transfection varied, but typically we recovered 10–100 G418-resistant colonies.

L and PC12 cells were transfected using calcium phosphate precipitation (Graham and van der Eb, 1973). P815 cells were transfected using a calcium phosphate precipitation method adapted to mastocytoma cells (van Pel *et al.*, 1985).

## B. Cell Culture

### 1. Solutions

Complete RPMI medium: RPMI1640 medium (Seromed), supplemented with 1% Nutridoma (Sigma, St. Louis, MO), 4 m*M* glutamine, 1 m*M* $\mu$*M* folic acid, and 10 m*M* HEPES

Collagen solution, obtained by mixing 8 volumes of Vitrogen® (collagen type I; Collagen Corporation) with 1 volume of M199 (10×) (GIBCO) and 1 volume of 0.12 *M* NaOH.

### 2. Hybridoma Cells

MB2/B7 polymeric IgA-producing hybridoma cells were cultivated as described (Weltzin *et al.*, 1989). For cocultures, hybridoma cells were harvested by centrifugation at 400 *g* for 10 min, washed once with phosphate-buffered saline (PBS, pH 7.4), and resuspended in a collagen solution (~$3 \times 10^6$ cells per ml collagen mix). The cell–collagen mix (1.5 ml) was then added to the lower chamber of a Transwell unit, and incubated for 60 min at 37°C to allow collagen polymerization. After washing and equilibration with fresh RPMI medium, the cells were cultivated for up to 5 days in complete RPMI medium.

### 3. MDCK Cells

Transfected and control MDCK cells grown to confluence in plastic flasks (75-cm$^2$) were trypsinized, washed, and suspended in DMEM at $5 \times 10^5$ cells/ml. Cell suspensions (2 ml) were added to the upper chamber of the

Transwell filters and cultivated for up to 5 days in complete DMEM. At day 5, the hybridoma cells and the MDCK cells were brought together and the cocultures were cultivated in complete RPMI medium. Induction of poly-Ig receptor expression was obtained by incubating the cultures with $10^{-6}$ *M* dexamethasone.

## VII. Conclusions

We have described the properties, the advantages, and the limitations of a glucocorticoid-inducible expression vector, the pLK-neo vector, that contains a new hormone-inducible LTR promoter derived from an MMTV of kidney origin, a polylinker site for insertion of foreign genes or cDNAs, and selectable markers for bacterial and animal cells. Several cDNAs encoding epithelial enzymes, receptors, and membrane proteins known to require cell polarization for correct trafficking and targeting have been expressed to demonstrate that this vector is particularly suitable for epithelial cells, particularly for cells of kidney origin such as the extensively studied MDCK cells. In the absence of the inducer, the basal level of expression is very low in transfected cells, making the vector ideal for antisense mRNA strategies or for the analysis of cytotoxic gene products since the cells can proliferate and differentiate before the foreign gene product accumulates. The amount of protein that is synthesized following induction can be regulated by the hormone concentration. The variation in levels of expression and inducibility still requires the screening of many clones of transfected cells, but up to 200-fold increases in protein synthesis and accumulation can be obtained.

## Acknowledgments

We are indebted to Lilane Racine, Monique Reinhardt, Nathalie Jeanguenat, and Michel Bernard for valuable technical assistance. We thank Elaine Hughson for her help in immunocytochemistry and Odile Poulain-Godefroy for her collaboration and numerous discussion in some aspects of this work and for the rhodamine-conjugated goat anti-rabbit antibody. This work was supported by Swiss National Science Foundation Grant 31-26404.89 to J. P. Kraehenbuhl and Grant 31-26320.89 to N. Fasel.

## References

Brandli, A. W., Parton, R. G., and Simons, K. (1990). Transcytosis in MDCK cells: Identification of glycoproteins transported bidirectionally between both plasma membrane domains. *J. Cell Biol.* **111,** 2909–2921.

Buetti, E., and Kühnel, B. (1986). Distinct sequence elements involved in the glucocorticoid regulation of the mouse mammary tumor virus promoter identified by linker scanning mutagenesis. *J. Mol. Biol.* **190,** 379–389.

Burnstein, K. L., and Cidlowski, J. A. (1989). Regulation of gene expression by glucocorticoids. *Annu. Rev. Physiol.* **51,** 683–699.

Cai, H., Szeberenyi, J., and Cooper, G. M. (1990). Effect of a dominant inhibitory Ha-*ras* mutation on mitogenic signal transduction in NIH 3T3 cells. *Mol. Cell. Biol.* **10,** 5314–5323.

Cohen, J. J. (1993). Programmed cell death and apoptosis in lymphocyte development and function. *Chest* **103,** 99S–101S.

Colman, A. (1990). Antisense strategies in cell and developmental biology. *J. Cell Sci.* **97,** 399–409.

Dickson, C. (1987). Molecular aspects of mouse mammary tumor virus biology. *Int. Rev. Cytol.* **108,** 119–147.

Duperrex, H., Kenouch, S., Gaeggeler, H. P., Seckl, J. R., Edwards, C. R., Farman, N., and Rossier, B. C. (1993). Rat liver 11 beta-hydroxysteroid dehydrogenase complementary deoxyribonucleic acid encodes oxoreductase activity in a mineralocorticoid-responsive toad bladder cell line. *Endocrinology* **132,** 612–619.

Fasel, N. J., Bernard, M., Deglon, N., Rousseaux, M., Eisenberg, R. J., Bron, C., and Cohen, G. H. (1991). Isolation from mouse fibroblasts of a cDNA encoding a new form of the fibroblast growth factor receptor (flg). *Biochem. Biophys. Res. Commun.* **178,** 8–15.

Garoff, H. (1985). Using recombinant DNA techniques to study protein targeting in the eucaryotic cell. *Annu. Rev. Cell Biol.* **1,** 403–405.

Goodman, L. J., and Firestone, G. L. (1993). Glucocorticoid-regulated stability of a constitutively expressed mouse mammary tumor virus glycoprotein. *Mol. Endocrinol.* **7,** 94–103.

Graham, F. L., and Van der Eb, A. J. (1973). A new technique for the assay of infectivity of human adenovirus 5 DNA. *Virology* **52,** 456–467.

Green, P. J., Pines, O., and Inouye, M. (1986). The role of antisense RNA in gene regulation. *Annu. Rev. Biochem.* **55,** 569–597.

Haffar, O. K., Aponte, G. W., Bravo, D. A., John, N. J., Hess, R. T., and Firestone, G. L. (1988). Glucocorticoid-regulated localization of cell surface glycoproteins in rat hepatoma cells is mediated with Golgi complex. *J. Cell Biol.* **106,** 1463–1474.

Harder, T., and Gerke, V. (1993). The subcellular distribution of early endosomes is affected by the annexin $II_2p11_2$ complex. *J. Cell. Biol.* **123,** 1119–1132.

Hirt, R. P., Poulain-Godefroy, O., Billotte, J., Kraehenbuhl, J. P., and Fasel, N. (1992). Highly inducible synthesis of heterologous proteins in epithelial cells carrying a glucocorticoid-responsive vector. *Gene* 111, 199–206.

Hirt, R. P., Hughes, G. J., Frutiger, S., Michetti, P., Perregaux, C., Poulain-Godefroy, O., Jeanguenat, N., Neutra, M. R., and Kraehenbuhl, J. P. (1993). Transcytosis of the polymeric Ig receptor requires phosphorylation of serine 664 in the absence but not the presence of dimeric IgA. *Cell* **74,** 245–255.

Hunziker, W., Male, P., and Mellman, I. (1990). Differential microtubule requirements for transcytosis in MDCK cells. *EMBO J.* **9,** 3515–3525.

Izant, J. G., and Weintraub, H. (1985). Constitutive and conditional suppression of exogenous and endogenous genes by anti-sence RNA. *Science* **299,** 345–352.

Kaufman, R. J. (1987). High level production of proteins in mammalian cells. *In* "Genetic Engineering: Principles and Methods" (J. K. Setlow, ed), Vol. 9, pp. 155–198. Plenum Press, New York.

Kaufman, R. J. (1990). Vectors used for expression in mammalian cells. *Meth. Enzymol.* **185,** 487–511.

Kawai, S., and Nishizawa, M. (1984). New procedure for DNA transfection with polycation and dimethyl sulfoxide. *Mol. Cell. Biol.* **4,** 1345–1347.

Keown, W. A., Cambell, C. R., and Kucherplata, R. S. (1990). Methods for introducing DNA into mammalian cells. *Meth. Enzymol.* **185,** 527–537.

Klessig, D. F., Brough, D. E., and Cleghon, V. (1984). Introduction, stable integration, and controlled expression of a chimeric adenovirus gene whose product is toxic to the recipient human cell. *Mol. Cell. Biol.* **4,** 1354–1362.

Kraehenbuhl, J. P., and Neutra, M. R. (1992). Transepithelial transport and mucosal defence II: Secretion of IgA. *Trends Cell Biol.* **2,** 170–179.

Kriegler, M. (1990). Assembly of enhancers, promoters, and splice signals to control expression of transferred genes. *Meth. Enzymol.* **185,** 512–527.

Lee, F., Mulligan, R., Berg, P., and Ringold, G. (1981). Glucocorticoids regulate expression of dihydrofolate reductase cDNA in mouse mammary tumour virus chimaeric plasmids. *Nature (London)* **294,** 228–232.

Lisanti, M. P., Le Bivic, A., Saltiel, A. R., and Rodriguez-Boulan, E. (1990). Preferred apical distribution of glycosyl-phosphatidylinositol (GPI) anchored proteins: A highly conserved feature of the polarized epithelial cell phenotype. *J. Membr. Biol.* **113,** 155–167.

Lledo, P. M., Vernier, P., Vincent, J. D., Mason, W. T., and Zorec, R. (1993). Inhibiton of Rab3B expression attenuates Ca($^{2+}$)-dependent exocytosis in rat anterior pituitary cells. *Nature* (London) **364,** 540–544.

Meulia, T., and Diggelmann, H. (1990). Tissue-specific factors and glucocorticoid receptors present in nuclear extracts bind next to each other in the promoter region of mouse mammary tumor virus. *J. Mol. Biol.* **216,** 859–872.

Migliorati, G., Pagliacci, C., Moraca, R., Crocicchio, F., Nicoletti, I., and Riccardi, C. (1992). Glucocorticoid-induced apoptosis of natural killer cells and cytotoxic T lymphocytes. *Pharmacol. Res.* **26** (*suppl 2*) 26–27.

Mok, E., Golovkina, T. V., and Ross, S. R. (1992). A mouse mammary tumor virus mammary gland enhancer confers tissue-specific but not lactation-dependent expression in transgenic mice. *J. Virol.* **66,** 7529–7532.

Osen-Sand, A., Catsicas, M., Staple, J. K., Jones, K. A., Ayala, G., Knowles, J., Grenningloh, G., and Catsicas, S. (1993). Inhibition of axonal growth by SNAP-25 antisence oligonucleotides *in vitro* and *in vivo*. *Nature* (*London*) **364,** 445–448.

Renegar, K. B., and Small, P. A. J. (1991). Passive transfer of local immunity to influenza virus infection by IgA antibody. *J. Immunol.* **146,** 1972–1978.

Rollini, P., Billotte, J., Kolb, E., and Diggelmann, H. (1992). Expression pattern of mouse mammary tumor virus in transgenic mice carrying exogenous proviruses of different origins. *J. Virol.* **66,** 4580–4586.

Roman, L. M., and Garoff, H. (1986). Alteration of the cytoplasmic domain of the membrane-spanning glycoprotein p62 in Semliki Forest virus does not affect its polar distribution in established lines of Madin–Darby canine kidney cells. *J. Cell Biol.* **103,** 2607–2618.

Roth, M. G. (1989). Molecular biological approaches to protein sorting. *Annu. Rev. Physiol.* 51, 797–810.

Samanta, H., Engel, D. A., Chao, H. M., Thakur, A., Garcia-Blanco, M. A., and Lengyel, P. (1986). Interferons as gene activators. Cloning of the 5′ terminus and the control segment of an interferon activated gene. *J. Biol. Chem.* **261,** 11849–11858.

Sambrook, J., Fritsch, E. F., and Maniatis, T. (1989). "Molecular Cloning. A Laboratory Manual." Cold Spring Harbor Laboratory Press, Cold Spring Harbor, New York.

Schaerer, E., Verrey, F., Racine, L., Tallichet, C., Reinhardt, M., and Kraehenbuhl, J. P. (1990). Polarized transport of the polymeric immunoglobulin receptor in transfected rabbit mammary epithelial cells. *J. Cell Biol.* **110,** 987–998.

Simons, K., and Wandinger-Ness, A. (1990). Polarized sorting in epithelia. *Cell* **62,** 207–210.

Tsubura, A., Inaba, M., Imai, S., Murakami, A., Oyaizu, N., Yasumizu, R., Ohnishi, Y., Tanaka, H., Morii, S., and Ikehara, S. (1988). Intervention of T-cells in transportation of mouse mammary tumor virus (milk factor) to mammary gland cells in vivo. *Cancer Res.* **48,** 6555–6559.

Van Pel, A., De Plaen, E., and Boon, T. (1985). Selection of highly transfectable variant from mouse mastocytoma P815. *Somatic Cell Mol. Genet.* **11,** 467–475.

Wahli, W., and Martinez, E. (1991). Superfamily of steroid nuclear receptors: Positive and negative regulators of gene expression. *FASEB J.* **5,** 2243–2249.

Wellinger, R. J., Garcia, M., Vessaz, A., and Diggelmann, H. (1986). Exogenous mouse mammary tumor virus proviral DNA isolated from a kidney adenocarcinoma cell line contains alterations in the U3 region of the long terminal repeat. *J. Virol.* **60,** 1–11.

Weltzin, R., Lucia-Jandris, P., Michetti, P., Fields, B. N., Kraehenbuhl, J. P., and Neutra, M. R. (1989). Binding and transepithelial transport of immunoglobulins by intestinal M cells: Demonstration using monoclonal IgA antibodies against enteric viral proteins. *J. Cell Biol.* **108,** 1673–1685.

Wessels, H. P., Hansen, G. H., Fuhrer, C., Look, A. T., Sjostrom, H., Noren, O., and Spiess, M. (1990). Aminopeptidase N is directly sorted to the apical domain in MDCK Cells. *J. Cell Biol.* **111,** 2923–2930.

Wilson, J. M., Fasel, N., and Kraehenbuhl, J. P. (1990). Polarity of endogenous and exogenous glycosyl-phosphatidylinositol-anchored membrane proteins in Madin–Darby canine kidney cells. *J. Cell Sci.* **96,** 143–149.

Wurm, F. M., Gwinn, K. A., and Kingston, R. E. (1986). Inducible overproduction of the mouse c-*myc* protein in mammalian cells. *Proc. Natl. Acad. Sci. USA* **83,** 5414–5418.

Zettl, K. S., Sjaastad, M. D., Riskin, P. M., Parry, G., Machen, T. E., and Firestone, G. L. (1992). Glucocorticoid-induced formation of tight junctions in mouse mammary epithelial cells in vitro. *Proc. Natl. Acad. Sci. USA* **89,** 9069–9073.

# CHAPTER 13

# Expression of Exogenous Proteins in Cells with Regulated Secretory Pathways

**R. A. Chavez, Y.-T. Chen, W. K. Schmidt, L. Carnell, and Hsiao-Ping Moore**

Department of Molecular and Cell Biology
University of California, Berkeley
Berkeley, California 94720



## I. Introduction

Proteins destined for secretion must be efficiently identified and transported to secretory vesicles that fuse with the plasma membrane. In some cells, these proteins are sorted between two distinct secretory pathways: constitutive and regulated (Burgess and Kelly, 1987; Miller and Moore, 1990). The constitutive pathway is responsible for the homeostatic maintenance of the cell. Constitutively secreted proteins are quickly released from the cell once targeted to this pathway, and their secretion is independent of extracellular signals. This pathway is common to all cells. In contrast, researchers generally believe that the regulated secretory pathway is found only in specialized secretory cell types such as endocrine, neuronal, exocrine, and some gastrointestinal tract cells. Regulated secretory proteins, such as hormones, are stored intracellularly within vesicles whose fusion with the plasma membrane is dependent on specific extracellular signals. Often these proteins are synthesized as inactive precursors that are cleaved into their mature functional forms during their transport through this pathway. Since cells that have a regulated secretory pathway are also capable of constitutive secretion, regulated secretory cells must be able to distinguish between both types of secretory proteins and must sort them efficiently. During the last 10 years, much progress has been made in the study of protein sorting, processing, and transport by expressing exogenous proteins in cells capable of regulated secretion. This approach was based on the initial finding that tumor pituitary corticotrophs in culture can correctly package, process, and target an exogenous hormone on transfection (Moore *et al.*, 1983b), thus providing a nice surrogate system for dissection of the cellular mechanisms underlying these processes.

Three major questions have been addressed by this approach with regard to regulated secretion. The first of these concerns the mechanism by which secretory proteins are sorted into the correct pathway in cells that are capable of both types of secretion. Early transfection studies demonstrated that exogenously regulated proteins such as insulin, growth hormone, and trypsinogen can be correctly sorted in the pituitary adrenocorticotropic hormone (ACTH)-secreting cell line AtT-20 (Moore *et al.*, 1983; Burgess and Kelly, 1984; Moore and Kelly, 1985). More recently, transfected cells have been used to define sorting signals on various soluble and membrane proteins destined for regulated secretory vesicles [e.g., growth hormone, Moore and Kelly, 1986; von Willebrand factor, Sporn *et al.*, 1986; insulin, Powell *et al.*, 1988; Grodsky *et al.*, 1992; somatostatin, Stoller and Shields, 1989; a basic proline-rich granule protein, Castle *et al.*, 1992; synaptophysin, Linstedt and Kelly, 1991; peptidylglycine alpha-amidating monooxygenase (PAM), Milgram *et al.*, 1993]. Transfection has also been used to distinguish between the two prevalent models of the sorting mechanism for regulated secretory proteins: aggregation of regulated secretory products with subsequent sorting away from other proteins or the more classical sorting signal–receptor model (Gross *et al.*, 1989; Chanat and Huttner, 1991; Quinn *et al.*, 1991). Second, the transfection approach has been used to investigate secretory protein maturation in regulated secretory cells. For many peptide hormones such as insulin and ACTH, the species that is initially sorted into the regulated pathway is not identical to the product that is secreted. Typically, these hormones are synthesized as large inactive precursors that must be proteolytically processed into their mature active forms. The question of how this processing occurs has been investigated using cells that are incapable of proper processing, typically by coexpression of a candidate endopeptidase with the hormone of interest (Thomas *et al.*, 1988; Bennett *et al.*, 1992; Ohagi *et al.*, 1992; Benjannet *et al.*, 1993). Finally, what proteins constitute the machinery responsible for the regulation of secretory processes? Advances in the molecular analysis of regulated secretion have identified many regulatory proteins involved in these processes (for reviews, see Rothman and Orci, 1992; Pryer *et al.*, 1992). Functional analysis of these proteins requires their study in cells capable of regulated secretion. For example, the function of Rab proteins, small GTP-binding proteins implicated in vesicular trafficking, has been studied by transfection of wild-type and dominant-negative forms into regulated secretory cells (Ngsee *et al.*, 1993).

Most of these advances have used four model cell lines: the mouse pituitary corticotroph line AtT-20, the rat pituitary growth-hormone secreting lines $GH_3$/$GH_4C_1$, the insulin-secreting $\beta$TC cells derived from transgenic mice expressing SV40 T antigen (Radvanyi *et al.*, 1993), and the rat pheochromocytoma cell line PC12. This chapter focuses on AtT-20 and PC12 cells, the two lines in which we have performed our investigations. For use of the $GH_3$/$GH_4C_1$ lines and the $\beta$TC cells, refer to Benjannet *et al.* (1993). Studies of the AtT-20 line by Gumbiner, Moore, and Kelly provided the initial evidence for the existence of two distinct secretory pathways (Gumbiner and Kelly, 1982; Moore *et al.*, 1983a; Moore and Kelly, 1985). The characteristic secretory product of this cell line is the polypeptide hormone ACTH, a primary regulator of glucocorticoid secretion from the adrenal cortex. The large ACTH precursor, proopiomelanocortin (POMC), is processed into ACTH and a variety of other small polypeptide hormones of varying actions. Mature ACTH is stored for secretion in vesicles that are electron-dense in electron micrographs, leading to their designation as dense core granules. PC12 cells are derived from a rat pheochromocytoma (Greene and Tischler, 1976), and thus possess some of the secretory characteristics of the cells in the adrenal medulla. They contain large (~115 nm) chromaffin granules that contain the neurotransmitters dopamine and norepinephrine as well as secretogranins. Regulated secretion of these transmitters can be stimulated by membrane depolarization (Rosa *et al.*, 1985).

AtT-20 and PC12 cells can be contrasted in several ways. First, PC12 cells contain an additional type of vesicle that mediates secretion in neither the constitutive nor the regulated biosynthetic pathway. These vesicles, known as synaptic or recycling microvesicles, are thought to be recruited from the plasma membrane via an endosomal intermediate and refilled with their secretory cargo (i.e., neurotransmitters) for another cycle of exocytosis. AtT-20 cells possess very low levels of the markers for these vesicles (e.g., synaptophysin). If these vesicles are in AtT-20 cells, they are likely to be a low-abundance component of the secretory apparatus. Therefore, AtT-20 cells can be used to study the biosynthetic constitutive and regulated (dense-core) secretory pathways with relatively little interference from the recycling regulated secretory pathway. In contrast, PC12 cells contain all three pathways for secretion, and provide a better system for studying the recycling, synaptic vesicle-like regulated secretory pathway. In addition, these cells regulate the sorting of secretory proteins more tightly than AtT-20 cells; over 95% of the protein secretogranin II is sorted into the regulated pathway in PC12 cells (Rosa *et al.*, 1985; Gerdes *et al.*, 1989) whereas AtT-20 cells typically store less than 50% of newly synthesized hormones intracellularly (Moore and Kelly, 1985). However, AtT-20 cells contain a number of enzymes that are required for efficient processing of prohormones, including PC1 for endoproteolytic processing and PAM for amidation (Moore *et al.*, 1983; Moore and Kelly, 1985; Milgram *et al.*, 1993). Thus, AtT-20 is a better cell line for studies involving processing. Finally, AtT-20 cells exhibit a number of neuronal properties such as stereotyped morphology, whereas undifferentiated PC12 cells are essentially round, nondescript cells. AtT-20 cells are spindle-shaped cells that occasionally exhibit one to three distinct neurites. When grown on laminin-coated substrate, nearly all the cells in an AtT-20 culture will extend neurites. The tips of these neurites contain vesicles filled with ACTH (Rivas and Moore, 1989). Although PC12 cells can extend neurites when grown in the presence of nerve growth factor (Greene and Tischler, 1976), it is generally better to use AtT-20 cells for these studies because of their relatively large size, clearly defined processes, and easy maintenance without the need for special inductive growth factors.

These two cell lines do share some features, however. Each expresses proteins that are sulfated *in vivo*. In AtT-20 cells, POMC and the prohormone convertase PC1 (Moore *et al.*, 1983; Benjannet *et al.*, 1993) are both sulfated; in PC12 cells, two granular proteins, secretogranins I and II, are sulfated (Rosa *et al.*, 1985). Since sulfation occurs in the *trans*-Golgi or the *trans*-Golgi network (Gerdes *et al.*, 1989), compartments in which constitutive and regulated secretory products first become segregated, the sorting and transport of these proteins beyond the *trans*-Golgi network can be monitored after pulse-labeling with radioactive sulfate (see Section V,A). Second, both cell lines are quite stable with respect to their functional characteristics. With other cell lines, characteristics such as levels of protein expression and secretion may be lost with increased passages. No differences are detected in the sorting efficiencies or secretion responses in either of these two cell lines with continued growth of the cells. Finally we have found that, although several commonly used transfection procedures can be applied to these two cell types, the exact procedures require optimization of important parameters (see subsequent discussion).

Other cell lines have been isolated that provide alternatives for investigators who prefer to study nonneuronal or nonneuroendocrine function. In particular, Hanahan and co-workers have isolated pancreatic $\beta$ cells from transgenic mice at an early stage in islet cell tumorigenesis (Radvanyi *et al.*, 1993). These cells preserve the insulin secretory response to glucose, as well as the surface expression of specific antigens and other features found in normal $\beta$ cells. Although the AtT-20 and PC12 cell lines behave differently from each other, the combinations of all their features make either of them an excellent first choice as a model system in which to address questions about regulated transport and secretion. In the following sections, we provide methods for the expression of exogenous proteins in both PC12 and AtT-20 cells, as well as selected methods to study the targeting and secretion of these proteins.

## II. Considerations for the Choice of Expression Systems

### A. Stable versus Transient Transfection

When contemplating the use of cell lines with regulated secretory pathways, the desirability of stable versus transient transfection must be considered. In a stable transfection, the entire population of selected cells expresses the gene of interest so the behavior of cells expressing the transfected protein can be directly compared with that of the parent untransfected cell line. This feature is especially useful when assaying the effects of a transfected gene on a given cellular process. For instance, the stable expression system is suitable for studying the effect of overexpressing wild-type or mutant small GTP-binding proteins (Rabs or Arfs) on secretion. In addition, different clones may express the gene of interest at different levels, providing data on the effects of protein expression over a range of expression levels. However, this process is time-consuming, especially for AtT-20 cells which tend to grow slowly at low density; a selection would typically require 3–4 wk for propagation and screening. Thus, for studies that do not require the entire cell population to express the transfected gene, it is more convenient to use a transient expression system. We have found that sorting of a transfected protein between the constitutive and the regulated secretory pathways in AtT-20 cells is comparable when studied by transient or stable transfection. The transient method is also useful in studying post-translational modifications such as protein folding, proteolytic processing, glycosylation, sulfation, and other processing events. In this case, a complete experiment requires only 3–5 days. However, in transient transfection the average expression is generally lower than in stable transfection; depending on the transfection method, usually only 10–30% of the cell population will express the gene of interest. This restriction makes certain studies difficult because of the background behavior of the untransfected cells. If techniques exist for the identification and analysis of single transfected cells, and if the gene of interest is not expressed in the host cell line, transient transfections are an obvious recourse. The unique morphology of AtT-20 cells makes them an ideal system for the analysis of transiently transfected genes at the single-cell level by immunofluorescence localization. Consider developing stably transfected lines in the following situations: (1) when the gene product and its effects on cellular processes cannot be detected above the background of untransfected cells (>70%) present in transient transfections or (2) when the gene product saturates the cellular machinery at high levels of expression. In the latter case, selecting a stable cell line with the proper expression level will be important. In general, it is difficult to control the exact level of a transfected gene in each cell in transient transfections.

We describe two different plasmid-based transfection methods—a calcium phosphate ($Ca/P_i$) precipitate method and a liposome-mediated transfection method. The $Ca/PO_i$ procedure is the most commonly used method, with wide applicability to a variety of cell lines. It is also considerably less expensive than the liposome-mediated transfection method because of the cost of the commercial liposome reagent (Lipofectin™; GIBCOBRL, Grand Island, NY). However, in AtT-20 cells the $Ca/P_i$ procedure requires the use of considerably more plasmid DNA for both the gene of interest and the selectable marker plasmid (12-fold more required for each). Both methods have proven successful with the AtT-20/F2 cell line, whereas PC12 cells are more easily transfected with the Lipofectin™-based method (referred to as "lipofection"). We have found that DEAE–dextran-based transfection methods (Seed and Aruffo, 1987) are unsuitable for transfections of either cell line. However, we have routinely used this method for transient transfection of COS 7 cells as a test for the expression of specific constructs before undertaking the more lengthy stable transfection procedures.

## B. Vaccinia Virus Infection

A possible approach to avoid the lengthy stable transfection procedure and to achieve high uniform expression is vaccinia virus-based expression. Vaccinia virus, a member of the family Poxviridae and the agent used in the vaccination against smallpox, has a wide host range. Expression in both AtT-20 and PC-12 cells can be achieved by this method. Two different techniques have been developed for the use of vaccinia virus in directing the expression of exogenous proteins in mammalian cells. The first technique involves the creation of a recombinant vaccinia virus, by homologous recombination, in which the gene of interest is inserted into the viral genome and is under the control of a viral promoter. Cells are infected with the recombinant virus, and the gene of interest is expressed along with the other viral genes. A major drawback to this method is the time required to prepare and clone the recombinant virus strain for each gene of interest. In contrast, the second technique combines transfection and infection to promote the expression of a cloned gene. The first step in this method is the infection of cells with a recombinant vaccinia virus strain that expresses the gene for T7 RNA polymerase [vTF7-3, VR-2153, American Type Culture Collection (ATCC), Rockville, MD], and the infection is followed by the liposome-mediated transfection of a plasmid vector, containing the gene of interest inserted downstream from a bacteriophage T7 RNA polymerase promoter, into the infected cell line (Fuerst *et al.*, 1987). Expression of the T7

RNA polymerase from the viral infection results in transcription of the gene of interest from the transfected plasmid, which in turn drives the expression of the desired protein. In AtT-20 and PC12 cells, however, this latter method produces only relatively low frequency expression of the desired product. For experiments in which high frequency expression is desired, a recombinant virus approach is thus recommended.

Depending on the nature of the experiment and the cell type used, several features of viral infection must be considered. The lytic nature of the vaccinia virus infection, as well as the virus-induced shut-down of host cell protein synthesis, restricts the use of this method to transient expression. The time course of host protein synthesis inhibition is paralleled by an increase in the level of expression of viral proteins. Brion *et al.* (1992) showed that proper packaging and storage of regulated secretory products in AtT-20 cells require continued new protein synthesis, and that inhibition of protein synthesis interferes with normal transport through the regulated secretory pathway. Transport through the constitutive pathway is not affected by the same conditions. Therefore, the time course of protein synthesis inhibition as a result of viral infection must be considered in the experimental design for studying packaging, sorting, and transport of proteins through the regulated pathway. The ability of vaccinia virus to infect a variety of mammalian cell lines allows the rapid expression of functional proteins that are correctly post-translationally modified. The multiplicity of infection (MOI) is an important parameter in recombinant virus expression. If the MOI is too high, vaccinia infection results in rapid cell lysis; if too low, the cytopathic effects will occur with an asynchronous onset across the cell population. The optimal MOI provides a frequency of transient expression much greater than that observed with either plasmid-based transfection method, because the efficiency of the infection process is quite high.

## III. Plasmid-Based Expression

### A. Culture Conditions

We have used the AtT-20/F2 cell line (CRL 1795; ATCC, Rockville, MD) for our transfections. AtT-20/F2 cells are maintained in Dulbecco's modified Eagle's medium (DMEM; Whittaker Bioproducts) containing 10% fetal bovine serum (FBS; Sigma, St. Louis, MO) at 37°C under 15% $CO_2$. We have found that AtT-20 cells are very sensitive to the serum conditions; it is advisable to screen several lots of FBS from various vendors. These cells also do not survive in medium supplemented with heat-inactivated serum. Thus, extensive warming of the media at 37°C should be avoided. At different levels of $CO_2$, the growth and selection conditions for AtT-20/F2 cells may vary. AtT-20/F2 cells require treatment with trypsin solution (STV; Whittaker Bioproducts) for replating. Extensive trituration is also required to prevent clusters of cells from plating together. These cells tend to grow atop one another, and this type of growth decreases the efficiency of transfection. Therefore when plating AtT-20/F2 cells for transfection, make sure that the cells are evenly dispersed on the dish for maximal transfection efficiency. We routinely split AtT-20/F2 cells at 1:5 dilution into fresh dishes, and grow them for 3–5 days before a subsequent split. Several other subclones of the original AtT-20 cell line are also available; in general, they do not differ in their characteristics or suitability for transfection. For instance, we have found no differences between the AtT-20/F2 subclone and another subclone designated AtT-20/NIH in transient transfections (W. K. Schmidt and H.-P. H. Moore, unpublished observations). However, the selection conditions for generating stably transfected cell lines differ with respect to the two cell lines. In this chapter, we describe conditions only for AtT-20/F2 cells.

For PC12 cells (CRL 1721; ATCC), normal growth medium is composed of DMEM containing 5% enriched calf serum (Gemini Bioproducts, Calabasas, CA) and 5% horse serum (Summit Biotech, Ft. Collins, CO). The cells are grown at 37°C under 10% $CO_2$. Unlike AtT-20/F2, PC12 cells can be split by trituration without STV; they are not as firmly attached to the substrate as AtT-20 cells. PC12 cells can be split at 1:5 dilution, with growth checked every 4–5 days.

## B. Expression Vectors

### 1. Promoters

We have used several different viral promoters active in mammalian cells for expression in transfected AtT-20 and PC-12 cells, including simian virus 40 (SV40) early, Rous sarcoma virus (RSV), and cytomegalovirus (CMV) early. Earlier studies have shown that the SV40 early promoter has an extremely low activity in AtT-20 cells relative to the RSV promoter as measured by chloramphenicol acetyltransferase (CAT) gene expression (less than 200-fold; see Moore and Kelly, 1985), and is therefore unsuitable for the expression of nonselectable markers in this cell line (but see subsequent discussion). Both RSV and CMV are strong promoters in AtT-20 cells, and have been used for routine transfection procedures. Muller *et al.* (1990) reported that RSV-driven expression of foreign genes in PC12 cells achieves significant levels only when liposome-mediated transfection rather than the $Ca/P_i$ method is used. However, we have found significant levels of expression of RSV-driven cDNAs in PC12 transient transfections using $Ca/P_i$-based procedures (L. Carnell, W. K. Schmidt, and H.-P. H. Moore, unpublished observations). More recently, we have settled on CMV promoter-driven expression as our method of choice. For this purpose, we use pCDM8, a commercially available vector (Invitrogen, San Diego, CA) originally developed by Seed (Seed and Aruffo, 1987). This vector possesses several features of interest. First, it contains a CMV early promoter and enhancer capable of driving expression of cloned genes in mammalian cell lines. Second, if mutants of the gene of interest are desired, pCDM8 contains an M13 phage origin of replication that allows single-stranded DNA recovery for oligonucleotide-directed mutagenesis. In addition, this vector contains a polylinker with six unique restriction sites. Finally, this vector contains a T7 RNA polymerase promoter for the *in vitro* synthesis of mRNA, as well as the expression of cloned genes after infection with T7 RNA polymerase recombinant vaccinia virus (see Section II,B; see also Fig. 1). These features make pCDM8 a flexible and attractive option when choosing a vector for either plasmid-based or vaccinia virus-based expression.

## 2. Epitope Tagging

If the protein to be expressed is a homolog or a mutant form of an endogenous protein, the detection of expression becomes an important factor. A useful method is the generation of a recombinant gene encoding a chimeric protein

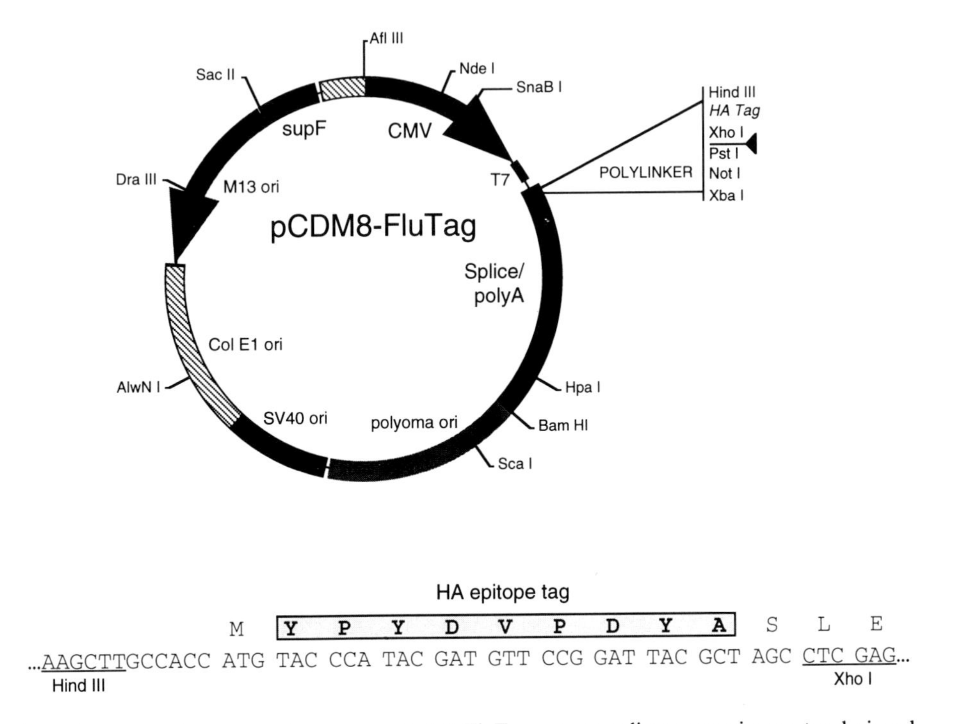


**Fig. 1** Map of the expression vector pCDM8-FluTag, a mammalian expression vector designed for amino-terminal epitope tagging of cytoplasmic proteins. This map contains selected unique restriction sites as well as the polylinker region of the plasmid. The directionality of the M13 origin and of the CMV promoter sequence is indicated. Below the plasmid map is an expansion of the sequence between the *Hin*d III and *Xho* I sites, with the nucleotide sequence encoding the influenza hemagglutinin epitope tag. The 5′ end of the gene of interest should be modified to place an *Xho* I site at the codon for the second amino acid to allow inframe insertion of the gene.

containing an epitope that distinguishes the expressed protein from its endogenous relative. We have used this method of epitope tagging to analyze the localization and function of several Rab proteins (Chen *et al.*, 1993; Yu *et al.*, 1993; R. Chavez and H.-P. Moore, unpublished experiments). In collaboration with Seed, we have developed a variant of the pCDM8 vector, pCDM8-FluTag (Fig. 1), that has been used in a variety of experiments. A synthetic nucleotide sequence encoding an 11-amino-acid sequence from the influenza virus hemagglutinin (HA) protein is inserted into the pCDM8 polylinker so the HA-derived sequence is fused in frame at the 5′ end of the gene of interest. The resultant chimeric protein can be recognized by the mouse monoclonal antibody 12CA5 (BAbCO, Berkeley, CA), which is effective in immunofluorescence and immunoblotting assays (Wilson *et al.*, 1984). Chen *et al.* (1993) demonstrated that HA epitope-tagged Sec4p, a small GTP-binding protein involved in secretion, is still functional and can complement a temperature-sensitive *sec*4 mutant strain. The gene of interest can be cloned into the vector at the *Xho* I site at the 5′ end and at one of the sites in the polylinker region (*Pst* I, *Not* I, or *Xba* I) at the 3′ end. Obviously, the HA epitope can be used in other positions along the primary sequence of a specific protein, as long as its insertion does not disrupt protein function. Other epitope tags have been described including c-*myc* (Evan *et al.*, 1985), FLAG (Kodak-IBI, New Haven, CT), and novel tags that allow the purification and detection of the tagged protein without reliance on antibody–antigen interactions, for example, 6xHIS (QIAGEN, Chatsworth, CA).

## C. Transfection Procedures: AtT-20

### 1. Ca/$P_i$ Transfection

We have used the following method for Ca/$P_i$-based transfections of AtT-20 cells, modified from that of Graham and van der Eb (1973). This method involves the preparation of a DNA–Ca/$P_i$ precipitate that is added to the cells for a specified time, followed by a standard glycerol shock treatment.

***Solutions required***

2× HeBS: 274 m*M* NaCl, 10 m*M* KCl, 11 m*M* glucose, 1.6 m*M* $Na_2HPO_4$, 42 m*M* HEPES, pH 7.05 (filter-sterilized, stored at −80°C)

2 *M* $CaCl_2$: 11.1 g $CaCl_2$ (anhydrous) in 50 ml distilled $H_2O$ ($dH_2O$) (filter-sterilized)

DMEM, supplemented with 2% FBS

25% glycerol in DMEM (filter-sterilized)

***Procedure***

1. Split AtT-20/F2 cells at 1–1.5 × $10^6$ cells per 10-cm dish 16–18 hr prior to the transfection, and place in normal AtT-20 growth medium (DMEM supplemented with 10% FBS).

2. Prepare $Ca/P_i$ transfection mixture by combining the 2× HeBS solution and a DNA-containing $CaCl_2$ solution to form a DNA–$Ca/P_i$ precipitate. For a 10-cm dish, combine 94 μl 2 *M* $CaCl_2$, 120 μg plasmid of interest (for transient transfection, 60–70 μg), 24 μl plasmid containing a selectable marker (e.g., pSV2-neo) if stable transfectants are desired, and sterile $dH_2O$ to a final volume of 750 μl in a sterile snap-cap polystyrene tube.

*Note:* The high amount of the plasmid of interest in contrast to that used in typical $Ca/P_i$ transfection methods, is vital to high-level expression in transfections of AtT-20/F2 cells. In transfections containing only 20 μg plasmid of interest, the level of expression achieved is less than 25% of the level attained with 120 μg; the level of expression decreases when the amount of the plasmid of interest is increased above 120 μg (H.-P. H. Moore, unpublished observations).

3. Form the DNA–$CaP_i$ precipitate by gently bubbling air through the 750 μl HeBS (2–3 bubbles/sec) using a pasteur pipet attached to a portable pipet-aid and adding the DNA–$CaCl_2$ solution dropwise (1 drop/sec). After all the DNA–$CaCl_2$ solution has been added, the combined solution should be slightly milky.

4. Allow the DNA–$Ca/P_i$ precipitate to incubate without disturbance at 20°C for an additional 40 min.

5. At the end of the 40-min period, wash the cells twice with 1x HeBS.

6. Gently resuspend the transfection mixture and add directly to the cells. Inspection of the cells using an inverted microscope should disclose small particles of the precipitate dispersed throughout the culture. Incubate the cells for 30 min at 20°C. In both this step and the incubation of the DNA–$Ca/P_i$ precipitate, we have found that the temperature requirement is quite strict. The transfection efficiency and level of expression is impaired when these steps are carried out at temperatures higher than 20°C.

7. Remove the transfection mixture by aspiration, and add 12 ml DMEM supplemented with 2% FBS to the cells. Place the cells in a tissue culture incubator for 7 hr at 37°C under 15% $CO_2$.

8. Aspirate the medium from the cells, and add 1 ml DMEM containing 25% glycerol (filter-sterilized) for 1 min at room temperature.

10. Aspirate the glycerol solution, rinse the cells 3 times with DMEM, and add normal growth medium. Return the cells to the tissue culture incubator.

## 2. Lipofection

We have used the following method for liposome-based transfections of AtT-20/F2 cells, using the Lipofectin™ reagent, with some modifications from the manufacturer's recommendations.

1. Plate cells 16–18 hr before the transfection at 1.5 × $10^6$ cells per 60-mm dish (approximately 75–80% confluent).

2. Prepare the transfection mixture by combining 50 $\mu$g Lipofectin™ reagent, 20 $\mu$g plasmid of interest (include 4 $\mu$g plasmid with a selectable marker, if desired), and sterile $dH_2O$ to a final volume of 200 $\mu$l.

3. Incubate this mixture at room temperature for 15 min. Just before adding this solution to the cells, add 1.8 ml DMEM and mix the resulting solution by inversion.

4. During the 15-min incubation, wash the cells 3 times with either DMEM or phosphate-buttered saline (PBS).

5. Add the transfection mixture to the culture dish, and incubate for 6–8 hr in the tissue culture incubator (37°C, 15% $CO_2$). If desired, this incubation period may be extended for 8–12 hr if an equal volume of normal AtT-20 growth medium is added to the cells.

6. At the end of the transfection period, the cells will be less readily attached to the substrate. Remove the cells from the plate by trituration.

7. Pellet cells in a clinical centrifuge to remove the residual Lipofectin™, and replate them in a new dish of the same size with normal AtT-20 growth medium. Even after replating, the cells should not become too confluent, since many of the cells will not replate after lipofection.

8. The cells are then grown as desired at 37°C in 15% $CO_2$ (see subsequent discussion).

## D. Transfection Procedures: PC12

### 1. Ca/$P_i$ Transfection

We have used a Ca/$P_i$-based transfection protocol for PC12 cells that is similar to the transfection procedure for AtT-20/F2 cells, with the following exceptions. First, $5 \times 10^6$ cells are plated per 10-cm dish. Second, we have found essentially no difference in expression levels for transient transfections that contain between 40 and 120 $\mu$g/ml of the plasmid of interest (L. Carnell, W. K. Schmidt, and H.-P. H. Moore, unpublished observations). In addition, the glycerol shock solution used in this procedure contains only 10% glycerol, not 25% glycerol as given in the AtT-20 transfection protocol.

### 2. Lipofection

For PC12 cells, we have used a liposome-based method adapted from that described by Muller et al. (1990).

1. For stable transfection, cells are plated at $1–1.5 \times 10^6$ cells per 10-cm dish, 12–24 hr before the transfection, on plates coated with 50 $\mu$g/ml poly-D-lysine (1 hr minimum) prior to use. Poly-D-lysine is used to pretreat the dish to prevent excessive loss of PC12 cells during the transfection procedure. For transient transfections, plate three times that number of cells.

2. For a transfection in a 10-cm dish, combine the DNA and DMEM to a final volume of 1.5 ml. The plasmid of interest should be 20 $\mu$g/ml in the final transfection mixture, and the plasmid with the selectable marker, if needed, should be 2.5 $\mu$g/ml.

*Note:* Muller *et al.* (1990) found maximal expression from the RSV promoter at plasmid concentrations greater than 10 $\mu$g/ml.

3. Prepare a second solution containing Lipofectin™ (final concentration in the transfection: 15 $\mu$g/ml) and DMEM in the same volume.

4. Wash the cells with DMEM, and aspirate any remaining medium.

5. Combine the DNA/DMEM solution and the Lipofectin™/DMEM solution, and add the transfection mixture to the cells.

6. Incubate the cells in the tissue culture incubator (37°C, 10% $CO_2$) for 3–5 hr.

7. At the end of the incubation, add an equal volume of normal PC12 medium to the transfection mixture, and allow the cells to grow overnight.

After the transfection, PC12 cells will be even less adherent than they normally are because of the effects of residual Lipofectin™ reagent. Growing the cells on plates treated with poly-D-lysine will help resolve this problem. The cells may be reseeded by trituration, pelleting, and replating as described for AtT-20 cells (see preceding protocol).

## E. Selection of Stable Cell Lines

If a plasmid containing a selectable marker was included in the transfection, stable transfectants can be isolated by growth of the transfected culture in the presence of selective media, followed by isolation, growth, and expansion of the resultant colonies. We have optimized the selection of AtT-20/F2 cells transfected with pSV2-neo, a plasmid that contains the neomycin resistance gene under the control of an early SV40 promoter, using geneticin sulfate (G418) (GIBCO/BRL) to supplement the selection medium. The level of expression of the neomycin resistance gene from the pSV2-neo plasmid is quite low. This plasmid is used, therefore, to increase further the probability that G418-resistant colonies also express the gene of interest. The optimum concentration for selection of AtT-20/F2 cells is 250 $\mu$g/ml of the active form of G418. For other subclones of AtT-20 cells, the optimal G418 concentration must be determined empirically; for example, we have found that 250 $\mu$g/ml G418 was not sufficient for selection of stable lines from an AtT-20/NIH subclone. We routinely make up a 10x solution of G418 in DMEM (either with or without 10% FBS), and dilute it as needed during the selection process. The selection procedures described here have been used successfully with cells transfected by either the Ca/$P_i$- or the Lipofectin™-based method.

## 1. Selection and Cloning of AtT-20/F2 Cells

1. The cells are split 1:2 into a 10-cm dish (or dishes) using STV 48 hr after the transfection. It is convenient to split them into a 10-cm dish to allow for greater ease during subsequent manipulations.

2. Selective medium (250 $\mu$g/ml active G418 in DMEM supplemented with 10% FBS) is added to the cells 12–16 hr after the split and is replenished every 3–4 days.

3. After 2 wk, G418-resistant colonies should be visible (1–2 mm in diameter). To isolate colonies, we use autoclaved glass cloning rings (8 mm high, 5 mm inner diameter) that have been inserted into a thin layer of silicone vacuum grease (Dow-Corning, Midland, MI).

4. Use 100 $\mu$l prewarmed STV to trypsinize the colony.

5. Transfer the contents of the cloning ring well to a multiwell cluster containing selective medium. We usually use 48- or 96-well dishes for the initial growth of the colony.

6. The colonies should be split into larger wells as they become confluent under continued selection with G418. If specific antibodies that recognize the gene of interest are available, or if the gene of interest has been modified by the addition of an epitope tag (see previous discussion), the selected colonies can be screened for positive expression by immunoblotting at the 24- or 12-well stage.

7. Growth after expansion can be maintained in selective medium that contains half the G418 concentration used in the initial selection. Positive colonies can be frozen at the 60-mm dish stage, using DMEM supplemented with 10% dimethylsulfoxide (DMSO) and 50% FBS.

## 2. Selection and Cloning of PC12 Cells

Selection and isolation of stably transfected colonies of transfected PC12 cells is performed essentially as described for AtT-20/F2 cells, with the following differences.

1. Selective medium is first applied 3 days following the transfection. Therefore, transfected PC12 cells should be split 1:2 2 days after the transfection, with selective medium added 24 hr later. If a selectable marker such as the neomycin resistance gene from pSV2-neo is used, the concentration of active G418 used for selection is 300–350 $\mu$g/ml.

2. No STV is needed for replating the cells or picking colonies, since the cells should detach readily by trituration. Colonies should be frozen in medium containing 25% each horse serum and enriched calf serum, with 10% glycerol, in DMEM.

# IV. Vaccinia Virus-Based Expression

## A. Expression Using Recombinant Virus

Since production of recombinant vaccinia virus and infection of mammalian cells have been described elsewhere (see Chapter 7) and the detailed procedures for the infection of AtT-20 have been described (Hruby *et al.*, 1986), these methods will not be discussed here. However, an important factor to consider when these methods are used to study traffic through the regulated secretory pathway is the shut-down of host protein synthesis after viral infection (see Section II,B). We have determined the time courses of exogenous protein expression and host cell protein synthesis inhibition in both AtT-20 and PC12 cells to facilitate the design of experiments for examining the regulated secretory pathway in these two cell lines.

To examine the time course of viral protein expression, a recombinant vaccinia virus containing the cloned gene for HA epitope-tagged human Rab8, a low molecular weight GTP-binding protein, was used to infect AtT-20 and PC12 cells at an MOI of 20 plaque-forming units (pfu)/cell. Cells were harvested at specific times and the time course of expression was determined by examining the expression of HA epitope-tagged Rab8 by Western blotting with a monoclonal antibody specific for the HA epitope tag (Fig. 2). In both cell lines, the expression of HA-tagged Rab8 is activated as early as 2 hr postinfection, but the expression reaches a plateau at 4–6 hr postinfection. To examine the host

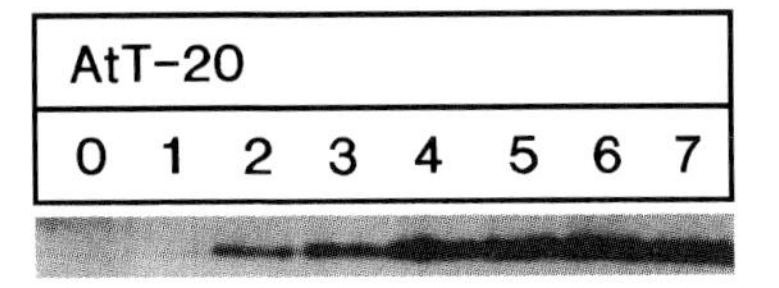


**Fig. 2** Time course of expression driven by the vaccinia 7.5 promoter in cultured neuroendocrine cells. A recombinant vaccinia virus carrying an influenza virus hemagglutinin (HA)-tagged small G protein (rab8) under the control of the p7.5 promoter was used to infect AtT-20 and PC12 cells. At the indicated times (hr postinfection), cell lysates were collected and resolved by SDS–PAGE (12.5% gel) and probed with the anti-HA monoclonal antibody 12CA5, after transferring to nitrocellulose membrane. The immunoblot was exposed with the ECL detection system (Amersham, Arlington Heights, IL).

cell protein synthesis profile during the course of viral infection, both AtT-20 and PC12 cells were infected with the recombinant virus used in this experiment. The infected cells were metabolically labeled, and whole cell lysates were collected and resolved by SDS–PAGE (Fig. 3). This analysis demonstrates that host protein synthesis diminishes significantly between 2 and 4 hr postinfection (see open arrows in Fig. 3), whereas viral products become clearly visible during this time period (filled arrows, Fig. 3). This expression time course agrees with cytoplasmic changes seen in infected cells (Moss, 1985; Sodeik *et al.*, 1993). Therefore, to study transport through the regulated secretory pathway, the optimal time window for the expression of the desired gene with minimal complications from the shut-down of host protein synthesis is between 2 and 4 hr postinfection.

We performed pulse–chase experiments to examine if the regulated secretory pathway is impaired during the 2- to 4-hr period after vaccinia virus infection. To assay the efficiency of the regulated secretory pathway in vaccinia virus-infected cells, Western Reserve strain virus (VR-119; ATCC) was used to infect AtT-20/F2 cells; the cells were labeled with [$^{35}$S]sulfate and chased in unlabeled

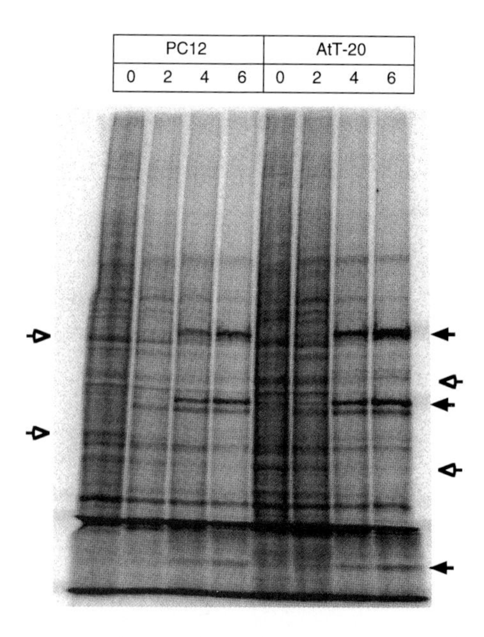


**Fig. 3** Time course of host and viral protein synthesis after vaccinia virus infection. AtT-20 and PC12 cells were infected with a recombinant vaccinia virus expressing HA-tagged rab8 and, at various time points, were starved in methionine-free medium for 30 min and labeled with [$^{35}$S]Met (1 mCi/ml) for 20 min. Whole-cell lysates were collected, resolved by SDS–PAGE (10% gel), and analyzed using a PhosphorImager (Molecular Dynamics, Sunnyvale, CA). Expression of viral products was evident by 2 hr postinfection (filled arrows), and the synthesis of several host proteins (open arrows) was drastically reduced between 2 and 4 hr after infection.

media for 45 min to allow unregulated secretion. The cells were then chased for an additional 45 min in the presence or absence of 8-bromo cyclic AMP, an ACTH secretagogue (see Fig. 4A for the two labeling protocols). We have found that POMC, the precursor of ACTH, is sulfated in infected cells and its processing products can be labeled by [$^{35}$S]sulfation (Fig. 4B). The processing of POMC into mature ACTH was determined by immunoprecipitation from the tissue culture medium and cell extracts, followed by resolution by SDS–PAGE (Fig. 4B). The labeling of POMC as well as the secretory response of AtT-20 cells to 8-bromo cyclic AMP was not significantly affected within 4 hr of infection with vaccinia virus. Collectively, these results indicate that the regulated secretory pathway is not impaired by vaccinia virus infection up to 4 hr postinfection, overlapping the time window in which virally driven protein expression has reached a high level (2–4 hr, see Figs. 2 and 3). This protocol is thus suitable for high-efficiency transient expression for studying various aspects of regulated secretion.

## V. Measurement of Transport through the Regulated Pathway

A major use of the transfection systems is to study features on transported proteins that direct them to the regulated secretory pathway. An important consideration in these studies is that sorting is not absolute. Depending on the cell line, a variable fraction (20–95%) of newly synthesized secretory proteins becomes sorted into regulated secretory granules. Thus, sorting of modified proteins must be measured quantitatively and must be compared with the unmodified proteins. We describe two protocols for measuring sorting efficiencies. These methods can be used to study the effects of mutations in the structural genes of secretory proteins on transport following transfection, or the consequences of expressing proteins that interfere with the endogenous sorting and secretory apparatus.

### A. Pulse–Chase Secretion Assay

Pulse–chase labeling can be used to study the sorting and processing of secretory proteins. To study secretion via the constitutive pathway, cells can be grown either in monolayer culture or in suspension culture after lifting them from the plates with EGTA during the pulse–chase period. However, protein packaging and processing in the regulated secretory pathway do not occur efficiently in suspended cells; transport through this pathway should therefore be studied with attached cells. Since PC12 cells may lift off the dish during frequent medium changes, they can be seeded onto poly-D-lysine-coated dishes for these experiments.

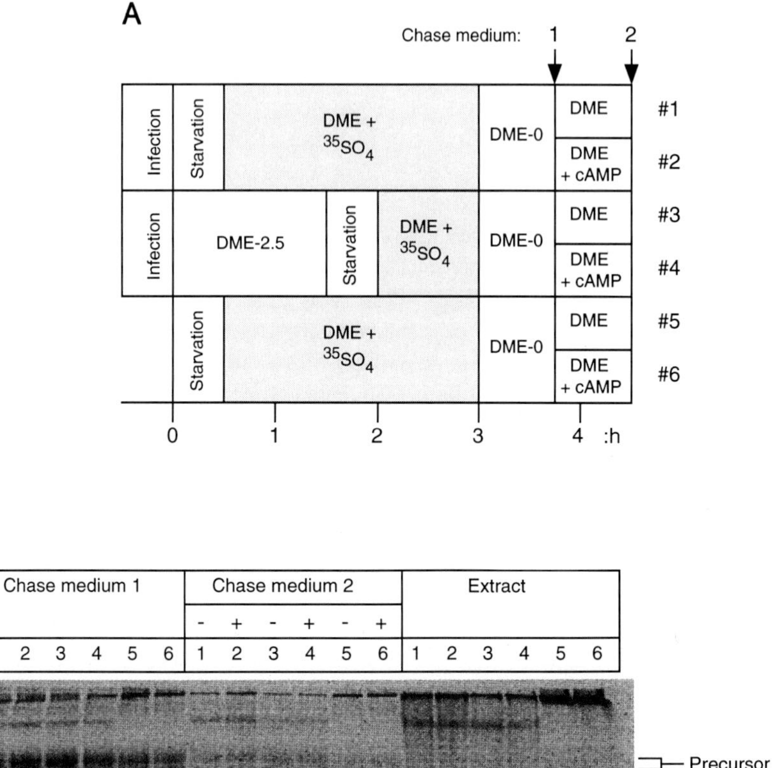


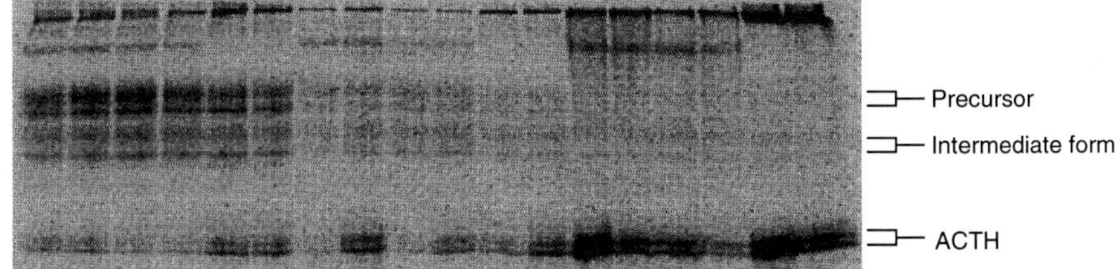


**Fig. 4** Effect of vaccinia virus infection on secretion of ACTH-related products from AtT-20 cells. (A) Label–chase protocols. Six dishes of AtT-20 cells seeded at equal densities were infected with vaccinia virus strain WR (ATCC# VR-119) at MOI of 20 pfu/cell for 30 min; then two different time courses of $[^{35}S]O_4$ labeling and chases were followed. For the first time course (dishes 1 and 2), immediately after the infection the cells were starved in sulfate-free DMEM for 30 min and labeled with DMEM containing $[^{35}S]O_4$ (0.5 mCi/ml) for 150 min. For the second time course (dishes 3 and 4), cells were kept in DMEM containing 2.5% FCS (DME-2.5) for 1.5 hr postinfection, and then were starved for 30 min in sulfate-free DMEM and labeled with DMEM containing $[^{35}S]O_4$ (0.5 mCi/ml) for 60 min. A parallel uninfected culture was labeled by the first method (dishes 5 and 6). The labeling was terminated by washing the cells with DMEM containing 5 m*M* unlabeled sodium sulfate. The labeled cells were then chased for 45 min in DMEM lacking FCS (DME-0) (chase medium 1), followed by a 45-min incubation in DMEM with or without 5 m*M* 8-bromo cyclic AMP (chase medium 2). At the end of the chases, the cells were extracted and both secreted materials and cell extracts were processed for analysis. (B) Regulated and nonstimulated release of sulfate-labeled POMC products from virally infected cells. Labeled POMC (precursor), intermediate cleavage products (intermediate form), and mature ACTH released into the medium, as well as that collected from cell extracts, were recovered by immunoprecipitation (Moore and Kelly, 1986) using an affinity purified rabbit antiserum against ACTH. Immunoprecipitates were resolved by SDS–PAGE (18% gel) and analyzed with a PhosphorImager.

### 1. Short-Term Pulse–Chase Protocol

The simplest method for measuring sorting efficiencies is to pulse-label the transported protein for a short time (5–15 min), and then to chase the fate of the labeled protein during periods up to 2–6 hr. One can take advantage of the differential turnover times in the constitutive and regulated pathways: constitutive secretion has a half-time of about 10 min between *trans*-Golgi network and the plasma membrane whereas proteins in the regulated secretory pathway have turnover rates of >10 hr. Thus, sorting between these two pathways can be determined by chasing the labeled cells for 2–6 hr, during which time proteins entering the constitutive pathway are secreted and those entering the regulated pathway are stored within the cells. To measure the amount actually stored in the regulated pathway, one can stimulate the cells with a secretagogue at the end of the chase period to induce secretion from the regulated secretory granules. The ratio of secretagogue-induced protein release and constitutive protein secretion during the early chase times then provides a measure of sorting efficiencies.

For studying transport through the entire secretory pathway, radiolabeled amino acids (e.g., [$^3$H]- and [$^{14}$C]-labeled amino acids, [$^{35}$S]methionine and -cysteine) can be used. If the protein of interest is modified by post-translational modification, incorporation of radiolabel into the modified moiety is often a convenient method to monitor a segment of transport events beyond the processing stage. For example, if the protein of interest is sulfated (e.g., POMC in AtT-20 cells), $Na_2[^{35}S]O_4$ can be used as the radiolabel. Most sulfation occurs in the TGN *trans*-Golgi network (Baeuerle and Huttner, 1987; Niehrs and Huttner, 1990); labeling at this point allows the monitoring of processes (such as sorting and proteolytic processing) that occur late in the secretory pathway without interference of early transport steps (see subsequent discussion). In addition, since few proteins are sulfated, [$^{35}$S]sulfate provides highly specific labeling; samples can often be loaded directly on gels without further purification by immunoprecipitation. This particular label can also be used for PC12 cells to examine sorting of secretogranins (see Section I). A typical pulse–chase protocol in AtT-20 or PC-12 cells follows.

1. The cells are first rinsed with PBS, and starved for 15–30 min in medium deficient in the appropriate radiolabeling moiety (e.g., amino acids, sulfate). For experiments with [$^{35}$S]sulfate labeling, we use a sulfate starvation medium, SBM (10 m*M* $MgCl_2$, 5.55 m*M* glucose, 110 m*M* NaCl, 5.4 m*M* KCl, 0.9 m*M* $Na_2HPO_4$, 2 m*M* $CaCl_2$, 20 m*M* Na-HEPES, pH 7.2). For sulfation starvation medium, this gives much higher labeling efficiencies than DMEM lacking sulfate.

2. The cells are then pulse-labeled at 37°C in a minimum volume of pulse medium (starvation medium supplemented with 0.2–2 mCi label/ml). For short pulse periods (1–15 min depending on the level of expression and the label used), this is best achieved by floating the dish on a 37°C water bath.

3. The labeled cells are then chased in unlabeled medium at 37°C for two consecutive 45-min periods. At the end of each chase interval, the chase media are collected and analyzed to quantify the amount of protein secreted by the constitutive pathway. Many proteins (especially genetically altered mutant proteins) exit the endoplasmic reticulum (ER) slowly because of inefficient folding. In these cases, the chase time may be increased to up to 6 hr.

4. The cells are then stimulated with an appropriate secretagogue for 30 min to 3 hr. For AtT-20 cells, the most effective secretagogue is 8-bromo cAMP dissolved in DMEM at a concentration of 5 m*M*.

5. Media samples are collected, and the cells are extracted with NDET buffer (1% Nonidet P-40, 0.4% deoxycholate, 66 m*M* EDTA, 10 m*M* Tris, pH 7.4).

6. Particulate materials are removed by centrifugation in a microfuge at 10,000 rpm for 10 min.

7. ACTH or the proteins of interest from the media and extract samples can be purified by single or double immunoprecipitation as described elsewhere (Moore and Kelly, 1985,1986).

#### *a. Modifications for PC12 cells*

For PC-12 cells, the chase medium in Step 3 (low K chase medium) consists of 127 m*M* NaCl, 5 m*M* KCl, 0.33 m*M* $Na_2HPO_4$, 0.44 m*M* $KH_2PO_4$, 4.2 m*M* $NaHCO_3$, 20 m*M* HEPES, pH 7.2, 2.2 m*M* $CaCl_2$, 5.6 m*M* glucose. For stimulation by depolarization (Step 4), the concentration of KCl is raised to 55 m*M* and the concentration of NaCl is decreased to 77 m*M*. Since depolarization occurs quickly, the stimulation period can be as short as 10 min.

### 2. Long-Term Label–Chase Protocol

The preceding protocol is suitable for situations in which the amount of label incorporated into the proteins in a 15-min pulse period is high enough for detection. In many cases, however, the level of the transfected gene products is low because of protein instability or low transcription or translation efficiencies. In these cases, a longer labeling time is needed. We have used 16-hr labeling followed by three consecutive 3-hr chases (last one with secretagogue) to measure sorting of proteins expressed at low levels. For these long labeling periods, the labeling medium should be supplemented with 1–2% FBS so the cells remain healthy. Long-term labeling may also deplete the amino acids or other precursors used for the label. Thus, the labeling medium may be supplemented with a small amount of normal DMEM (5–10% of the total volume).

The long-term labeling protocol presents a slight complication in the measurement of sorting efficiencies. Since the half-life of proteins stored in the regulated pathway is in excess of 10 hr during the 16-hr labeling period incorporation of label into regulated secretory proteins will increase with time. In contrast, incorporation of label into most proteins in the constitutive pathway will reach a plateau within a few hours. Thus, the "sorting index" (measured by the ratio of labeled proteins secreted in the presence of secretagogue and those secreted constitutively during the first two chases) is biased in favor of the regulated pathway in the long-term labeling protocol. Therefore, the exact numbers obtained by this method should not be directly compared with those obtained by the short-term label protocol. Furthermore, since the ER exit rates differ significantly among different proteins, the ER pool sizes also differ significantly. In a long-term label, for instance, a protein that exists the ER slowly will exhibit a higher ratio of constitutive to regulated secretion than rapidly exiting protein because of its larger labeled ER pool. This will produce an apparently lower sorting efficiency (for the regulated pathway).

We have devised a protocol for long-term label that minimizes the problem of different ER pool sizes (Moore and Kelly, 1985). In brief, cells are labeled to steady state for 16 hr. To measure constitutive secretion, the labeling medium is removed and replaced with an aliquot of fresh labeling medium and incubated at 37°C for 1 hr. Labeled proteins released into the medium during this hour are collected. This medium, rather than the products of secretion collected during chases, is then used to quantify constitutive release. Since secretion is measured in the continuous presence of the label, it represents the flux rate rather than the pool size. The labeled cells are then chased for two 3-hr periods and the media are discarded. Regulated release is then measured by stimulating the cells for 3 hr. The secretagogue-induced secretion during the last 3-hr chase period is divided by constitutive secretion during the 1-hr labeling period to yield an index for sorting.

## B. Fractionation of Secretory Granules

Another method for measuring sorting efficiencies is to isolate mature secretory granules from transfected cells. The amount of protein targeted to the granules can then be quantified directly. To purify secretory granules from these cells, we have used a method modified from that reported by Gumbiner and Kelly (1981). Using a combination linear and step gradient of $D_2O$ and Ficoll as the separation medium, fractions containing secretory granules can be isolated from cell lysates by equilibrium centrifugation. Typically, the recovery of mature ACTH in the purified fractions is on the order of 1–4% of the total ACTH in the initial homogenate.

The most critical factor in the purification of secretory granules is the homogenization procedure. We use an EMBL cell homogenizer (European Molecular Biology Laboratory, Heidelberg, Germany) composed of a stainless steel cylindrical chamber with Luer ports at either end and containing a ball bearing. The cell suspension is forced through the chamber and homogenization is effected by the shear forces created within the small clearance between the bearing and the chamber wall. Bearings of various sizes are provided with the homogenizer when purchased, providing flexibility in the amount of clearance. However, because of the variability of manufacture, it is necessary to empirically determine the ball bearing that gives the correct degree of homogenization for this procedure. Conditions for each homogenizer (e.g., cell density in the suspension, bearing size, number of passages through the apparatus) should be optimized to provide 80–90% intact free nuclei.

## 1. Solutions Needed

5X salts: 1.25 *M* sucrose, 50 m*M* HEPES, 100 m*M* KCl, 5 m*M* EDTA (pH 7.4); adjust pH with 10 *N* NaOH in $D_2O$

9.5% (w/v) Ficoll:40% $F_2O$: 20 ml 5X salts, 9.5 g Ficoll, 20 ml $D_2O$, $H_2O$ to 100 ml

12% (w/v) Ficoll:100% $D_2O$: 20 ml 5X salts, 12 g Ficoll, $D_2O$ to 100 ml

20% (w/v) Ficoll:100% $D_2O$: 20 ml 5X salts, 20 g Ficoll, $D_2O$ to 100 ml

PBS–EGTA: $Ca^{2+}/Mg^{2+}$-free PBS containing 4 m*M* EGTA

Homogenization buffer: 250 m*M* sucrose, 2m*M* EGTA, 1 m*M* EDTA, 10 mM HEPES, pH7.0.

## 2. Procedure

1. Grow cells in two 15-cm dishes to 80–90% confluence. The yield of granules depends critically on the state of the cells; cells that grow upon each other to form clumps tend to give low yields.

2. Wash the cells twice with PBS-EGTA; then add 10 ml PBS-EGTA to each dish.

3. Incubate 10 min in tissue culture incubator to detach the cells from the dish. Transfer the cells from the dish to a 50-ml conical tube (Falcon or equivalent) on ice.

4. Pellet the cell suspension using a Beckman GPR centrifuge (or other low-speed centrifuge) at 400 *g* for 5 min at 4°C and remove the supernatant.

5. Gently resuspend the pellet by swirling in 10 ml ice-cold homogenization buffer and centrifuge a second time. Finally, thoroughly resuspend the cells in 15 ml cold homogenization buffer.

6. Prechill the homogenization apparatus on ice, and rinse once with cold homogenization buffer.

7. Withdraw 5 ml cell suspension into a 5-ml syringe and attach the syringe to a Luer port on the apparatus; attach an empty syringe to the second Luer port. Pass the cell suspension through the homogenizer for 6–8 strokes (i.e., each direction is one stroke).

8. Check the homogenate for proper breakage of the cells (see preceding discussion).

9. Transfer the homogenate into a pre-chilled Sorvall SS34 centrifuge tube. Repeat the homogenization procedure until all the cells have been homogenized.

Pellet the combined homogenates in a Sorvall RC5B centrifuge (SS34 rotor) at 6,300 *g* (9,000 rpm) at 4°C for 4 min, with brake.

10. Transfer the postnuclear supernatant into another SS34 tube, and recentrifuge the supernatant at 20,000 *g* (16,000 rpm) at 4°C for 40 min.

11. Resuspend the resultant pellet (P16k) in 1 ml homogenization buffer by trituration with a pipetman.

### 3. Gradient Preparation

The $D_2O$/Ficoll gradient used in this granule isolation procedure consists of a linear gradient (9.5% Ficoll:40% $D_2O$–12% Ficoll:100% $D_2O$) layered above a Ficoll:100% $D_2O$ step gradient. The step gradient varies from 12 to 20% Ficoll in 2% increments. The 14 to 18% solutions are prepared from the 12 and 20% solutions.

1. At the bottom of a polyallomer centrifuge tube (Beckman, Fullerton, CA; 25 × 89 mm) on ice, the step gradient is layered as follows: 1 ml 20% (w/v) Ficoll:100% $D_2O$; 2 ml each (successively) 18% (w/v) Ficoll:100% $D_2O$, 16% (w/v) Ficoll:100% $D_2O$, and 14% (w/v) Ficoll:100% $D_2O$; 1 ml 12% (w/v) Ficoll:100% $D_2O$.
2. The linear gradient is formed using a linear gradient maker and a peristaltic pump with equal volumes (14.5 ml each) of the 9.5% (w/v) Ficoll:40% $D_2O$ and the 12% (w/v) Ficoll:100% $D_2O$ solutions.
3. The resuspended P16k solution is layered atop the gradient, and centrifuged in a Beckman SW28 rotor at 25,000 rpm (82,740 *g*) for 12–16 hr at 6°C.
4. After the run, a diffuse cloudy band can be detected near the middle of the gradient by visual inspection. This band, high in protein content, is composed of ER and Golgi membranes, and other membranes that contain immature forms of ACTH (i.e., POMC and other ACTH processing intermediates). Collect 1.7-ml fractions from the bottom by puncturing the centrifuge tube with a 20-gauge needle.
5. The mature granules containing only the mature forms of ACTH should be recovered in the bottom third of the gradient, between fractions 4 and 6. This can be detected by Western blotting with an affinity-purified anti-ACTH antibody or by an ACTH radioimmunoassay (Moore *et al.*, 1983). Fractions can then be assayed as desired for the transfected gene product. The membranes in these fractions can be pelleted by combining 1 volume of the fraction with 2 volumes cold PBS, and centrifuging for 1 hr at 100,000 *g* at 4°C.

## VI. Overview and Conclusions

During the last 10 years, transfection studies have yielded important data about the cellular processes and mechanisms governing protein sorting, transport, and processing through the regulated secretory pathway. These include substrate specificities of endoproteases, cellular environment and cell type specificity for these enzymes, structural features for granular localization, factors controlling protein sorting, and budding and fusion of secretory granules. Undoubtedly these systems will continue to be valuable tools for such studies, as well as for other unexplored issues such as the control of the volume and number of secretory granules produced, and maintenance of the sizes of intracellular organelles and the plasma membrane. We hope that the ability to manipulate cells by expressing foreign genes will make these inquiries possible in the near future.

## Acknowledgments

The authors thank members of the Moore laboratory for critical reading of the manuscript. This work was supported by grants from the Public Health Service (GM 35239), the American Cancer Society (CD-497, CB-89A), the Lucille P. Markey Charitable Fund, and the Muscular Dystrophy Association to H.-P. Moore, a Howard Hughes Predoctoral Fellowship to W. K. Schmidt, an NSF Predoctoral Fellowship to L. Carnell, and an NSF Postdoctoral Fellowship to R. A. Chavez.

## References

Baeuerle, P. A., and Huttner, W. B. (1987). Tyrosine sulfation is a trans-Golgi-specific protein modification. *J. Cell Biol.* **105,** 2655–2664.

Benjannet, S., Rondeau, N., Paquet, L., Boudreault, A., Lazure, C., Chretien, M., and Seidah, N. G. (1993). Comparative biosynthesis, covalent post-translational modifications and efficiency of prosegment cleavage of the prohormone convertases PC1 and PC2: Glycosylation, sulphation and identification of the intracellular site of prosegment cleavage of PC1 and PC2. *Biochem. J.* **294,** 735–743.

Bennett, D. L., Bailyes, E. M., Nielsen, E., Guest, P. C., Rutherford, N. G., Arden, S. D., and Hutton, J. C. (1992). Identification of the type 2 proinsulin processing endopeptidase as PC2, a member of the eukaryote subtilisin family. *J. Biol. Chem.* **267,** 15229–15236.

Brion, C., Miller, S., and Moore, H.-P. H. (1992). Regulated and constitutive secretion: Differential effects of protein synthesis arrest on transport of glycosaminoglycan chains to the two secretory pathways. *J. Biol. Chem.* **267,** 1477–1483.

Burgess, T. L., and Kelly, R. B. (1984). Sorting and secretion of adrenocorticotropin in a pituitary tumor cell line after perturbation of the level of a secretory granule-specific proteoglycan. *J. Cell Biol.* **99,** 2223–2230.

Burgess, T. L., and Kelly, R. B. (1987). Constitutive and regulated secretion. *Annu. Rev. Cell Biol.* **3,** 243.

Castle, A. M., Stahl, L. E., and Castle, J. D. (1992). A 13-amino acid N-terminal domain of a basic proline-rich protein is necessary for storage in secretory granules and facilitates exit from the endoplasmic reticulum. *J. Biol. Chem.* **267,** 13093–13100.

Chanat, E., and Huttner, W. B. (1991). Milieu-induced, selective aggregation of regulated secretory proteins in the trans-Golgi network. *J. Cell Biol.* **115,** 1505–1519.

Chen, Y.-T., Holcomb, C., and Moore, H.-P. (1993). Expression and localization of two small-molecular weight GTP-binding proteins, rab8 and rab10, by epitope tag. *Proc. Natl. Acad. Sci. USA* **90,** 6508–6512.

Evan, G. I., Lewis, G. K., Ramsay, G., and Bishop, M. J. (1985). Isolation of monoclonal antibodies specific for human c-*myc* proto-oncogene product. *Mol. Cell Biol.* **5,** 3610–3616.

Fuerst, T. R., Earl, P. L., and Moss, B. (1987). Use of a hybrid vaccinia virus-T7 RNA polymerase system for expression of target genes. *Mol. Cell Biol.* **7,** 2538–2544.

Gerdes, H. H., Rosa, P., Phillips, E., Baeuerle, P. A., Frank, R., Argos, P., and Huttner, W. B. (1989). The primary structure of human secretogranin II, a widespread tyrosine-sulfated secretory granule protein that exhibits low pH- and calcium-induced aggregation. *J. Biol. Chem.* **264,** 12009–12015.

Graham, F., and van der Eb, A. (1973). A new technique for the assay of infectivity of human adenovirus 5 DNA. *Virology* **52,** 456–467.

Greene, L. A., and Tischler, A. S. (1976). Establishment of a noradrenergic clonal line of rat adrenal pheochromocytoma cells which respond to nerve growth factor. *Proc. Natl. Acad. Sci. USA* **73,** 2424–2428.

Grodsky, G. M., Ma, Y. H., Cullen, B., and Sarvetnick, N. (1992). Effect on insulin production, sorting and secretion by major histocompatibility complex class II gene expression in the pancreatic beta-cell of transgenic mice. *Endocrinology* **131,** 933–938.

Gross, D. J., Halban, P. A., Kahn, C. R., Weir, G. C., and Villa-Kormaroff, L. (1989). Partial diversion of a mutant proinsulin (B10 aspartic acid) from the regulated to the constitutive secretory pathway in transfected AtT-20 cells. *Proc. Natl. Acad. Sci. USA* **86,** 4107–4111.

Gumbiner, B., and Kelly, R. B. (1981). Secretory granules of an anterior pituitary cell line, AtT-20, contain only mature forms of corticotropin and beta-lipotropin. *Proc. Natl. Acad. Sci. USA* **78,** 318–322.

Gumbiner, B., and Kelly, R. B. (1982). Two distinct intracellular pathways transport secretory and membrane glycoproteins to the surface of pituitary tumor cells. *Cell* **28,** 51–59.

Hruby, D. E., Thomas, G., Herbert, E., and Franke, C. A. (1986). Use of vaccinia virus as a neuropeptide expression vector. *Meth. Enzymol.* **124,** 295–309.

Linstedt, A. D., and Kelly, R. B. (1991). Synaptophysin is sorted from endocytotic markers in neuroendocrine PC12 cells but not transfected fibroblasts. *Neuron* **7,** 309–317.

Milgram, S. L., Mains, R. E., and Eipper, B. A. (1993). COOH-terminal signals mediate the trafficking of a peptide processing enzyme in endocrine cells. *J. Cell Biol.* **121**(1), 23–36.

Miller, S. G., and Moore, H.-P. H. (1990). Regulated secretion. *Curr. Opin. Cell Biol.* **2,** 642–647.

Moore, H.-P. H., and Kelly, R. B. (1985). Secretory protein targeting in a pituitary cell line: Differential transport of foreign secretory proteins to distinct secretory pathways. *J. Cell Biol.* **101,** 1773–1781.

Moore, H.-P. H., and Kelly, R. B. (1986). Rerouting of a secretory protein by fusion with human growth hormone sequences. *Nature (London)* **321,** 443–446.

Moore, H.-P. H., Gumbiner, B., and Kelly, R. B. (1983a). A subclass of proteins and sulfated macromolecule secreted by AtT-20 (mouse pituitary tumor) cells is sorted with adrenocorticotropin into dense secretory granules. *J. Cell Biol.* **97,** 810–817.

Moore, H.-P. H., Walker, M., Lee, F., and Kelly, R. B. (1983b). Expressing a human proinsulin cDNA in a mouse ACTH secretory cell. Intracellular storage, proteolytic processing and secretion on stimulation. *Cell* **35,** 531–538.

Moss, B. (1985). Replication of poxviruses. *In* "Virology" (B. N. Fields, ed.), pp. 685–698. Raven Press, New York.

Muller, S. R., Sullivan, P. D., Clegg, D. O., and Feinstein, S. C. (1990). Efficient transfection and expression of heterologous genes in PC12 cells. *DNA Cell Biol.* **9,** 221–229.

Ngsee, J. K., Fleming, A. M., and Scheller, R. H. (1993). A rab protein regulates the localization of secretory granules in AtT-20 cells. *Mol. Biol. Cell* **4,** 747–756.

Niehrs, C., and Huttner, W. B. (1990). Purification and characterization of tryosylprotein sulfotransferase. *EMBO J.* **9,** 35–42.

Ohagi, S., LaMendola, J., LeBeau, M. M., Espinosa, R., Takeda, J., Smeekens, S. P., Chan, S. J., and Steiner, D. F. (1992). Identification and analysis of the gene encoding human PC2, a prohormone convertase expressed in neuroendocrine tissues. *Proc. Natl. Acad. Sci. USA* **89,** 4977–4981.

Powell, S. K., Orci, L., Craik, C. S., and Moore, H.-P. H. (1988). Efficient targeting to storage granules of human proinsulins with altered propeptide domain. *J. Cell Biol.* **106,** 1843–1851.

Pryer, N. K., Wuestehube, L. J., and Schekman, R. (1992). Vesicle-mediated protein sorting. *Annu. Rev. Biochem.* **61,** 471–516.

Quinn, D., Orci, L., Ravazzola, M., and Moore, H.-P. H. (1991). Intracellular transport and sorting of mutant human proinsulins that fail to form hexamers. *J. Cell Biol.* **113,** 987–996.

Radvanyi, F., Christgau, S., Baekkeskov, S., Jolicoeur, C., and Hanahan, D. (1993). Pancreatic beta cells cultured from individual preneoplastic foci in a multistage tumorigenesis pathway: A potentially general technique for isolating physiologically representative cell lines. *Mol. Cell Biol.* **13,** 4223–4232.

Rivas, R. R., and Moore, H.-P. H. (1989). Spatial segregation of the regulated and constitutive secretory pathways. *J. Cell Biol.* **109,** 51–60.

Rosa, P., Hille, A., Lee, R. W. H., Zanini, A., De Camilli, P., and Huttner, W. B. (1985). Secretogranins I and II: Two tyrosine-sulfated secretory proteins common to a variety of cells secreting peptides by the regulated pathway. *J. Cell Biol.* **101,** 1999–2011.

Rothman, J. E., and Orci, L. (1992). Molecular dissection of the secretory pathway. *Nature (London)* **355,** 409–415.

Seed, B., and Aruffo, A. (1987). Molecular cloning of the CD2 antigen, the T-cell erythrocyte receptor, by a rapid immunoselection procedure. *Proc. Natl. Acad. Sci. USA* **84,** 3365–3369.

Sodeik, B., Coms, R. W., Ericsson, M., Hiller, G., Machamer, C. E., van't Hof, W., van Meer, G., Moss, B., and Griffiths, G. (1993). Assembly of vaccinia virus: Role of the intermediate compartment between the endoplasmic reticulum and the golgi stacks. *J. Cell Biol.* **121,** 521–541.

Sporn, L. A., Marder, V. J., and Wagner, D. D. (1986). Inducible secretion of large biologically potent von Willebrand factor multimers. *Cell* **46,** 185–190.

Stoller, T. J., and Shields, D. (1989). The propeptide of preprosomatostatin mediates intracellular transport and secretion of alphaglobin from mammalian cells. *J. Cell Biol.* **108,** 1647–1655.

Thomas, G., Thorne, B., Thomas, L., Allen, R., Hruby, D., Fuller, R., and Thorner, J. (1988). Yeast KEX2 endopeptidase correctly cleaves a neuroendocrine prohormone in mammalian cells. *Science* **241,** 226–230.

Wilson, I. A., Niman, H. L., Houghten, R. A., Cherenson, A. R., Connolly, M. L., and Lerner, R. A. (1984). The structure of an antigenic determinant in a protein. *Cell* **37,** 767–778.

Yu, H., Leaf, D. S., and Moore, H.-P. H. (1993). Gene cloning and characterization of a Rab protein from mouse pituitary AtT-20 cells. *Gene* **132,** 273–278.

# CHAPTER 14

# Expression of Foreign Proteins in a Human Neuronal System

**David G. Cook, Virginia M.-Y. Lee, and Robert W. Doms**

Department of Pathology and Laboratory Medicine
University of Pennsylvania Medical Center
Philadelphia, Pennsylvania 19104



## I. Introduction

### A. Background

As is evident from other chapters in this volume, there are numerous ways to introduce foreign genes into cells both *in vitro* and *in vivo,* including standard procedures such as calcium phosphate-, or liposome-mediated transfection, viral vectors of many sorts, and, more recently, gene gun transfection. In addition, once a gene is introduced into a cell, it may be expressed transiently,

METHODS IN CELL BIOLOGY, VOL. 43

stably, or under some sort of inducible promoter that provides a degree of regulation. Promoters can also be chosen to give a wide range of expression levels. However, the options are more limited when attempting to express a foreign protein in neurons. First, fully differentiated neurons are not amenable to transfection by standard techniques (Meichsner *et al.*, 1993). Second, obtaining a source of pure completely differentiated neurons is difficult. Culturing primary neurons is expensive and time consuming. Even under the best conditions they are heterogeneous, and it is difficult to obtain sufficient numbers of cells for biochemical experiments. As a result, studies of neuronal physiology, protein targeting, and polarity have been severely hampered by these limitations. Much of what is known about fundamental neuronal cell biology comes largely from morphological studies. In this chapter, we will describe a recently developed model human neuronal system that makes it possible to obtain unlimited quantities of fully polarized, postmitotic neurons, and approaches that can be taken to express exogenous proteins transiently as well as to obtain stable cell lines. Thus, this system offers a way to study dendritic and axonal transport, membrane recycling, and protein targeting at the genetic and biochemical levels.

### B. Neuronal Systems

Mature human neurons do not divide. As a result, studies on how proteins are processed in the central nervous system (CNS) have been largely limited to primary neuronal cultures and clonal cell lines that exhibit some neuronal characteristics, such as PC12 cells. These cell lines, however, more closely resemble embryonic neurons or neuroblasts than mature neurons. For example, they do not establish polarized neurites with axonal or dendritic characteristics. As a result, primary cultures must be used to study processes such as axonal transport.

Like polarized epithelial cells, neurons have at least two distinct plasma membrane domains—axonal and dendritic. Thus, neurons must use specific mechanisms to sort and maintain proteins in the right domain accurately. Relatively little is known about protein sorting in neurons because of the lack of a neuronal cell line that establishes polarity. However, evidence has been obtained that suggests that neurons may utilize at least some of the same mechanisms used by polarized epithelial cells. In polarized epithelial cells, the influenza hemagglutinin (HA) protein is normally delivered to the apical surface whereas the vesicular stomatitis virus (VSV) G protein is sorted to the basolateral surface (Rodriguez-Boulan and Nelson, 1989). Indeed, much of what is known about protein sorting in polarized epithelial cells is derived from studies on these two proteins. Using morphological techniques, Dotti and Simons found that G protein was sorted to dendrites whereas HA was preferentially transported to axons in primary rat hippocampal neurons (Dotti and Simons 1990; Dotti *et al.*, 1991). Furthermore, sorting appeared to occur prior to delivery to the cell surface. A diffusion barrier that exists at the axonal hillock appears to maintain the compositional differences between these domains (Kobayashi *et al.,* 1992). These studies were hampered, however, by an inability to perform biochemical experiments because of the low number of cells and contaminating cell types, and illustrates the limitations of using primary neuronal cultures to address fundamental cell biological questions.

## C. Attempts to Express Foreign Genes in Neurons

Even after a primary neuronal culture is obtained, it is difficult to introduce foreign genes with any degree of efficiency. Standard transfection procedures, such as calcium phosphate precipitation or liposome-mediated transfection, work best with mitotic cell lines and thus are inefficient when used on differentiated cells. Meichsner *et al.* (1993), for example, expressed an epitope-tagged form of MAP2c in cultured rat hippocampal neurons by calcium phosphate transfection. However, the frequency of transfection was low, with only 1 in 1000 cells expressing detectable levels of the epitope-tagged protein. Nevertheless, this was sufficient for morphological experiments. Clearly, much higher efficiencies must be obtained to contemplate biochemical experiments. A number of viral vectors have also been tried with some success. Retroviral vectors require mitotic cells for integration, and so are not useful for mature postmitotic neurons. However, vectors based on herpes simplex virus type 1 (HSV-1), which does not require integration, have been more successful. A variety of HSV-1 vectors have been employed to introduce genes into neurons *in vivo,* and may prove to be useful for gene therapy of neuronal diseases (Geller, 1991a,b; Ho *et al.,* 1993). A replication-deficient adenovirus vector has also been used to transfer a reporter gene into rat nerve cells both *in vivo* and *in vitro* (Le Eal La Salle *et al.,* 1993). In addition to the HSV-1 and adenovirus vectors tried to date, vaccinia virus vectors have also been used to express proteins in primary neuronal cultures *in vitro* (Karschin *et al.,* 1991). As we describe below, this vector provides an efficient way to express proteins in the NT2N model neuronal system.

## D. The NT2N System

To obtain efficient expression of a foreign gene in neurons, two limitations must be overcome: a convenient source of postmitotic, fully polarized neurons must be found and a way to introduce foreign genes that will result in efficient expression must be developed. In this section, we will describe the properties of the NT2N cell system. NT2N cells, derived from the NT2 human teratocarcinoma cell line by retinoic acid treatment, exhibit many of the features of fully polarized, postmitotic human CNS neurons (Pleasure *et al.,* 1992). Because they can be obtained in large quantities, biochemical studies can easily be performed. In addition, because they are derived from a mitotically active cell line, a variety of approaches that would otherwise be unavailable when working with primary neuronal cultures may be taken to obtain expression of foreign proteins.

## II. NT2N Cells: A Model Human Neuronal System

### A. Properties of NT2N Cells

NT2N cells are derived from Ntera 2/D1 (NT2) cells, a human teratocarcinoma cell line that can be committed to a CNS neuronal lineage by retinoic acid treatment (Andrews, 1984; Pleasure *et al.*, 1992; Pleasure and Lee, 1993). Induction of the mature neuronal phenotype takes approximately 4–5 wk. The resulting NT2N cells are postmitotic and can be made >99% pure by sequential replating over the course of an additional 4 wk. NT2N cells share many features associated with primary neuronal cultures. They are postmitotic and maintain a stable phenotype for long periods. Because they express alpha-internexin but not peripherin or a high molecular weight form of tau associated with the peripheral nervous system (PNS), their characteristics most closely resemble those of CNS rather than PNS neurons (Pleasure *et al.*, 1992). Both *N*-methyl-D-aspartate (NMDA) and non-NMDA glutamate-receptor channels are expressed by NT2N cells and, like primary cultured neurons, they exhibit glutamate-induced neurotoxicity (Younkin *et al.*, 1993). Finally, NT2N cells express almost exclusively APP695, the CNS isoform of the amyloid precursor protein from which $\beta$-amyloid is derived (Wertkin *et al.*, 1993). The stable neuronal phenotype of NT2N cells and the fact that they are derived from NT2 cells, which can be easily grown in large quantities, will make it possible to study the transport, distribution, structure, and function of neuronal proteins at both the biochemical and the morphological level.

One of the most striking manifestations of the mature neuronal phenotype is the development of polarized processes, typically a single axon and multiple dendrites. Axons appear as long untapering processes whereas dendrites are thick at the cell base and grow progressively thinner. During the course of retinoic acid treatment, NT2N cells exhibit progressive development of an extensive neuritic network composed of processes that are morphologically similar to axons and dendrites. The distribution of several markers has shown that the processes of NT2N cells have many of the properties of differentiated axons and dendrites. For example, processes with an axonal morphology are positive for highly phosphorylated neuro-filament (middle molecular weight) protein (NF-M) whereas the dendritic processes are largely negative for this marker. The dendritic processes are positive for hypophosphorylated forms of NF-M and MAP2 whereas the axonal processes are negative for these markers. In addition, ribosomes can be seen in processes with dendritic morphology but are excluded from processes with axonal morphology. Thus, NT2N cells express axons and dendrites as judged by several criteria, making it possible for the first time to study the development and consequences of neurite polarity without the technical limitations imposed by primary neuronal cultures.

A particularly interesting feature of NT2N cells is that the growth of polarized neurites can be controlled. When essentially pure cultures of NT2N cells are enzymatically detached and replated, they once again elaborate axons and dendrites. The growth of these polarized processes is dependent on the tissue culture substrate used. Processes grow rapidly on poly-D-lysine plus laminin or on Matrigel (Becton Dickinson, Bedford, MA), but not on poly-D-lysine alone. Thus, by replating NT2N cells on the appropriate substrate, it is possible to synchronize axonal and dendritic development. NT2N cells can therefore be used as models to study the development of polarized neurites as well as to study how proteins are differentially sorted to axons, dendrites, and the plasma membrane of the cell body.

## B. Advantages and Disadvantages

The most significant advantage of NT2N cells over other model neuronal systems is that nearly pure, fully polarized, postmitotic neurons can be generated from an undifferentiated teratocarcinoma cell line. As a consequence, essentially pure neurons can be obtained in nearly unlimited quantities, making it possible to perform biochemical experiments. Previous studies on protein sorting in neurons, for example, have had to rely almost exclusively on morphological techniques because of inadequate cell numbers, coupled with the presence of contaminating nonneuronal cells. Because fully polarized NT2N cells are derived from retinoic acid treatment of an undifferentiated precursor (NT2 cells), transfection and selection procedures (described in the first part of this volume) can be used to generate a stable cell line expressing a protein of interest prior to neuronal differentiation, making it possible to obtain neurons that stably express a foreign protein. In addition, transient expression systems using viral vectors can be used. A further advantage of NT2N cells is that they can be grafted into the brains of nude mice where they survive for months. The fact that no tumors develop following microinjection is further evidence that the NT2N cells are postmitotic.

The most significant drawback of the NT2N system is the amount of time required to obtain fully differentiated, postmitotic neurons of sufficient purity—generally 6–8 wk following the beginning of induction—and the large amount of tissue culture work involved as a result. Nevertheless, by inducing a new lot of cells each week, a steady supply of neurons can be obtained. Furthermore, once fully differentiated, the NT2N cells are quite stable and can be kept in culture for several weeks prior to use. In addition, the final purification step can be omitted for all but the most critical experiments, with a significant savings in time.

## C. Growth and Culture

NT2N cells are derived from NT2 cells by retinoic acid treatment and purified from nonneuronal cells by three sequential replating steps. Basically, NT2 cells are treated with retinoic acid for 5 wk, after which they are transferred to a new flask (Replate 1). The neuronal cells (NT2N) grow atop the more tightly adherent nonneuronal cell layer, making it possible to enrich for NT2N cells selectively by two additional replating steps. The procedures are described here and photomicrographs of NT2 and NT2N cells at different stages are shown in Fig. 1.

### *Special Materials*

Retinoic acid: 1000-fold stock in dimethylsulfoxide (DMSO), stored at −80°C (1000× stock), made monthly

5-Fluoro-2′-deoxyuridine: 1 m*M* stock in $dH_2O$, stored at −80°C (100× stock)

Uridine: 1 m*M* stock in $dH_2O$, stored at −80°C (100× stock)

Cytosine-β-D-arabinofuranoside: 1 m*M* stock in $dH_2O$, stored at −80°C (1000× stock)

Poly-D-lysine: 1 mg/ml in $dH_2O$, stored at −20°C (100× stock)

Matrigel (Becton Dickinson, Cockeysville, MD): Comes in 10-ml bottles; thaw on ice and make 0.5-ml aliquots that are stored at −20°C; when needed, thaw overnight at 4°C

All of these materials, except Matrigel, are from Sigma (St. Louis, MO).

### *NT2 Cells*

1. NT2 cells may be obtained from the American Type Culture Collection (CRL1973; ATCC, Rockville, MD).
2. Maintain NT2 cells in Dulbecco's modified Eagle's medium (DMEM) (high glucose; DMEMHG) with 10% fetal calf serum (FCS).
3. To split, wash with phosphate-buffered saline (PBS) before adding 1 ml trypsin–EDTA.
4. Split 1 : 3.

*Note:* Undifferentiated NT2 cells look like normal epithelial cells (Fig. 1). We do not carry these cells more than 75 passages. Also, we have used bovine calf serum in place of FCS with good success.

### *Differentiation*

To induce differentiation:

1. Plate 2–2.5 × $10^6$ NT2 cells into a 75-$cm^2$ flask.
2. Allow cells to grow for 1 day.
3. Replace media with DMEMHG, 10% FCS containing 10 μ*M* retinoic acid.
4. Change media twice a week, always with retinoic acid.

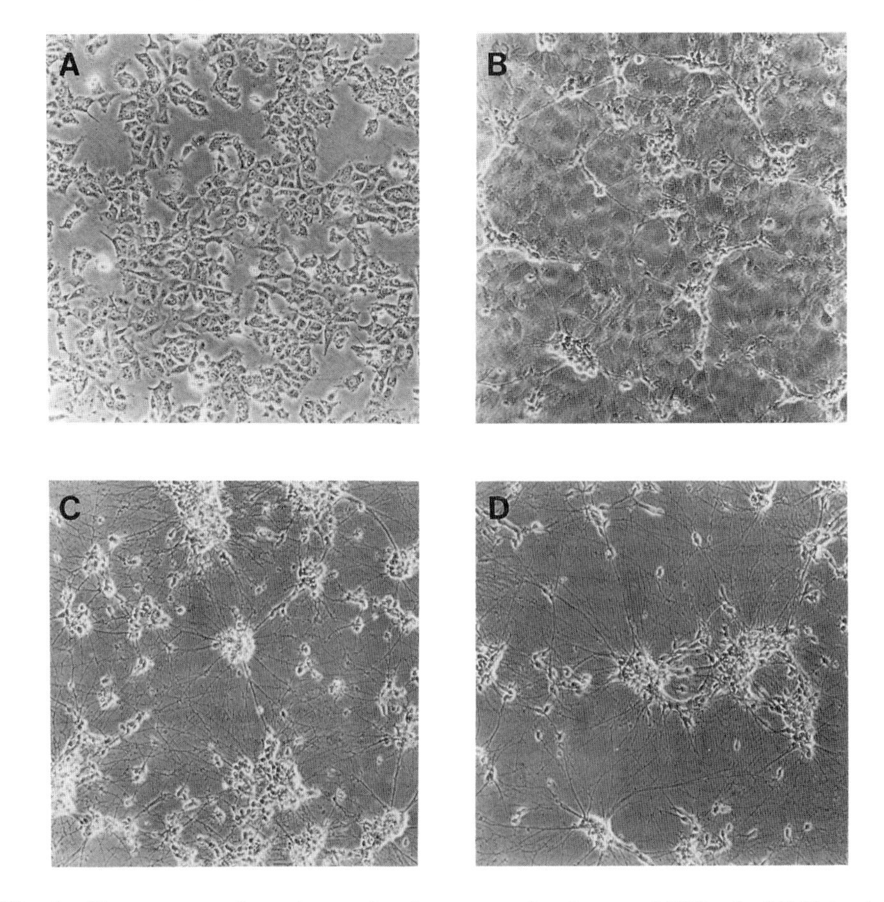


**Fig. 1** Phase contrast photomicrographs of nonneuronal and neuronal NT2 cells. (A) Untreated nonneuronal NT2 cells. (B) Mixed "Replate II" culture. Neurons are phase bright cells growing on monolayer of nonneuronal cells. (C) Neuronal "Replate III" culture from which virtually all nonneuronal cells have been removed by the replating procedure and treatment with mitotic inhibitors. (D) Replate III neuronal culture 4 days after infection with recombinant vaccinia virus vPE16 (Earl *et al.*, 1990) expressing human immunodeficiency virus type 1 (HIV-1) envelope glycoprotein gp160. Postmitotic NT2N cells display a very limited cytopathic response to vaccinia virus infection.

5. Allow to grow and differentiate for 5 wk (a total of 10 retinoic acid treatments).

*Note:* The media should begin to look acidic by the second or third retinoic acid treatment. The cells begin to stack up (see Fig. 1). Generally, we start treating 1–2 T75 flasks of NT2 cell with retinoic acid per week.

***Replate 1 Procedure***

1. Wash NT2 cells (in a 75-$cm^2$ tissue culture flask) with PBS.
2. Add 2 ml trypsin–EDTA.
3. Knock flask to dislodge cells.
4. Add 8 ml DMEMHG, 10% FCS with PBS and resuspend by pipetting.
5. Place 5 ml resuspended cells into each of two 225-$cm^2$ flasks; add additional 45 ml media.
6. Allow cells to grow for 2 days.

*Note:* After 2 days, it will be evident that neuronal NT2N cells are growing atop a confluent monolayer of nonneuronal, epithelial-like cells (Fig. 1). This will make it possible to enrich for neuronal cells by the sequential replating steps described next.

***Replate 2 Procedure***

1. Remove media from the 225-$cm^2$ flask (save this conditioned media because it will be used later to feed Replate 3 cells).
2. Rinse cells with PBS (10 ml per flask).
3. Add 4 ml trypsin–EDTA for 1–5 min at room temperature.
4. Knock flask to dislodge neuronal cells, checking often to be sure that the underlying, more tightly adherent, nonneuronal cell layer is not lifting off (see *Note* below). When most neuronal cells have been dislodged, stop the digestion by adding 8 ml DMEMHG, 10% FCS.
5. Remove medium and pellet cells 5 min at 1000 rpm.

*Optional:* Add an additional 10 ml medium to the flask to collect additional cells, but process separately since they usually contain a greater proportion of nonneuronal cells.

6. Resuspend cells in 10 ml medium containing 10 $\mu M$ 5-fluoro-2′-deoxyuridine (FUDR), 10 $\mu M$ uridine, 1 $\mu M$ cytosine-$\beta$-D-arabinoside (Ara C). These inhibitors are kept in the medium through all subsequent steps.
7. Count cells and seed as desired:
   6-cm dish: $2.5 \times 10^6$
   10-cm dish: $9–10 \times 10^6$
   T75 flask: $25 \times 10^6$
   T225 flask: $70–80 \times 10^6$
8. Feed cells twice a week with media containing the inhibitors, but using 5% FCS rather than 10%.

*Note:* After several weeks, the NT2N cells are seen growing individually and in clumps, with processes extending out for great distances (Fig. 1). Nonneuronal cells appear as flat epithelial-like cells. In a poor Replate 2 preparation, a nearly intact monolayer of nonneuronal cells will be present, whereas in a good preparation only occasional epithelial cells are seen. These preparations can be used for most experiments. Obviously, the most critical parameter in the production of NT2N cells is the extent of trypsinization and ''knocking'' employed in generating the Replate 2 cells. By dislodging the cells less vigorously, more pure preparations of NT2N cells can be obtained. With some practice, cultures containing approximately 95% NT2N cells can be obtained routinely. These cells can be maintained in culture for at least 10 wk. After 3 wk in culture, the cells can be maintained in 50% DMEMHG, 5% FCS and 50% conditioned medium from the Replate 1 cells. This conditioned medium should first be clarified by centrifugation and filter sterilized. Inhibitors can be omitted from the medium once the culture is free of nonneuronal cells.

***Replate 3 Procedure***

1. Grow Replate 2 cells for 3–4 wk or until the culture is composed largely of NT2N cells.
2. Wash twice with PBS.
3. Add 1 ml trypsin–EDTA (for a T75) or 2 ml (for a T225).
4. Knock flask occasionally to dislodge neuronal cells (as for making Replate 2 cells). Watch carefully to make sure nonneuronal cells are not being dislodged.
5. Stand flask on end for 1 min to allow cells to drain to bottom.
6. Add 8 ml medium with 5% FCS.
7. Pellet cells 5 min at 1000 rpm.
8. Remove supernatant and resuspend cells in 1 ml medium with inhibitors. Carefully break up the cell pellet.
9. Because the Replate 2 cells typically grow in clumps, cells are generally not counted at this stage. Instead, one T75 flask of Replate 2 cells can be used to generate two 6-cm dishes of Replate 3 cells or approximately 24 microscope cover slips in a 24-well plate.

*Note:* We use Replate 3 cells only for the most critical experiments, in which the purest NT2N preparations are required (Fig. 1). These cells can be maintained for 2 mo, or they can be frozen and used at a later date. Neurite growth can be controlled as described next.

***Matrigel and Poly-D-Lysine***

Neurite growth is controlled by altering the matrix on which the Replate 3 cells are placed. If the cells are replated only on poly-D-lysine, neurites do not grow to any appreciable extent. However, if cells are replated on Matrigel, extensive neurite growth is evident over the course of several days. The following description is for a 6-cm dish of cells.

1. Add 3 ml poly-D-lysine at a final concentration of 10 $\mu$g/ml in water.
2. Let sit overnight in the hood with the lids on the plates.
3. Remove excess and allow to dry.
4. When dry, add 1–3 ml Matrigel (Matrigel is first diluted 1 : 36 in DMEMHG without serum) (enough to cover the dish after spreading). Remove excess and allow the dish to dry.

### D. Monitoring Neuronal Differentiation

Retinoic acid treatment of NT2 cells causes them to differentiate into terminally differentiated postmitotic neurons, although a complement of undifferentiated cells remains that retains the original NT2 phenotype. This resembles a stem cell mode of division that results in nondifferentiated cells that retain the original phenotype and a population of differentiated neurons. Cells with other phenotypes are never observed. Thus, we do not routinely monitor neuronal differentiation of NT2 cells undergoing retinoic acid treatment. This can, however, be accomplished by monitoring the expression of a panel of cytoskeletal proteins, either by immunofluorescence or by Northern blot analysis as described in detail by Pleasure and Lee (1993).

## III. Expression of Foreign Genes in NT2N Cells

We have found that fully differentiated NT2N cells, like primary neurons, are refractory to standard transfection procedures. Therefore, to obtain efficient expression of a foreign gene, two approaches are available: the use of viral vectors to transiently express a gene and establishing a permanent NT2 cell line prior to retinoic acid differentiation. Both methods are discussed here.

### A. Transient Expression Using Viral Vectors

Because of the length of time required to establish permanent NT2N cell lines, it may be advisable to employ a transient expression system first to determine if establishment of a permanent line is worthwhile. This is particularly true when expressing mutant forms of proteins; transient expression makes it possible to determine whether a mutant protein exhibits an interesting phenotype. A further advantage of transiently expressing a foreign protein in NT2N cells is that expression in NT2 or other nonneuronal cells may not accurately reflect what happens in the differentiated phenotype. Because transfection of NT2N cells in inefficient, recombinant viruses can be used to efficiently express foreign genes in the fully differentiated NT2N cells. We have found that recombinant vaccinia viruses, which have been used successfully to express foreign proteins in primary neuronal cultures (Karschin *et al.*, 1991), infect NT2N cells efficiently. Other recombinant virus vectors such as those based on Semliki Forest virus (Chapter 2), Sindbis virus (Chapter 3), or HSV-1 (Chapters 9 and 10) may work as well. Because a detailed discussion of vaccinia virus vectors is found in Chapter 7, only a brief overview is given here.

## 1. Recombinant Vaccinia Virus Vectors

There are a number of advantages to using recombinant vaccinia virus vectors. Vaccinia has a wide host range, has a large capacity for inserts, and infects cells efficiently (Earl and Moss, 1991). Detectable levels of protein synthesis occur 3–4 hr after infection and a number of strong promoters are available, making it possible to obtain high levels of protein expression. Performing double or triple infections to coexpress several foreign proteins can also be done efficiently. Vaccinia virus replicates in the cytoplasm of the host cell; it does not integrate into the host genome. Thus, vaccinia can be used to express protein in nonmitotic cells such as primary neuron cultures and NT2N cells. With most cell types, vaccinia has significant cytopathic effects but there is, nevertheless, an experimental window from 4 to 12 hr postinfection in which biochemical experiments can be performed. During this time period, the biosynthesis and processing of a variety of foreign proteins has been shown to faithfully reflect native conditions (Braakman *et al.,* 1991,1992; Earl *et al.,* 1991). As a result, vaccinia virus can be useful for quickly assessing the effects of mutations on protein processing. Mutants of particular interest can then be stably transfected and studied in greater detail. For gene products that may be toxic to cells, an inducible expression system that uses the regulatory elements of the *Escherichia coli lac* operon in the vaccinia virus expression system is available (Zhang and Moss, 1991). A recombinant vaccinia virus, vlacI, expresses high levels of the lac repressor. The gene of interest is cloned into a plasmid just downstream from the *lac* operator and the vaccinia virus 7.5K promoter. A recombinant virus is then made by standard techniques (Earl and Moss, 1991). In the absence of isopropyl-$\beta$-thiogalactoside, expression of the gene is suppressed by approximately 95%. Thus, expression can be modulated by adding isopropyl-$\beta$-thiogalactoside.

## 2. Infection of NT2N Cells

We have found that fully differentiated NT2N cells are remarkably resistant to vaccinia virus-induced cytopathic effects. Even 24 hr after infection, the cells show little obvious sign of being infected: the cells do not round up, and their processes are intact. NT2N cells are infected as usual. At a multiplicity of infection of 50, approximately 80% of the cells are infected. Infection efficiency can be judged by immunofluorescence using antibodies against vaccinia virus structural proteins or against the foreign protein expressed by the recombinant virus. However, if a recombinant virus was made using a plasmid containing the $\beta$-galactosidase gene for color selection, then infection efficiency can be quickly and easily monitored by *in situ* staining for this enzyme using X-gal.

#### *Solutions*

Fixative: 2% paraformaldehyde in PBS

X-gal: 40 mg/ml stock solution in dimethylformamide (stored at $-20°C$)

Stain: 5 m*M* potassium ferricyanide, 5 m*M* potassium ferrocyanide, and 2 m*M* magnesium chloride in $dH_2O$

#### *Staining Procedure*

1. Infect NT2N cells as usual.
2. At 16 hr postinfection, wash the monolayer twice with PBS.
3. Remove PBS and add fixative for 5 min at room temperature.
4. Wash the monolayer twice with PBS.
5. Make a 1:40 dilution of X-gal in the stain solution; equilibrate to 37°C.
6. Remove PBS and add stain solution. Infected cells turn blue generally within 1 hr, although the reaction can be allowed to go overnight.

## 3. Construction of Recombinant Vaccinia Virus Vectors

Recombinant vaccinia viruses are made by transfecting virus-infected cells with a plasmid transfer vector that contains an expression cassette flanked by vaccinia virus DNA, allowing for homologous recombination (Earl and Moss, 1991). Vectors are often designed so recombination occurs within the viral thymidine kinase gene, which is thereby inactivated, allowing selection for recombinants by infecting a thymidine kinase negative cell line in the presence of 5-bromodeoxyuridine. Phosphorylation of 5-bromodeoxyuridine by the viral thymidine kinase carried by viruses that have not recombined allows incorporation of this toxic nucleotide into the viral DNA. As a consequence, thymidine kinase positive wild-type virus is selected against (Earl and Moss, 1991). The most commonly used transfer vectors contain additional components that allow for color or antibiotic selection of recombinant viruses. A variety of transfer vectors are available that contain different types of vaccinia virus promoters; cellular or other viral promoters are not used because they are not recognized by the vaccinia transcriptional machinery. The three basic types of vaccinia promoters are associated with early, intermediate, or late genes. Naturally occurring compound early–late promoters also exist that allow for expression throughout the course of infection. Genes driven by early promoters are usually not expressed after 6 hr of infection, whereas late genes are generally expressed by 4 hr postinfection. Early transcripts terminate after the sequence TTTTTNT. Thus, coding sequences should be scanned for the presence of this motif and should be altered by mutagenesis if an early promoter is used. Strong late promoters, such as the vaccinia virus 11K promoter, give the highest levels of expression. In our studies with NT2N cells, we routinely utilize a synthetic early–late promoter in the p65 transfer vector that gives high levels of expression late in infection. Adequate levels of expression are also attained early after infection. An advantage of having a strong early promoter is that early genes are expressed in the presence of Ara c, which prevents expression of late gene products. As a result, vaccinia replication and the resulting cytopathic effects of the virus are greatly reduced. This approach is useful if morphological techniques, such as immunofluorescence, are to be employed.

When working with vaccinia virus, class I or II biological safety cabinets should be used. At present, the Centers for Disease Control recommend that all laboratory personnel who work directly with the virus be vaccinated once every 10 years. However, the efficacy of vaccination has been questioned, so the institutional biosafety officer should be contacted to determine local policy regarding vaccination and containment.

## B. Establishing Permanent Cell Lines

Although NT2N cells can be transfected only inefficiently, the precursor NT2 cells can be transfected by a variety of standard techniques including calcium phosphate precipitation, DEAE–dextran, and liposome-mediated transfection. As a result, making stable NT2 cell lines is no different than making stable lines of other cells such as Chinese hamster ovary (CHO) and Madin-Darby canine kidney (MDCK) cells. Detailed procedures for several of these techniques are discussed in Chapters 11 and 13 of this volume. We have expressed foreign proteins in NT2N cells by transfecting NT2 cells with the $\beta$-gal expression plasmid SPUD1. The NT2 cells were co-transfected with psVneo, and resistant cells were selected with the aminoglycoside G418. After selection for 3 wk, a large fraction of the NT2 cells expressed $\beta$-gal. At this point, two approaches can be taken. The NT2 cells selected for resistance to G418 can be subcloned by limiting dilution, and colonies can be grown up in quantity. Induction with retinoic acid and enrichment of NT2N cells is then performed as usual, except that G418 is included in all steps. Alternatively, the cloning step can be omitted and the G418-selected cells can be induced directly. Again, G418 selection is maintained. Although this option this ultimately results in a more heterogeneous population of NT2N cells expressing the foreign gene, we have found that expression levels are generally sufficient for biochemical and morphological experiments. By omitting the cloning step, considerable time can be saved. Other selectable markers have not yet been tested in the NT2N system.

## IV. Summary

The NT2N cell system offers an attractive way to overcome some of the technical limitations inherent in working with primary neuronal cultures. In particular, it is possible to obtain large quantities of neurons with which to perform biochemical experiments, and the growth of neurites can be synchronized and controlled by varying the substrate on which the cells grow. In addition, because the differentiated NT2N neurons are derived from a mitotically active precursor cell line *in vitro,* it is possible to employ a variety of techniques, that are not otherwise available when working directly with postmitotic neurons, to obtain expression of foreign proteins. Because they are fully polarized, NT2N cells offer a way to study protein sorting to axons and dendrites at both the biochemical and the morphological level. Further characterization of NT2N cells is underway, and more efficient ways to obtain expression of foreign proteins will no doubt be found.

### References

Andrews, P. W. (1984). Retinoic acid induces neuronal differentiation of a cloned human embryonal carcinoma cell line *in vitro. Dev. Biol.* **103,** 285–293.

Braakman, I., Hoover-Litty, H., *et al.* (1991). Folding of influenza hemagglutinin in the endoplasmic reticulum. *J. Cell Biol.* **114,** 401–411.

Braakman, I., Helenius, J., *et al.* (1992). Manipulating disulfide bond formation and protein folding in the endoplasmic reticulum. *EMBO J.* **11,** 1717–1722.

Dotti, C. G., and Simons, K. (1990). Polarized sorting of viral glycoproteins to the axon and dendrites of hippocampal neurons. *Cell* **62,** 63–72.

Dotti, C. G., Parton, R. G., *et al.* (1991). Polarized sorting of gliated proteins in hippocampal neurons. *Nature (London)* **349,** 158–161.

Earl, P., and Moss, B. (1991). Expression of proteins in mammalian cells using vaccinia viral vectors. *Curr. Prot. Mol. Biol.* **16,** 15.1–18.10.

Earl, P., Hugin, A. W., and Moss, B. (1990). Removal of cryptic poxvirus transcription termination signals from the human immunodeficiency virus type 1 envelope gene enhance expression and immunogenicity of a recombinant vaccinia virus. *J. Virol.* **64,** 2448–2451.

Earl, P. L., Moss, B., *et al.* (1991). Folding, interaction with GRP78-BiP, assembly, and transport of the human immunodeficiency virus type 1 envelope protein. *J. Virol.* **65,** 2047–2055.

Geller, A. I. (1991a). Influence of the helper virus on expression of $\beta$-galactosidase from a defective HSV-1 vector, pHSVlac. *J. Virol. Meth.* **31,** 229–238.

Geller, A. I. (1991b). A system, using neural cell lines, to characterize HSV-1 vectors containing genes which affect neuronal physiology, or neuronal promoters. *J. Neurosci. Meth.* **36,** 91–103.

Ho, D. Y., Mocarski, E. S., *et al.* (1993). Altering central nervous system physiology with a defective herpes simplex virus vector expressing the glucose transporter gene. *Proc. Natl. Acad. Sci. USA* **90,** 3655–3659.

Karschin, A., Aiyar, J., *et al.* (1991). $K^+$ channel expression in primary cell cultures mediated by vaccinia virus. *FEBS* **278,** 229–233.

Kobayashi, T., Storrie, B., *et al.* (1992). A functional barrier to movement of lipids in polarized neurons. *Nature (London)* **359,** 647–650.

Le Gal La Salle, G., Robert, J. J., *et al.* (1993). An adenovirus vector for gene transfer into neurons and glia in the brain. *Science* **259,** 988–990.

Mcichsner, M., Doll, T., *et al.* (1993). The low molecular weight form of microtubule-associated protein 2 is transported into both axons and dendrites. *Neurosci.* **54,** 873–880.

Pleasure, S. J., and Lee, V. M.-Y. (1993). NTERA 2 cells: A human progenitor cell line committed to a CNS neuronal lineage. *J. Neurosci. Res.* **35,** 585–602.

Pleasure, S. J., Page, C., *et al.* (1992). Pure, postmitotic, polarized human neurons derived from NTera 2 cells provide a system for expressing exogenous proteins in terminally differentiated neurons. *J. Neurosci.* **12,** 1802–1815.

Rodriguez-Boulan, E., and Nelson, W. J. (1989). Morphogenesis of the polarized epithelial cell phenotype. *Science* **245,** 718–725.

Wertkin, A. W., Pleasure, S. J., *et al.* (1993). Neurons derived from a human cell line secrete $\beta$A4 and express a unique profile of potentially amyloidogenic carboxy terminal fragments. *Proc. Natl. Acad. Sci. USA* **90,** 9513–9517.

Younkin, D. P., Tang C-M., *et al.* (1993). Inducible expression of neuronal glutamate receptor channels in the NT2 human cell line. *Proc. Natl. Acad. Sci. USA* **90,** 2174–2178.

Zhang, Y., and Moss, B. (1991). Inducer-dependent conditional lethal mutant animal viruses. *Proc. Natl. Acad. Sci. USA* **88,** 1511–1515.

# CHAPTER 15

# Homologous Recombination for Gene Replacement in Mouse Cell Lines

**Thomas E. Willnow and Joachim Herz**

Department of Molecular Genetics
University of Texas Southwestern Medical Center
Dallas, Texas 75235



METHODS IN CELL BIOLOGY, VOL. 43

# I. Introduction

Much of our understanding of the physiological functions of specific genes has been derived from the study of organisms or cells in which a particular gene has been rendered dysfunctional by a genetic mutation. Traditionally, the route that led to the isolation and identification of the mutated gene began with the analysis of the mutant phenotype. The underlying defect is characterized by genetic, biochemical, morphological, or other criteria and is then traced back to the affected gene using a variety of different methods that largely depend on the organism or species in which the defect occurs. Powerful methods include the study of temperature-sensitive mutants in yeast and in *Drosophila,* complementation analysis in cultured cells, lineage analysis in *Caenorhabditis elegans,* and the positional cloning of inheritable disease genes in humans or mice. In some cases the gene defect may disturb a metabolic pathway, leading to the accumulation of a metabolite in the affected organism. The study of the pathway and the purification of the protein encoded by this particular gene then allow definition of the molecular basis of the gene defect by cDNA cloning of the defective allele or by analysis and comparison of the normal and mutated genomic loci.

As the number of cloned genes is continuously increasing and the genomic maps of several species, including human, are approaching high resolution, the large amount of nucleotide sequence information stored in generally accessible databases represents an invaluable research tool. Frequently now research efforts of separate groups meet at the level of the gene that was mapped as a disease locus by one group and was independently cloned by another group studying the protein produced by this particular gene (e.g., Cremers *et al.,* 1990; Seabra *et al.,* 1992). In other cases, the identified gene might share homology with other previously identified genes for which functional data are available, thus speeding the elucidation of the physiological function of the encoded protein (Martin *et al.,* 1990).

Researchers now recognize that many proteins belong to gene families whose members either are derived from a common ancestor or share significant sequence homology, suggesting that these proteins have related but not identical functions. For many of these related genes that are now being identified by homology cloning approaches, functional data are not available. However, the possession of cloned nucleotide sequences and the knowledge of the primary structure of the encoded proteins allows the derivation of tools with which the protein can be studied. These methods include hybridization probes, antibodies, and the ability to overexpress the cloned gene in bacteria, eukaryotic cells, or transgenic animals.

Another method that aims at the functional elimination of a gene rather than its overexpression makes use of the site-specific integration of homologous transfected DNA into the genome. The frequency of homologous recombination differs widely among species. This technique was initially mainly used as a tool by yeast geneticists, since site-specific rather than random integration of homologous DNA is the preferred mode of integration in this organism (Hinnen *et al.,* 1978; Orr-Weaver *et al.,* 1981). Homologous recombination also occurs at high frequency in the slime mold *Dictyostelium* (De Lozanne and Spudich, 1987). First considered to be a rare event in the mammalian system, the techniques for homologous integration of transfected DNA into the genome of mammalian cells have been perfected by the work of Capecchi and co-workers and by Smithies and colleagues (Smithies *et al.,* 1985; Doetschmann *et al.,* 1987; Thomas and Capecchi, 1987; Mansour *et al.,* 1988).

The ability to disrupt genes by homologous recombination in murine embryonic stem (ES) cells and transmission of the disrupted gene through the germ line of mice after reimplantation of the stem cells into early mouse embryos has opened a new dimension in mammalian genetics. This technique has proven particularly useful for the generation of animal models for human diseases (Snouwaert *et al.,* 1992; Ishibashi *et al.,* 1993) on which therapeutic approaches can be tested. Results obtained from such animals, in which a defined defect has been engineered into the genome, also frequently offer insight into specific roles that individual members of gene families play in certain cell types, whereas their functional absence may be of no discernible consequence in other tissues (Soriano *et al.,* 1991).

Positive/negative selection schemes (Mansour *et al.,* 1988) routinely yield frequencies of homologous versus random integration ranging from 1 in 2 to 1 in 500 stem cell colonies that have integrated the transfected DNA into their genome. These high frequencies make gene targeting a feasible and extremely valuable method of studying the physiological and cell biological consequences of the absence of the encoded protein. Homologous recombination also occurs in other established somatic mammalian cell lines including COS (Jasin and Berg, 1988), NIH 3T3 (Sedivy and Sharp, 1989), and Chinese hamster ovary (CHO) cells (Zheng and Wilson, 1990) but it is most widely used for the specific disruption or mutation of genes in ES cells.

The functional disruption of an autosomal gene in cultured cells *in vitro,* present in two copies in the genome of each cell, requires two separate targeting events. It is possible to use two different selectable markers, usually resistance to G418 and hygromycin B (te Riehle *et al.,* 1990), or, alternatively, to select for the double knock-out by increasing the G418 concentration in the culture medium (Mortensen *et al.,* 1992). In many cases it may be advantageous to disrupt only one copy of the gene of interest in ES cells and to establish a mouse strain that carries this defined genetic defect in its germ line. Heterozygous animals can then be bred to homozygosity, thus circumventing the need for the separate targeting of the second allele *in vitro.* Should the gene defect be nonviable in the homozygous state, it will often still be possible to isolate viable cell lines from homozygous embryos early during development, provided the gene is not essential for cell survival.

In this chapter we describe the procedures commonly employed in our laboratory to generate and analyze mice and ES cell lines in which the genes for endocytic cell surface receptors have been destroyed. Many of the detailed procedures have been adapted from previously described methods.

## II. Materials

### A. Cell Lines

Mouse ES cells were derived in our laboratory from delayed blastocysts of the 129 strain of mice or were kindly provided by Allan Bradley (Baylor College of Medicine, Houston; AB1 line). As feeder cells, STO fibroblasts transfected with an expression construct for LIF (leukemia inhibitory factor) were used (SNL 76 clone; A. Bradley).

### B. Media and Tissue Culture Reagents

STO fibroblasts and ES cells should be cultured in isotonic (290 mosmol) Dulbecco's modified Eagle's medium (DMEM). Early studies on the *in vitro* development of mouse embryos by Brinster (1965a–c) have demonstrated that the development of the mouse embryo is impaired in hypertonic medium whereas slightly hypotonic culture conditions are well tolerated. The osmolarity in the oviduct of a mouse was found to be 308 mosmol.

***Preparation of DMEM***

1. Dissolve 1 bag DMEM powder (13.4 g/bag, 4500 mg/liter glucose, with L-glutamine, without sodium pyruvate, without sodium bicarbonate; Cat. No. 12100-038; GIBCO/BRL, Grand Island, NY) in 1080 ml tissue culture grade water (e.g., Milli-Q filtration system; Millipore, Milford, MA).
2. Add 2.4 g sodium bicarbonate (cell culture tested; Cat. No. 895-1810IN; GIBCO/BRL) and stir slowly until dissolved completely.
3. Measure osmolarity and adjust to 290 mosmol. (This step is optional. We find that medium prepared by this protocol works well and reproducibly without further adjustment.)
4. Sterile filter the medium through a 0.22-$\mu$m cellulose acetate filter into sterile medium bottles.

There is no need to adjust the pH, because it will equilibrate to 7.4 in a humidified atmosphere containing 5% $CO_2$. It is convenient to make a batch of 5–6 liters of medium at a time and store it in 200-ml and 500-ml aliquots at 4°C. Prior to use, the medium is supplemented with 15% fetal calf serum (FCS, endotoxin tested; Cat. No. A-1115-L; HyClone Laboratories, Logan, UT), 0.1 m*M* nonessential amino acids (MEM nonessential amino acids solution; Cat. No. 320-1140PG; GIBCO/BRL), 2 m*M* L-glutamine (Cat. No. G7513; Sigma,

St. Louis, MO), and 0.1 m*M* β-mercaptoethanol (cell culture tested; Cat. No. M-7522; Sigma). Also, 100 U/ml penicillin and 100 μg/ml streptomycin can be added to the medium (penicillin/streptomycin mixture; Cat. No. 17-602E; BioWhittaker, Walkersville, MD). However, we prefer to culture the cells without antibiotics. This requires meticulous cell culture techniques to prevent contamination of the cultures with bacteria, yeast, or mycoplasma. Complete medium (containing β-mercaptoethanol) can be kept for up to 1 wk at 4°C. Longer storage is not recommended.

Mouse embryonic fibroblasts are grown in DMEM supplemented with 2 m*M* glutamine, 10% FCS, 0.1 m*M* β-mercaptoethanol, and penicillin/streptomycin.

Ligand degradation experiments are performed in DMEM (1000 mg glucose/liter, without L-glutamine; Cat. No. D-9036; Sigma), supplemented with 0.2% (w/v) bovine serum albumin (BSA, albumin fraction V; Cat. No. 10868; United States Biochemical, Cleveland, OH).

#### *Additional Reagents Required for Tissue Culture Procedures*

Phosphate buffered saline (PBS) (Dulbecco's phosphate buffered saline without $Ca^{2+}$ and $Mg^{2+}$; Cat. No. 59-32178P; JRH Biosciences, Lenexa, KS)

0.25% Trypsin–1 m*M* EDTA solution (Cat. No. 610-5200 AG; GIBCO/BRL)

0.05% Trypsin–EDTA solution (Cat. No. 610-5300PG; GIBCO/BRL)

0.1% Gelatin in PBS (autoclave to dissolve) (gelatin type I from swine skin; Cat. No. G-2500; Sigma)

### C. Chemicals, Enzymes, and Other Reagents

Proteinase K is from Merck (Darmstadt, Germany, Cat. No. 24568) and native *Taq* DNA polymerase is from Perkin Elmer (Cat. No. N 801-0046). Mitomycin C can be obtained from Boehringer Mannheim (Cat. No. 107 409; Indianapolis, IN) and asialofetuin is from Sigma (Cat. No. A-4781). Urokinase (uPA) was generously provided by Jack Henkin (Abbott Laboratories, Abbott Park, IL). Recombinant human PAI-1 was a gift from Ed Madison (Scripps Research Institute, LaJolla, CA). G418 (Geneticin®; Cat. No. 860-1811IJ; GIBCO/BRL Laboratories, Cat. No. 860-1811IJ) is prepared as a 100× stock (19 mg/ml active concentration) in PBS and sterile filtered through a 0.22-μm filter. It is stable at 4°C for several weeks. FIAU [1-(2-deoxy,2-fluoro-β-D-arabinofuranosyl)-5 iodouracil; Oclassen Pharmaceuticals, San Rafael, CA] is prepared as a 2 mg/ml stock in water, sterile filtered, and stored at −80°C. Recently, FIAU has been taken off the market. Although we only have direct experience with FIAU, other investigators successfully use ganciclovir (Syntex, Palo Alto, CA) as a substitute. *Pseudomonas aeruginosa* exotoxin A (Cat. No. P-0184; Sigma) is dissolved in PBS at a concentration of 100 μg/ml and stored at −80°C. $\alpha_2$-Macroglobulin was prepared from human plasma and activated by methylamine (Strickland *et al.*, 1984) and uPA/PAI-1 complexes were formed as described elsewhere (Herz *et al.*, 1992).

Animals are anesthetized with sodium pentobarbital (Nembutal®; Cat. No. 0074-3778-05; Abbott Laboratories).

$^{125}$I-Labeled ligands (LDL, asialofetuin, $\alpha_2$-macroglobulin, uPA/PAI-1) are iodinated using the iodogen (Fraker and Speck, 1978) or the iodomonochloride method (Goldstein *et al.,* 1983).

## D. Plastic Ware

### *For Cell Culture*

12-well dishes (Cat. No. 25815; Corning, New York, NY)

24- and 96-well dishes (Cat. No. 3524 and 3596; Costar, Cambridge, MA)

25-cm$^2$ and 75-cm$^2$ flasks (Cat. No. 3055 and 3375; Costar)

100-mm culture dishes (Cat. No. 25020-100; Corning)

1-, 5-, and 10-ml disposable sterile plastic pipets (Cat. No. 4011, 4051, 4101; Costar)

Sterile media bottles (Cat. No. 25628-500; Corning)

Sterile 15- and 50-ml tubes (Cat. No. 2070, 2001; Falcon, Lincolnpark, NJ)

0.22-$\mu$m sterile filters (MILLEX®-GS; Cat. No. SLGS025OS; Millipore)

Medium filtration units (Cat. No. 25932-200; Corning)

Disposable electroporation chambers (Cat. No. 1601AB; BRL)

Cryovials (Cat. No. 375418; Nunc, Naperville, IL)

### *For Ligand Turnover Studies*

Heparin treated glass capillaries (pasteur capillary pipets, 9″ length; Cat. No. 13-678-20C; Fisher Scientific, Pittsburgh, PA)

0.5-cc LO-DOSE® Insulin syringes (Cat. No. 9461; Becton Dickinson, Cockeysville, MD)

Microcuvettes for hematological determinations (0.75-ml Microvette® KE CB 1000; Cat. No. 17.455.100; Sarstedt, Newton, NC)

Disposable glass tubes (13 × 100 mm borosilicate tubes; Cat. No. 14-961-27; Fisher Scientific)

## E. Buffers and Solutions

TE: 10 m*M* Tris/HCl, 1 m*M* EDTA, pH 8

ES cell buffer: 10 m*M* HEPES, 2 m*M* $CaCl_2$, 2 m*M* $MgCl_2$, 140 m*M* NaCl, pH 7.4

Wash buffer: 50 m*M* Tris/HCl, 150 m*M* NaCl, pH 7.4

TBS: 10 m*M* Tris/HCl, 150 m*M* NaCl, 2 m*M* $CaCl_2$, 2 m*M* $MgCl_2$, pH 7.4

Saline: 150 m*M* NaCl

*Reagents for PCR Analysis*

10× PCR buffer: 166 m*M* ammonium sulfate, 670 m*M* Tris/HCl, 67 m*M* $MgCl_2$, 50 m*M* β-mercaptoethanol, 67 μ*M* EDTA, pH 8.8

Proteinase K-buffer: 1× PCR buffer, 1.7 μ*M* SDS, 50 μg/ml proteinase K

10 m*M* dNTP mix (made from ultrapure dNTP set; Cat. No. 27-2035-01; Pharmacia, Piscataway, NJ)

Dimethyl sulfoxide (DMSO) (Cat. No. D-2650; Sigma)

1.6 mg/ml BSA, molecular biology grade (Cat. No. 5561 UB; GIBCO/BRL)

100 μg/ml PCR primer solution in TE

All PCR buffers and solutions are stored in aliquots at −20°C.

## F. Instrumentation

*For Cell Culture*

37°C water bath

5% $CO_2$ incubator

Clinical centrifuge to sediment cells

Inverted phase contrast microscope

Laminar flow hood

Hemocytometer to determine cell numbers

Osmometer

*For Gene Targeting Experiments in Embryonic Stem Cells*

Irradiator (Mark 1; J. L. Shepherd & associates, San Fernando, CA)

Cell electroporation system (Cell Porator™; BRL)

Binocular dissecting microscope (e.g., Nikon SMZ-2B)

PCR machine (e.g. Thermal cycler; MJ Research, Watertown, MA)

Mouth controlled suction device: This device is made from a microcapillary (50-μl capillaries; Cat. No. 21-164-2G; Fisher Scientific) drawn out over a Bunsen flame to a very small diameter of about 0.2 mm. The capillaries are fit into a piece of polyethylene tubing, connected with a 0.22-μm sterile filter and a mouthpiece (white mouthpiece and tubing; Cat. No. 13-647-5; Fisher Scientific)

*For Mouse Surgery Procedures*

Blunt and toothed forceps

Iris and bone scissors

Wound clips and applier (Cat. No. 7630 and 7631; Clay Adams, Parsippany, NJ)

Slide warmer (serves as warming plate)

# III. Experimental Procedures: Gene Disruption and Derivation of Mice and Cells

## A. DNA Constructs

Vector constructs used for gene targeting are either of the replacement or of the insertion type and typically include a positive selectable marker to enrich for cells that have taken up the targeting vectors. Usually the *neo* gene is used, which confers resistance to the drug G418 on the cells. The differences between both vector types and their respective advantages and disadvantages have been investigated and discussed in detail (Capecchi, 1989a,b; Smithies, 1993). The frequency of homologous recombination for insertion type vectors is often higher than for replacement vectors. However, because homologous recombination by insertion leads to a linear duplication of genomic sequences, integrated DNA sequences can be lost again during subsequent cell divisions by the reverse process, resulting in the excision of the integrated DNA and restoration of the wild-type locus.

In this chapter we focus on the use of replacement type vectors which cannot be lost from the insertion site by homologous intragenic rearrangements. Furthermore, replacement vectors can be constructed to allow convenient identification of homologous recombination events by polymerase chain reaction (PCR) analysis of *neo*$^R$ clones (discussed in detail later).

The lower frequency of homologous integration of replacement vectors can be effectively compensated for by the introduction of a negative selection step that preferentially eliminates clones that arise from nonhomologous integration of the targeting vector. Mansour and colleagues used the herpes simplex virus thymidine kinase (HSV-TK) gene as a negative selectable marker. HSV-TK has a relaxed substrate specificity compared with mammalian thymidine kinases and converts the inactive nucleoside analogs FIAU and ganciclovir into chain terminators that can be incorporated into the DNA of replicating cells. During homologous integration of a replacement construct, the HSV-TK gene flanking the homologous DNA (Fig. 1) is excluded and generally does not integrate into the genome whereas it will remain intact in the majority of cases in which the incoming DNA has integrated randomly. The presence of FIAU or ganciclovir in the culture medium will therefore strongly select against G418-resistant cells which have integrated a functional copy of the HSV-TK gene, whereas clones that have undergone homologous recombination remain insensitive to the drugs.

### 1. Isolation of Genomic Sequences

The construction of a targeting vector requires the cloning of a stretch of genomic DNA, 10–15 kb in length, that covers the region of the gene in which the disruption is to take place. Ideally, the DNA should be isogenic, which means it should be derived from the strain of mouse that gave rise to the ES

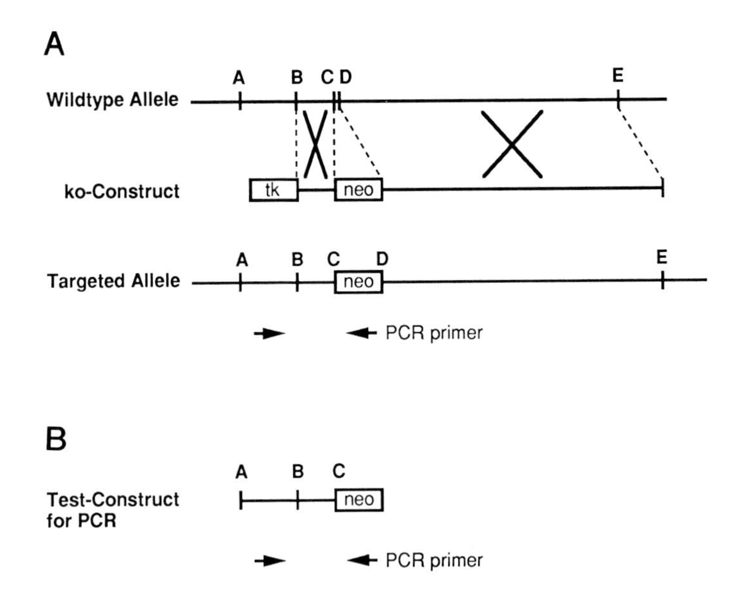


**Fig. 1** Construction of gene replacement vector and PCR test construct. (A) The replacement-type targeting vector is constructed from DNA segments BC and DE derived from the genomic locus of the wild-type allele. The short arm (fragment BC) should be 1–1.5 kb; the long arm (fragment DE) should be approximately 10 kb in length. Alternatively, the short arm may be cloned by PCR using oligonucleotides originating from sites B and C. This obviates the need for convenient restriction sites at the chosen positions. To facilitate cloning, suitable restriction sites may be included within the oligonucleotides. The neo expression cassette (neo) is cloned between sites C and D; the herpes simplex virus thymidine kinase expression cassette (tk) flanks the short arm immediately adjacent to site B. The structure of the targeted allele is shown in the third line. If, as shown in this schematic, sites C and D are not identical, gene sequences between C and D are deleted in the targeted allele. The targeted allele is detected by PCR using primers (indicated by the arrows) that are located within neo and between sites A and B. (B) A test construct on which the sensitivity of detection of the targeted allele by PCR can be determined. Fragment AC is a restriction fragment or can be cloned by PCR using primers derived from sites A and C.

cells. Ideally, DNA should be isolated from the ES cells used for the targeting experiments (te Riehle *et al.*, 1992). If a genomic library of ES cell DNA is not available, a suitable fragment may be directly cloned from a digest of genomic ES cell DNA. A detailed protocol of the possible cloning procedures is beyond the scope of this chapter. Please refer to a general cloning manual (e.g., Sambrook *et al.*, 1989) for this purpose. A possible strategy is briefly outlined here.

1. Digest genomic ES cell DNA with a panel of restriction enzymes. We usually use enzymes for which suitable cloning sites exist in commercially available bacteriophage lambda vectors (Lambda Dash, Lambda Fix; Stratagene, La Jolla, CA).
2. Analyze digested DNA by Southern blotting with a cDNA probe covering the region in the gene of interest that is to be disrupted.

3. Identify enzymes that produce hybridizing fragments of 10–20 kb.

4. Digest approximately 400 μg genomic DNA with the most suitable enzyme.

5. Enrich for the fragment of interest by size fractionating the digested DNA on sucrose density gradients (described in detail by Sambrook *et al.*, 1989).

6. Identify fractions containing the fragment by Southern blotting or by PCR.

7. Prepare a subgenomic library of the peak fraction in the lambda vector.

8. Screen the library by hybridization and isolate recombinant phages containing the desired fragment.

In our experience, size fractionation results in an at least 10-fold enrichment of clones in the subgenomic library containing the desired fragment over a library containing a random representation of the ES cell genome.

## 2. Design of a Replacement-Type Targeting Vector

The preferred design of a gene replacement vector is shown in Fig. 1A. The vector should fulfill the following requirements:

1. The *neo* gene should be flanked by a long (~10 kb) and a short (~1–1.5 kb) arm of linear genomic sequences homologous to the target gene. The *neo* cassette contains the tn5 phosphotransferase (*neo*) gene driven by a suitable promoter and followed by a polyadenylation signal. Suitable *neo* expression cassettes are the MC1 (Stratagene) or the pol2neobpA (Soriano *et al.*, 1991) cassettes.

2. The *neo* cassette should be inserted into an exon leading to the disruption of the reading frame of the gene and the premature termination of the translated protein. The orientation of the *neo* gene (transcription parallel or antiparallel to the target gene) is not important.

3. The short arm should be flanked by at least one copy of the HSV-TK expression cassette. We use a tandem array of two HSV-TK genes to insure that at least one copy remains intact in the case of random integration. We prefer the HSV-TK cassette to flank the short rather than the long arm of homology. This reduces the chance of the TK cassette becoming separated from the *neo* gene because of a break of the DNA strand during transfection, which would result in an increased number of ES cell colonies in which the targeting vector has integrated randomly into the genome.

4. The targeting vector must be linearized prior to transfection with a restriction enzyme that does not cut within the genomic sequences or the *neo* or TK cassettes.

Depending on the distribution of restriction sites in the isolated genomic DNA fragment, the construction of the targeting vector may be facilitated by PCR-assisted cloning to insert the *neo* cassette into an exon. Many possible strategies that depend on the structure of the gene can be employed to construct the final targeting vector and no specific recommendations can be given. However, the exon into which the *neo* cassette is to be inserted should be carefully chosen to insure that the functional disruption of the gene is complete.

## 3. Construction of a PCR Test Vector

Following transfection and selection of the ES cells, homologous recombination events in individual clones can be detected either by Southern blotting or by PCR analysis. If a replacement vector of the design just described was used, PCR is the preferable strategy because of its ease, the speed with which recombinant clones can be identified, and the number of clones that can be screened. Positive clones identified by PCR are then verified by Southern analysis. To insure that the PCR reaction is sensitive enough to detect low copy numbers of homologous recombination events, it is advantageous to construct a test vector that, like the knock-out vector, contains the *neo* gene and the short arm of homology and an additional few hundred bases of genomic DNA contiguous with the sequences contained in the short arm (Fig. 1B). This vector is used to test different sets of PCR primers beforehand.

The principle of detection of a homologous integration by PCR is depicted in Fig. 1A. PCR primers are derived from the *neo* gene and from an area in the genome located immediately adjacent to the sequences contained in the short arm of the targeting vector. Only in the case of a homologous recombination event will both primer sites be in close proximity to each other and result in the amplification of a DNA fragment of predictable size and sequence. Random integration will not generate a specific DNA product.

Not all primer pairs are equally sensitive. To reliably detect homologous recombination events in a few hundred cells, a given primer pair must yield an unambiguous amplification product with as little as 1 fg test vector mixed with 100 ng native genomic ES cell DNA. Examples for different sensitivities of primer pairs with the same test construct are shown in Fig. 2. To avoid false

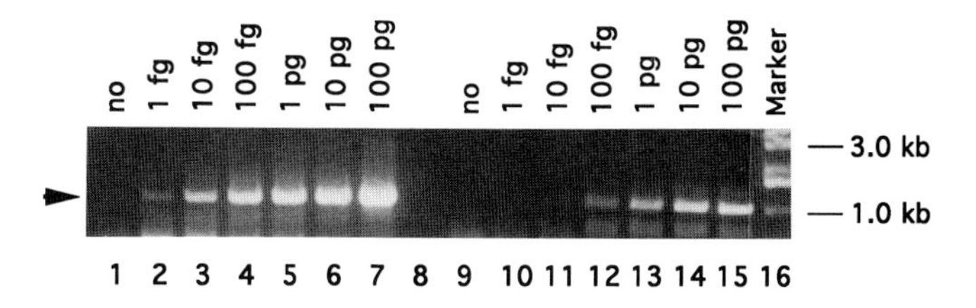


**Fig. 2** Determination of the sensitivity of primer pairs for PCR screening. In this assay, 100 ng mouse genomic DNA without (lanes 1 and 9) or with (lanes 2–7 and lanes 10–15) the indicated amount of PCR test construct were analyzed by PCR using the conditions stated in the text. Two different primer pairs were used (lanes 1–7 and 9–15, respectively). Primers were derived from the neo cassette and from sequences of the receptor-associated protein RAP (Herz *et al.*, 1991). The primer pair in lanes 1–7 allows the detection of approximately 100 times less DNA than the pair used in lanes 9–15. The positive PCR signal is indicated by the arrowhead. DNA fragments were separated on a 0.8% agarose gel and stained with ethidium bromide.

negative results, it is advisable to design and match primers to be used for the detection of the recombination events carefully. Although computer programs exist to help in primer design, they are no guarantee for the design of a successful primer set and several different primer combinations may have to be tested empirically.

## B. Cultivation of Cells

### 1. STO Fibroblasts

Mitotically inactivated STO fibroblasts expressing LIF are used to prepare feeder cell layers. They provide both secretory and matrix factors required for the successful propagation of ES cells *in vitro*. Stocks of STO cells are propagated in ES medium in a 5% $CO_2$ atmosphere and are useful as long as the cultured cells are morphologically homogeneous (usually 25–30 passages). Do not allow STO cells to become overconfluent because this will select for cells that have lost contact inhibition and therefore have become unsuitable to serve as feeder cells.

It is important to plan the need for feeder layers ahead of time. A sufficient supply of STO cells is guaranteed by continuously keeping 2–3 75-$cm^2$ flasks of STO cells in culture during the course of a gene targeting experiment in ES cells. Once a culture has reached 90% confluence, the cells are split and further propagated, frozen for storage, or used for the preparation of feeder layers, as described next.

### 2. Routine Passage of a Nearly Confluent Culture of STO Cells in 75-$cm^2$ Flasks

1. Aspirate the medium and rinse the cells once with 6 ml PBS.
2. Add 1.5 ml 0.25% trypsin–EDTA solution and incubate at room temperature.
3. When the cells start to detach from the bottom of the flask, vigorously tap the flask several times to shake all cells off the bottom and inactivate the trypsin by adding 6 ml ES medium.
4. Resuspend the cells in the flask by repeated pipetting.
5. Reseed cells into fresh culture vessels at the desired density (usually 1:10 ratio). Do not exceed a splitting ratio of 1:20.

For 25-$cm^2$ flasks, volumes are scaled down accordingly.

### 3. Preparation of Mitotically Inactivated STO Feeder Layers

Plan the numbers of plates and flasks required to propagate the ES cells for a particular experiment! Mitotically inactivated feeder layers are good for up to 2 wk, but are preferably used within 1 wk. As a rule, a nearly confluent 75-

$cm^2$ flask of STO cells will yield 1.5 100-mm plates, 4–5 25-$cm^2$ flasks, or 1.5 24-well plates containing feeder layers. It is not necessary to count the cells. Feeder layers are prepared in gelatin-coated culture vessels to insure optimal attachment of the feeder layers.

*Gelatin Coating*

1. Autoclave PBS containing 0.1% gelatin (w/v).
2. Flood the bottom of the culture vessels with gelatin solution and let sit for at least 1 hr at room temperature.
3. Completely remove gelatin solution before plating feeder layers.

STO cells can be mitotically inactivated by treatment with mitomycin C or, alternatively, by $\gamma$ irradiation. Although we prefer $\gamma$ irradiation because it is cheaper, easier and faster than mitomycin C treatment, a radiation source may not always be readily available.

*Mitomycin C Inactivation*

1. Grow STO cells to 90% confluence.
2. Replace medium with fresh medium containing 10 $\mu$g/ml mitomycin C.
3. Incubate in 5% $CO_2$ incubator for 2 hr.
4. Remove medium and thoroughly rinse cells 3 times in PBS.
5. Trypsinize cells as described earlier.
6. Resuspend cells in ES medium and dispense into gelatin-coated flasks, wells, or plates.

*Inactivation by $\gamma$ Irradiation*

1. Grow STO cells to 90% confluence.
2. Trypsinize cells as described earlier. Resuspend in 5–15 ml ES medium and transfer cell suspension to a 15-ml sterile plastic tube.
3. Place cell suspension into irradiator and expose to a dose of 10,000 rad.
4. Bring cells up in desired final volume and seed into gelatin-coated culture vessels.

For the SNL 76 clone of STO cells, a dose of 10,000 rad is required to insure that all cells are inactivated while maintaining a viability of >99%. The irradiation dose may vary with feeder cells of different origin and should be tested before ES cells are plated onto feeder layers prepared with this method. If the dose was insufficient, overgrowth of the feeder layers will be apparent within a few days. Once the correct dose has been established, we have found this method to be highly reproducible for the preparation of high quality feeder layers.

After mitotic inactivation, it takes about 6 hr until all cells have attached to the dish and spread out. Feeder layers are then ready to use.

## 4. Isolation and Propagation of Mouse Embryonic Stem Cells

ES cells are derived from the inner cell mass of the early mouse embryo at the blastocyst stage. Many independently derived ES cell lines are described in the literature (Evans and Kaufmann, 1981; Bradley *et al.*, 1984). In our laboratory we have used the AB-1 line of ES cells. We have also, in collaboration with Robert E. Hammer and David Clouthier, derived several lines of ES cells using essentially the procedures described by Robertson (1987). The AB-1 line and the stem cell lines derived in our laboratory have given rise to germ line chimeric founder animals with high frequency. Several factors are important for the maintenance of stem cells and the successful generation of germ line chimeras:

1. Passage number: Low-passage stem cells are more likely to have retained a euploid set of chromosomes. Aneuploid cells that will inevitably accumulate within a stem cell population with prolonged culture will not contribute to the germ line.
2. Optimal culture conditions: Stem cells should always be cultured at relatively high densities to insure a fast and uniform growth rate. Strenuous splitting of a stem cell culture will retard cell growth and select for abnormal cells. As a rule, a stem cell culture should become confluent within 3 days of seeding. Although this requirement has not been tested rigorously for the culture of ES cells, it is probably advisable to supply the cells with culture conditions that match as closely as possible the conditions found in the oviduct and the uterus. These include an isotonic osmolarity of the culture medium and a final $CO_2$ concentration of 5% in the incubator ($P_{CO_2}$ = 38 mm Hg).
3. High quality uniform feeder layers to provide uniform attachment and conditioning of the medium.

## 5. Routine Passage of ES Cells

For each targeting experiment, we start with a fresh vial of low passage stem cells.

1. Thaw vial as described in Section III,B,7.
2. Change medium on feeder layer in a 25-cm$^2$ flask and plate cell suspension (day 0). Grow culture at 37°C in a humidified 5% $CO_2$ atmosphere.
3. On day 1, change medium and assess viability of stem cells. Viability should be >90%.
4. On day 2, change medium again. If cells are confluent, trypsinize cells and replate onto fresh feeder layers at a ratio of 1:5 to 1:10.

A good indicator for the confluence of ES cells is the rate at which the medium is consumed. A color change of the medium from a healthy red to a light orange within 12 hr of refeeding indicates that the cells are approaching confluence. Figure 3 shows a subconfluent ES cell culture 2 days after splitting.

## 6. Trypsinization of ES Cells

1. Refeed cells with fresh medium 2–4 hr prior to trypsinization to insure high cell viability.

2. Remove medium from a confluent ES cell culture and rinse once with PBS.

3. Add 0.25% trypsin–EDTA solution to cover the cell layer. Use 1 ml for a 25-$cm^2$ flask and 1.5–2 ml for a 75-$cm^2$ flask. Adjust volume to surface area when using other culture dishes. A confluent 25-$cm^2$ flask will contain approximately $1–2 \times 10^7$ ES cells.

4. Return the culture to the incubator and incubate at 37°C for at least 5 min. This is necessary to insure complete trypsinization of the ES cells, which grow in tightly packed cell clusters.

5. Remove culture vessels from incubator and tap several times as hard as possible to shake cells loose and to disrupt clumps.

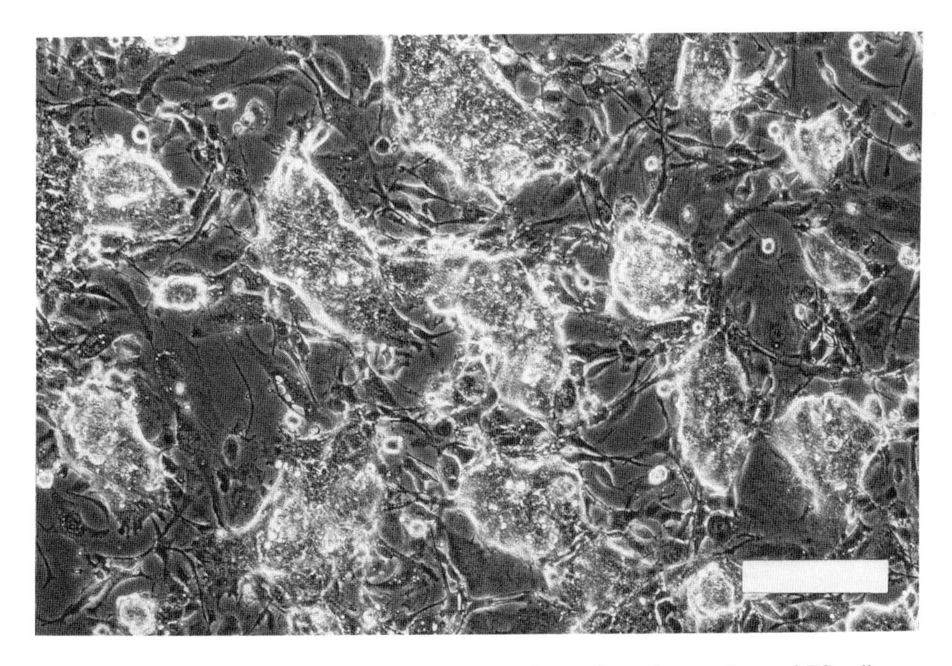

**Fig. 3** Phase-contrast image of ES cells on feeder layer. A confluent culture of ES cells was split at a ratio of 1 : 10 and plated onto a fresh feeder layer. The picture of this subconfluent ES cell culture was taken 48 hr after replating the cells. Notice ES cells growing as compact clumps on top of a uniform fibroblast feeder layer. Bar, 200 $\mu$m.

6. Add ES medium (4 ml for a 25-$cm^2$ flask, 8 ml for a 75-$cm^2$ flask) and forcibly pipet cells up and down the pipet while pressing the pipet mouth against the bottom of the flask or the plate. Partial occlusion of the pipet mouth increases shear forces and helps disrupt stem cell clumps. It is very important at this stage that a single cell suspension is obtained, since clumps of ES cells will invariably differentiate on replating. If obtaining a single cell suspension proves repeatedly difficult it might be necessary to replace the trypsin solution or to extend the trypsinization time.

7. Transfer cells to a sterile 15-ml plastic tube and sediment cells (1000 rpm, 1–2 min).

8. Cells can now be resuspended and replated for further expansion, injected into blastocysts, frozen for storage as detailed in Section III,B,7, or used for electroporation.

### 7. Freezing and Thawing of Cells

***Cell Freezing***

1. Harvest cells by trypsinization, as just described. Usually one confluent 75-$cm^2$ flask will yield enough STO cells to freeze 2–3 vials and sufficient ES cells for 5–6 vials (final cell density $5 \times 10^6$–$1 \times 10^7$ cells/ml).

2. Transfer the cell suspension into a 15-ml sterile plastic centrifuge tube and spin for 1–2 min at 1000 rpm in a clinical centrifuge.

3. Resuspend cells in complete ES medium (volume for STO cells, 1–1.5 ml; for ES cells, 2.5–3 ml). Count cells and adjust cell density accordingly, if desired. We have found that the formula just given works well and results in high viability of the cells after thawing without the prior need to count the cell density.

4. Slowly add an equal volume of 2× freezing medium (20% DMSO, 40% FCS, 40% DMEM). Mix carefully and aliquot suspension into cryovials (1 ml/vial).

5. Place vials in a styrofoam or cardboard box and place at −80°C for 2–24 hr.

6. Long-term store frozen cells in liquid nitrogen storage containers.

***Cell Thawing***

1. Move cryovial from the liquid nitrogen directly to a 37°C water bath.

2. Transfer cell suspension (1 ml) to a sterile 15-ml centrifuge tube and add 5 ml complete ES medium.

3. Pellet the cells by centrifuging for 1–2 min at 1000 rpm.

4. Aspirate the medium and resuspend cells in complete ES medium.

5. Plate cells at desired density. We usually reseed 1 vial of STO cells into 2 75-$cm^2$ flasks. ES cells are replated in 25-$cm^2$ flasks containing a layer of irradiated feeder cells (1 flask/vial).

## C. Gene Targeting in Mouse Embryonic Stem Cells

The following section describes a protocol for the introduction of the targeting vector into the ES cells by electroporation, as well as positive and negative selection procedures.

### 1. Electroporation of ES Cells with a Gene Targeting Construct

1. Trypsinize $3 \times 10^7$ ES cells (approximately 1.5 25-$cm^2$ flasks) as described and sediment the cell suspension in a 15-ml sterile plastic centrifuge tube at 1000 rpm for 1–2 min.
2. Resuspend the cell pellet in 10 ml PBS and recentrifuge. Repeat this step twice.
3. Resuspend the cells in 1 ml PBS.
4. Dilute 50–100 $\mu$l DNA solution (1 mg linearized DNA/ml TE) in 1 ml PBS and combine with cell suspension. Mix well.
5. Pipet cells into disposable electroporation chambers and electroporate using the Cell Porator™ (BRL) set at 330 $\mu$F, 275 V, low $\Omega$.
6. Add electroporated cell suspension to 60 ml ES medium and mix well.
7. Pipet 15 ml each to 4 100-mm dishes containing freshly prepared feeder layers and place dishes in 5% $CO_2$ incubator.
8. Designate the day of electroporation as day 0 of the experiment.

### 2. Positive/Negative Selection

A two-step selection process is employed to enrich the electroporated ES cell culture for clones that have undergone homologous recombination. In the first step, ES cells that have incorporated the targeting vector are selected for by culture in the presence of G418 (positive selection). In the second step, *neo*-resistant clones that still contain the HSV-TK gene are eliminated by including the drug FIAU in the culture medium (negative selection).

1. Replace ES medium with fresh medium containing 190 $\mu$g/ml G418 (active concentration) 24 hr after the electroporation (day 1). Change medium every day.
2. On day 5, switch medium to ES medium containing G418 and 0.25 $\mu M$ FIAU. Depending on the metabolic activity of the cells, it may not be necessary to change the medium on day 6.
3. On day 7, refeed cells with ES medium containing G418 without FIAU.
4. On day 9, refeed cells with regular ES medium (without G418 or FIAU). At this point, discrete ES cell colonies should be clearly visible. For an average electroporation experiment, 100–200 colonies/plate can be expected. Between day 10 and day 12, colonies are usually large enough to be analyzed.

## D. Identification of Targeted ES Cell Clones by PCR

Analysis by PCR is a very efficient method for screening large numbers of ES cell colonies in a short time for the homologous recombination event. Two different methods are described. The first method is technically less demanding but more labor intensive and thus does not allow the screening of large numbers of colonies.

### 1. Method 1

1. Remove medium from plate containing ES cell colonies and rinse once with PBS. Flood dish with 10 ml ES cell buffer.
2. Using a drawn-out pasteur pipet or glass capillary controlled by mouth suction, dislodge ES cell colony from feeder layer and pull into the orifice of the pipet or capillary. Be careful to take up as little buffer as possible. This step is best done under a stereomicroscope and, with reasonable care, can be performed outside the hood. Instead of a drawn-out pipet or capillary, an Eppendorf pipet equipped with a small tip may be used to remove the colony from the plate.
3. Transfer colony with as little buffer as possible to one well of a 96-well plate containing ~50 μl 0.25% trypsin–EDTA (1 drop).
4. Repeat Steps 2 and 3 until 10–20 colonies have been collected.
5. Place plate into incubator for 5 min.
6. Remove plate from incubator and add 2 drops ES cell medium/well.
7. Using an Eppendorf pipet set at 200 μl, disrupt colonies by rapidly pipetting cell suspension up and down (~10 times).
8. Transfer 50 μl cell suspension to microfuge tube for PCR analysis. Transfer remainder of cell suspension to fresh feeder layers in 24-well plates.
9. Return 24-well plates to incubator and expand cells.

This protocol allows the analysis of ~200 colonies in 1 day. Once all colonies have been picked proceed with PCR analysis.

10. Pellet cells in microfuge tube by brief centrifugation. Remove 45 μl supernatant.
11. Resuspend cell pellet in 15 μl proteinase K buffer.
12. Proceed as detailed in Section III,C,3.

### 2. Method 2

1. Aspirate the medium from one plate. Invert plate and mark and number colonies on bottom of plate using a fine felt tip pen.
2. Overlay cells with 5 ml ES cell buffer.

3. Pick a portion (~1/4) of each of the numbered colonies under a stereomicroscope using a finely drawn out pasteur pipet or capillary controlled by mouth suction. Take care to aspirate colony fragment in a minimum volume of buffer since the cells will easily stick to the glass and be lost if drawn up too far into the capillary.

4. Transfer the piece of the colony into a 0.5-ml microfuge tube containing 12 $\mu$l proteinase K buffer.

5. Use a fresh capillary for each colony to avoid contamination of the PCR samples.

6. After picking between 50 and 100 colonies, replace ES cell buffer with medium and return plate to $CO_2$ incubator. To control possible contamination of the plate during sample collection, include penicillin and streptomycin in the medium.

If preliminary experiments indicate that the targeting frequency for the gene of interest is low, a modification of Method 2 can be used to avoid an excessive number of individual PCR reactions.

1. Proceed with Steps 1–3 as given.

2. Instead of transferring the colony fragment into proteinase K buffer, transfer cells to a 1.5-ml microfuge tube containing 20 $\mu$l PBS. Do so for a pool of 10–25 colonies.

3. Centrifuge cells, remove supernatant, and resuspend cells in 15 $\mu$l proteinase K buffer.

4. Proceed to PCR reaction.

Screening colonies in pools requires a second round of sampling to identify the targeted colony within an identified positive pool. This increases the risk of contamination of the plate, resulting in the subsequent loss of the culture. Whenever the targeting frequency allows, we prefer to perform single colony PCRs. Instead of numbering individual colonies on the plate, colonies contained within one pool may be color coded for the first round of PCR screening and numbered during the rescreen.

## 3. PCR Reaction

Proceed with treatment of colony samples in proteinase K buffer obtained by Method 1 or 2.

1. Incubate samples in proteinase K buffer for 1 hr at 37°C.

2. Inactivate the proteinase by incubation for 12 min at 85°C.

3. Prepare a PCR mix using 11.2 $\mu$l $H_2O$, 2 $\mu$l 10× PCR buffer, 2.5 $\mu$l dNTP mix, 2.5 $\mu$l DMSO, 1.25 $\mu$l BSA solution (1.6 mg/ml), 0.2 $\mu$l each primer (100 $\mu$g/ml), and 0.75 U *Taq* DNA polymerase for each reaction.

4. To 5 μl DNA sample (in inactivated proteinase K buffer), add 20 μl PCR mix and amplify. A typical PCR program is: [93°C, 2 min; 60°C, 1 min; 65°C, 5 min] × 1; [93°C, 30 sec; 60°C, 30 sec; 65°C, 3 min] × 40; [65°C, 7 min] × 1; [4°C] indefinitely. The proper annealing temperature (here 60°C) must be determined empirically for each primer pair.

For each PCR experiment, genomic ES cell DNA and the PCR control construct are used as negative and positive controls. If the PCR reaction has been optimized by proper selection of primers, targeted clones can usually be identified easily by running the PCR products on ethidium bromide-containing agarose gels (Fig. 4). Optionally, the gels can be blotted on nitrocellulose filters and the filters probed with an oligonucleotide complementary to a sequence contained within the expected PCR fragment to confirm the results.

## E. Expansion of Targeted ES Cell Clones

If targeted ES cell colonies were identified according to Method 1, ES cell clones can immediately be expanded from the cultures in the 24-well plate. If clones were identified by partial picking of the colony (Method 2), the positive clone must be retrieved from the plate. To do this, pick colony as detailed in Method 1, trypsinize, and seed onto feeder layer in 24-well dish.

ES cells in the wells should become confluent after 3–5 days. Passage clones into 25-cm$^2$ flasks.

1. Remove medium from confluent wells and rinse once with PBS.
2. Add 3 drops 0.25% trypsin–EDTA to wells and return plate to incubator for at least 5 min.
3. Add 1 ml medium to each trypsinized cell layer.

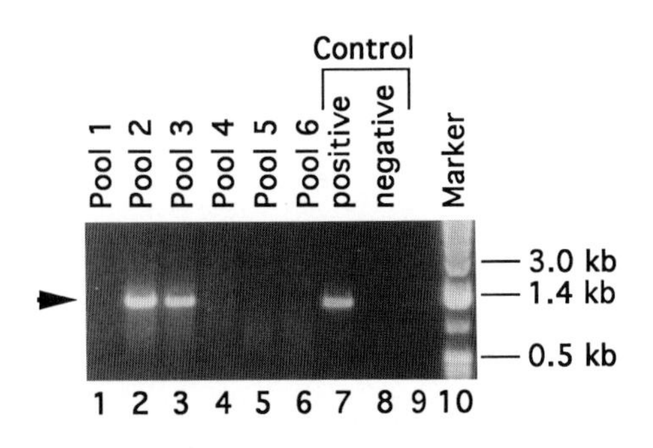


**Fig. 4** Identification of pools containing targeted ES cell colonies by PCR analysis. Pools of 10 ES cell clones/pool were analyzed by PCR for the disruption of the murine gp330 gene (lanes 1–6). Previously identified positive (lane 7) and negative (lane 8) ES cell clones were used as controls. The positive PCR signal is indicated by the arrowhead. DNA fragments were separated on a 0.8% agarose gel and stained with ethidium bromide.

4. Using a 1-ml Eppendorf pipet, vigorously and repeatedly pipet cells until a single-cell suspension is obtained. Assess successful trypsinization under the microscope.

5. Save 50 $\mu$l cell suspension to verify authenticity of clone by PCR. Transfer 100 $\mu$l to a gelatin-coated well without feeder layer, grow to confluence, and analyze genomic DNA by Southern blotting. Add remainder of cell suspension to 25-$cm^2$ flask containing fresh feeder layer.

6. The expanded culture should be confluent within another 3 days.

7. Freeze two aliquots from each targeted cell line.

## F. Generation of Germ Line Chimeras

The methodology of isolating early mouse enbryos, introducing ES cells into the blastocoel cavity, and reimplanting the blastocysts into the uterus of a pseudopregnant foster mother is too extensive to be covered in the space available. Excellent textbooks dealing with all aspects of this topic are available (Hogan *et al.,* 1986; Bradley, 1987).

## G. Isolation of Primary Embryonic Fibroblasts

Once ES cells carrying the mutated gene have given rise to chimeric animals that transmit the mutated allele through their germ line, a system has been established with which multiple aspects of the physiological function of this particular gene can be addressed. This can be done in the whole animal by comparing wild-type, heterozygous, and homozygous genetically deficient animals obtained by breeding heterozygous mice. This system also allows study of the function of the gene during development. Isolated cell cultures can be obtained from the adult animals or from embryos to answer specific questions about the function of this gene in specialized differentiated cell types including hematopoetic cells, immune system cells, nervous system cells, and hepatocytes (Pevny *et al.,* 1991; Thomas *et al.,* 1991; Rosahl *et al.,* 1993). If the gene defect is lethal in the homozygous state, embryos will often still provide an abundant source for the isolation of fibroblasts, which in turn may allow the study of the gene defect at the cellular level. Here we give a procedure for the isolation of embryonic fibroblasts from postimplantation embryos. Embryonic fibroblasts may be derived from embryos at almost any stage of pregnancy (usually between day 9.5 and 15), depending on the gene and its role in embryonic development.

1. Mate animals of the desired genotype (e.g., $+/- \times +/-$) and denote the morning on which the vaginal plug is observed as day 0.5 ($\sim$12 hr after the middle of the dark cycle).

2. Sacrifice the animal at the designated stage of pregnancy by cervical dislocation, wipe abdomen with a generous amount of 70% ethanol, and open abdominal cavity using scissors.

3. Pushing aside the gut, dissect out the uterine horns containing the embryos and remove them into a 60-mm petri dish containing 3 ml saline.

4. Open uterus with fine iris scissors, remove embryos, and dissect them from placenta and yolk sac.

5. Place each embryo separately into a sterile 60-mm petri dish containing 2–3 ml PBS.

6. If the embryos are older than day 12 of gestation, remove as much of the visceral organs (liver and heart) as possible.

7. Rinse the embryo generously with sterile PBS to wash away blood. In earlier embryos (day 9.5–12), this is not necessary.

8. Move embryo to a fresh sterile culture dish and tear it apart using forceps.

9. Transfer the tissue pieces to a sterile 50-ml plastic tube containing 1–5 ml (depending on size of embryo) ice-cold 0.05% trypsin–EDTA solution and incubate overnight at 4°C. Because of its reduced activity at 4°C, the trypsin will diffuse into the tissues during the overnight incubation without digesting the cells.

10. On the following day, remove excess trypsin solution without disturbing the sedimented embryonic tissue pellet.

11. Incubate samples at 37°C for 30 min.

12. Add 5–10 ml complete ES medium containing antibiotics and disaggregate tissue by vigorous pipetting.

13. Allow remaining clumps to sediment briefly.

14. Transfer the supernatant containing disaggregated cells to 75-$cm^2$ flasks or 100-mm culture dishes. Use 3 dishes (3 75-$cm^2$ flasks) per 15-day embryo, 2 dishes per 12–13 day embryo, and 1 dish for younger embryos. Incubate in a 5% $CO_2$ incubator until cells become confluent.

15. Freeze cells in aliquots as described earlier and denote passage number 1. Expand an aliquot of the cells for genotyping. An example of a Southern blot analysis of the genotypes of embryonic fibroblasts of 10.5-day embryos isolated from the mating of mice heterozygous for the disruption of the LDL receptor-related protein (LRP) gene (Herz *et al.*, 1993) is shown in Fig. 5. All

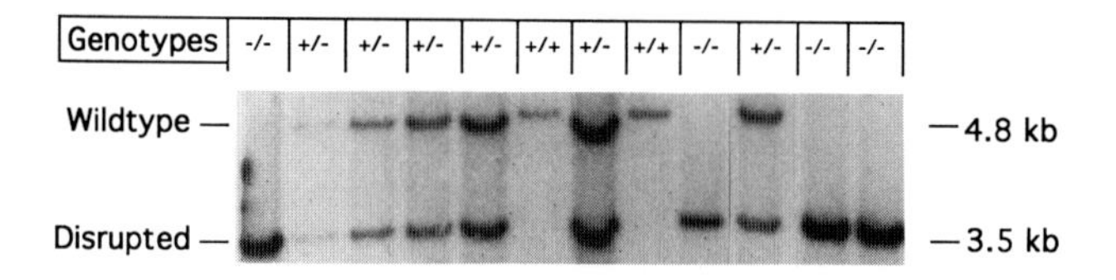


**Fig. 5** Southern blot analysis of primary mouse fibroblasts. Primary embryonic fibroblasts were isolated from mouse embryos resulting from the mating of mice heterozygous for the disruption of the LRP gene. Approximately 20 μg genomic DNA prepared from the fibroblasts was digested with *Xba* I and *Bam* HI and analyzed by Southern blotting using a genomic fragment of the LRP gene as a hybridization probe (Herz *et al.*, 1992). Hybridizing fragments representing the wild-type (4.8 kb) and the disrupted (3.5 kb) allele are indicated.

expected genotypes (wild-type, heterozygous, and homozygous deficient) are represented.

Late stage embryos (> day 13) yield abundant amounts of fibroblasts for experiments. Primary embryonic fibroblasts can be passaged several times before they become senescent. Early embryos may not yield enough cells to serve as a convenient steady source of fibroblasts. In this case, it may be of advantage to consider immortalization of the cells by transformation with SV40 T antigen (Livingston and Bradley, 1987).

## IV. Experimental Procedures: Analysis of Genetically Altered Cells and Mice

In this section, we describe three examples of how mutant cells and animals can be analyzed to characterize the phenotype caused by gene disruption. These methods have proven useful for the study of two endocytic receptors, the LDL receptor and the LRP. LDL receptor-deficient mice develop normally and the physiological phenotype can be studied in adult animals (Ishibashi *et al.*, 1993). LRP-deficient embryos die early during gestation (Herz *et al.*, 1992,1993). Embryonic fibroblasts were used to study the receptor defect *in vitro* (Willnow and Herz, 1994).

### A. Selection for Gene Conversion: Isolation of Toxin-Resistant Fibroblasts

LRP is a multifunctional endocytic receptor with functions in lipoprotein and protease metabolism (reviewed in Herz, 1993). In addition to its physiological ligands, LRP is also a receptor for *Pseudomonas aeruginosa* exotoxin A (PEA), which enters the cell through LRP, translocates across the endosomal membrane, and inactivates elongation factor 2 in the cytoplasm, thus inhibiting protein synthesis. PEA therefore can serve as a negative selectable marker against cells that express LRP. The following protocol was devised to select among a starting population of heterozygous (LRP+/−) embryonic fibroblasts for cells in which the wild-type allele has been converted to the disrupted allele (LRP−/−).

1. Grow LRP+/− fibroblasts to confluence (5 75-$cm^2$ flasks) in complete medium.
2. Add PEA (100 $\mu$g/ml stock in PBS) to final concentration of 30 ng/ml.
3. Incubate for 24 hr.
4. Replace medium with fresh fibroblast medium without PEA. Almost all cells in the flask will die within 48–72 hr.
5. Replace medium every 3 days. Between 10 and 20 colonies will become apparent within 1–2 wk.

6. Once the colonies have reached a diameter of approximately 4–5 mm, trypsinize each flask separately and reseed into a fresh flask.
7. Grow to confluence.
8. Add PEA (100 $\mu$l/ml stock in PBS) to final concentration of 30 ng/ml and incubate for 24 hr.
9. Replace medium with fresh fibroblast medium without PEA.
10. Trypsinize surviving cells and reseed onto a 100-mm culture dish at a density of approximately 500 cells/plate.
11. Once colonies are beginning to form, subject cells to a third round of toxin selection (PEA at 30 ng/ml for 24 hr).
12. Pick individual colonies 72 hr later and expand initially in 24-well plates.
13. Analyze genotype by Southern blotting.

Approximately 50% of the clones should be homozygous for the LRP disruption. The remainder of the PEA-resistant clones carry mutations in other genes (e.g., elongation factor 2) that render them toxin resistant.

## B. *In Vitro* Analysis of a Receptor-Defective Genotype: Degradation of Ligands by Cultured Cells

Both LRP and the LDL receptor are endocytic receptors that internalize ligands that bind to the receptors on the cell surface. $^{125}$I-Labeled ligands that have been endocytosed by either receptor are eventually degraded in the lysosomes, leading to the release of trichloroacetic acid (TCA)-soluble [$^{125}$I]monoiodotyrosine into the culture medium. The rate of accumulation of TCA-soluble radioactivity can be used as a measure for receptor activity. A detailed method for measuring the lysosomal degradation of $^{125}$I-labeled LDL by cultured cells has been described by Goldstein *et al.* (1983). This method has been adapted in the following time-course experiment to assay for the ability of wild-type, heterozygous, and homozygous LRP-deficient fibroblasts to degrade the $^{125}$I-labeled ligands $\alpha_2$-macroglobulin and complexes of urokinase with plasminogen activator inhibitor 1 (uPA/PAI-1). To determine the time course of ligand degradation by the cells, addition of the radiolabeled ligand to the cultures is staggered and degradation is terminated simultaneously in all wells.

1. On day 0, trypsinize stock cultures of cells of the different genotypes and seed into 12-well dishes at $2 \times 10^5$ cells/well. Grow overnight in fibroblast medium.

2. On day 1, aspirate medium and add 1 ml prewarmed degradation medium (DMEM without glutamine, with 0.2% BSA). Incubate for 20 min.

3. Remove medium from wells representing the longest time point and add 1 ml degradation mix ($^{125}$I-labeled ligand in degradation medium; specific activities approximately 1000 cpm/ng; final concentration 50–100 ng/ml for uPA/PAI-1, 1–10 $\mu$g/ml for $\alpha_2$-macroglobulin).

4. Return plate to incubator.

5. Proceed the same way with the other time points. For blank values, determine nonspecific degradation in wells that do not contain cells.

To determine the amount of TCA-soluble degradation products at the end of the time course:

6. Chill multiwell plates to 4°C.

7. Transfer 600 $\mu$l medium from each well to 1.5-ml microfuge tubes.

8. Add 300 $\mu$l 30% TCA (final concentration 10%), vortex, and incubate on ice for 30 min.

9. Spin down the precipitate in a microcentrifuge for 10 min at maximum speed.

10. Transfer 500 $\mu$l supernatant into disposable glass tubes (13 $\times$ 100 mm).

11. Add 5 $\mu$l 40% potassium iodide and let sit for 5 min at room temperature.

12. Add 20 $\mu$l 30% hydrogen peroxide; vortex. Incubate 5 min at room temperature.

13. Add 1 ml chloroform to each tube, vortex, and incubate for 5 min.

14. Remove 250 $\mu$l aqueous (upper) phase and determine radioactivity.

15. To calculate the specific degradation, subtract the no-cell blank from each value and multiply by 6 to obtain the total degraded counts per well.

Receptor-dependent cellular degradation of ligands is best expressed as amount of ligand degraded/mg cell protein/hr. To determine cellular protein content of the wells, proceed as follows.

16. Aspirate residual medium from the wells at 4°C.

17. Wash cell layers twice for 5 min with 1 ml wash buffer.

18. Remove wash solution and add 1 ml 0.1 *N* NaOH to each well.

19. Incubate at room temperature for 20 min with vigorous shaking on a rocker platform.

20. Determine protein concentration in the cell lysate (e.g., using the Pierce Coomassie Reagent, Cat. No. 23236, Pierce, Rockford, IL).

Figure 6 shows an example of a degradation experiment in which the ability of murine embryonic fibroblasts of different genotypes (LRP+/+, +/−, and −/−) to degrade $^{125}$I-labeled uPA/PAI-1 complexes was determined.

## C. $^{125}$I-Labeled Ligand Turnovers in Mice

An important method with which to study the physiological effects of the functional elimination of endocytic cell surface receptors is the ligand turnover experiment *in vivo* (Ishibashi *et al.*, 1993). $^{125}$I-Labeled ligands are prepared essentially the same way as for cell culture degradation experiments (Goldstein

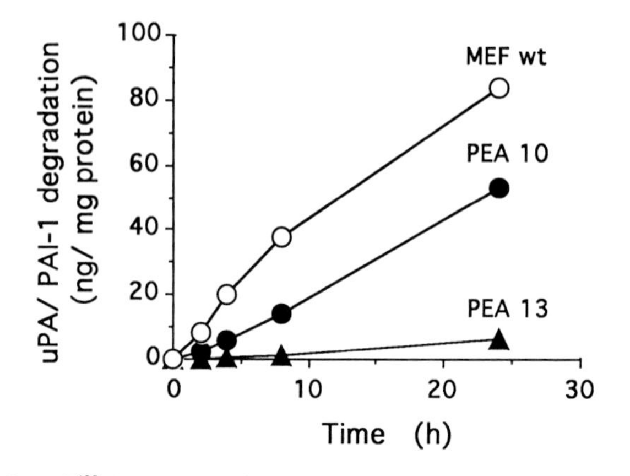


**Fig. 6** Degradation of $^{125}$I-labeled uPA/PAI-1 complex by mouse fibroblasts. Replicate monolayers of cells that were either wild-type (MEF wt, open circles), heterozygous (PEA 10, closed circles), or homozygous (PEA 13, closed triangles) for a disruption of the LRP gene locus received 1 ml DMEM (without glutamine) containing 0.2% BSA (w/v) and human $^{125}$I-labeled uPA/PAI-1 complex (50 ng/ml uPA, specific activity 7100 cpm/ng). After incubation at 37°C for the indicated periods of time, the amount of uPA degradation products secreted into the medium was determined. Each value represents the mean of duplicate incubations.

*et al.*, 1983; Herz *et al.*, 1991), except that they should be diluted in TBS containing 0.2% BSA prior to injection. Carefully plan the time course of the experiment and make sure that injection syringes and blood collection tubes are prepared in advance. Allow sufficient time for surgery, injections, wound closure, and blood collection. We find ligand turnovers to be highly reproducible among individual animals, provided injections and blood collections adhere strictly to the time course.

1. Anesthetize mouse by intraperitoneal injection of Nembutal® (12.5 mg/ml in PBS, 80–100 $\mu$g/g body weight).
2. Immobilize the animal by taping the limbs to a flat surface (e.g., kitchen tray). Make sure the ambient temperature is not too low (ideally between 25 and 30°C). Illumination with a 100 W light bulb prevents the body temperature from dropping during the injection.

Although it is possible to inject a mouse through the tail vein, we prefer the external jugular vein as the injection site. The main reason is that this vein can be easily exposed by a simple incision of the skin, allowing complete visual control over the injection process. This reduces the failure rate of injection and insures that no $^{125}$I-labeled ligand is lost into the perivascular space, leading to high variability of results among individual animals. Thus, statistically significant results can be obtained with fewer animals.

3. Select injection site (right or left side of neck). Turn the head of the mouse to the opposite side and extend the neck slightly. Wipe neck with a liberal quantity of 70% ethanol.

4. Using pointed scissors, cut skin in the midclavicular line from 2 mm below to 5–6 mm above the clavicle (collar bone). Cut through the fascia and spread apart subcutaneous fat and connective tissue using the tip of the scissors and/or blunt forceps until the vein is exposed.

5. To make the vein appear more prominent, lightly compress the thoracic wall just underneath the clavicle with the tip of a blunt forceps. Position the injection needle (bevel up) on the muscle but above the bone, approximately 0.5–1 mm caudal to the point at which the vein disappears behind the clavicle.

6. Piercing through a few muscle bundles, gently push the needle up toward the head into the vein. This technique greatly reduces bleeding from the injection site. Make certain that the needle is completely inside the vessel by moving the tip up against the vessel wall and down into the bloodstream. Inject $^{125}$I-labeled ligand (maximum volume, 250 $\mu$l) over a period of 10 sec.

7. Pull out needle and close wound with one metallic wound clip.

8. Exactly 1 min after the injection, collect 70–100 $\mu$l blood into a heparinized pasteur pipet from the retro-orbital venous plexus. Gently blow blood from the pipet into a microcuvette containing EDTA (Sarstedt).

9. Place animal onto a warming plate (e.g., slide warmer) preheated to 35–36°C. It is very important to prevent the animals from losing body heat because the resulting centralization of the circulation will adversely affect the reproducibility of the experiment.

10. Collect blood at designated time points. Keep samples on ice until the end of the experiment. After the last time point, kill animal by neck dislocation or by Nembutal® injection and collect organs for further analysis, if desired.

The amount of radioactive ligand injected, as well as the time course of the experiment, depends on the properties of the particular ligand studied. The systemic clearance of 5 $\mu$g $^{125}$I-labeled asialofetuin or $\alpha_2$-macroglobulin is very rapid in a normal animal (>90% cleared within 20 min). The clearance of LDL, on the other hand, is relatively slow even in the normal animal ($t_{1/2}$ ~5 hr for human LDL in the mouse). To determine the rate of clearance:

1. Centrifuge blood samples in a microcentrifuge for 2 min at 10,000 rpm.
2. Transfer 20 $\mu$l plasma to a 1.5-ml microfuge tube and add 1 ml 10% TCA. Vortex and incubate at room temperature for 5 min.
3. Spin down the precipitate in a microcentrifuge (10,000 rpm, 2 min).
4. Determine the TCA-soluble radioactivity in the supernatant and the TCA-precipitable counts in the pellet.

## V. Discussion

The methods described in this chapter have focused mainly on the application of homologous recombination to disrupt genes in ES cells with the goal to generate a strain of mice carrying a specific gene defect. However, the methodology is not strictly limited to the ES cell system but can be adapted to a variety of other cell types. Selected examples are given in the reference list. Although we have focused here on the use of this technology to study the properties of endocytic cell surface receptors, it is certainly not restricted to this narrow application but could be useful in many different aspects of cell biology. Given a suitable design of the targeting vector, it is also possible to generate mutant genes that may express a truncated version of the native protein, for example, lacking a regulatory domain. Likewise, it is possible to introduce subtle mutations into the coding region or to change promoter sequences.

Similar to the breakthrough caused by the advent of monoclonal antibodies, homologous recombination has revolutionized the field of mammalian genetics in only a few years. The ability to alter specifically the genetic make-up of whole animals has had significant impact on a variety of other areas including immunology, neurobiology, cell biology, and medicine in general. "Knocking out" specific genes may be expected to become an essential tool for the investigation of the functions of gene families.

## Acknowledgments

We thank Tom Wilkie and Mary Barnard for critical reading of the manuscript. Many of the experimental procedures and details in this chapter have been adapted from or been modified after countless helpful suggestions from our colleagues. We are particularly endebted to Phil Soriano, Bob Hammer, David Clouthier, Shun Ishibashi and Kathy Graves for sharing protocols and for their collaboration on projects in the course of which these methods have been tested and refined. This work was supported by National Institutes of Health grant HL-20948. T. E. W. is supported by a training grant from the Deutsche Forschungsgemeinschaft (DFG). J. H. is a Lucille P. Markey Scholar and is supported by the Syntex Scholar Program and the Perot Family Foundation.

## References

Bradley, A. (1987). Production and analysis of chimeric mice. *In* "Teratocarcinomas and Embryonic Stem Cells: A Practical Approach" (E. J. Robertson, ed.), pp. 113–151. IRL Press, Oxford.

Bradley, A., Evans, M., Kaufmann, M. H., and Robertson, E. (1984). Formation of germ-line chimaeras from embryo-derived teratocarcinoma cell lines. *Nature (London)* **309,** 255–256.

Brinster, R. L. (1965a). Studies on the development of mouse embryos in vitro. I. The effect of osmolarity and hydrogen ion concentration. *J. Exp. Zool.* **158,** 49–57.

Brinster, R. L. (1965b). Studies on the development of mouse embryos in vitro. II. The effect of energy source. *J. Exp. Zool.* **158,** 59–68.

Brinster, R. L. (1965c). Studies on the development of mouse embryos in vitro. III. The effect of fixed-nitrogen source. *J. Exp. Zool.* **158,** 69–77.

Capecchi, M. R. (1989a). Altering the genome by homologous recombination. *Science* **244,** 1288–1292.

Capecchi, M. R. (1989b). Altering the genome by gene targeting. *Trends Genet.* **5,** 70–76.

Cremers, F. P. M., van de Pol, D. J. R., van Kerkhoff, L. P. M., Wieringa, B., and Ropers, H.-H. (1990). Cloning of a gene that is rearranged in patients with choroideremia. *Nature (London)* **347,** 674–677.

De Lozanne, A., and Spudich, J. A. (1987). Disruption of the *Dictyostelium* myosin heavy chain gene by homologous recombination. *Science* **236,** 1086–1091.

Doetschmann, T., Gregg, R. G., Maeda, N., Hooper, M. L., Melton, D. W., Thompson, S., and Smithies, O. (1987). Targetted correction of a mutant HPRT gene in mouse embryonic stem cells. *Nature (London)* **330,** 576–578.

Evans, M. J., and Kaufmann, M. H. (1981). Establishment in culture of pluripotential cells from mouse embryos. *Nature (London)* **292,** 154–156.

Fraker, P. J., and Speck, J. C. (1978). Protein and cell membrane iodinations with a sparingly soluble chloroamide, 1,3,4,6-tetrachloro-3a,6a-diphenylglycoluril. *Biochem. Biophys. Res. Commun.* **80,** 849–857.

Goldstein, J. L., Basu, S. K., and Brown, M. S. (1983). Receptor-mediated endocytosis of LDL in cultured cells. *Meth. Enzymol.* **98,** 241–260.

Herz, J. (1993). The LDL receptor-related protein—Portrait of a multifunctional receptor. *Curr. Opin. Lipidol.* **4,** 107–113.

Herz, J., Goldstein, J. L., Strickland, D. K., Ho, Y. K., and Brown, M. S. (1991). 39-kDa protein modulates binding of ligands to low density lipoprotein receptor-related protein/$\alpha_2$-macroglobulin receptor. *J. Biol. Chem.* **266,** 21232–21238.

Herz, J., Clouthier, D. E., and Hammer, R. E. (1992). LDL receptor-related protein internalizes and degrades uPA/PAI-1 complexes and is essential for embryo implantation. *Cell* **71,** 411–421.

Herz, J., Clouthier, D. E., and Hammer, R. E. (1993). Correction: LDL receptor-related protein internalizes and degrades uPA-PAI-1 complexes and is essential for embryo implantation. *Cell* **73,** 428.

Hinnen, A., Hicks, J. B., and Fink, G. R. (1978). Transformation of yeast. *Proc. Natl. Acad. Sci. USA* **75,** 1929–1933.

Hogan, B., Constantini, F., and Lacy, E. (1986). "Manipulating the Mouse Embryo." Cold Spring Harbor Laboratory Press, Cold Spring Harbor, New York.

Ishibashi, S., Brown, M. S., Goldstein, J. L., Gerard, R. D., Hammer, R. E., and Herz, J. (1993). Hypercholesterolemia in LDL receptor knockout mice and its reversal by adenovirus-mediated gene delivery. *J. Clin. Invest.* **92,** 883–893.

Jasin, M., and Berg, P. (1988). Homologous integration in mammalian cells without target gene selection. *Genes Dev.* **2,** 1353–1363.

Livingston, D. M., and Bradley, M. K. (1987). The Simian Virus 40 large T antigen—A lot packed into a little. *Mol. Biol. Med.* **4,** 63–80.

Mansour, S. L., Thomas, K. R., and Capecchi, M. R. (1988). Disruption of the proto-oncogene *int*-2 in mouse embryo-derived stem cells: A general strategy for targeting mutations to non-selectable genes. *Nature (London)* **336,** 348–352.

Martin, G. A., Viskochil, D., Bollag, G., McCabe, P. C., Crosier, W. J., Haubruck, H., Conroy, L., Clark, R., O'Connell, P., Cawthon, R. M., Innis, M. A., and McCormick, F. (1990). The GAP-related domain of the neurofibromatosis type 1 gene product interacts with ras p21. *Cell* **63,** 843–849.

Mortensen, R. M., Conner, D. A., Chao, S., Geisterfer-Lowrance, A. A. T., and Seidman, J. G. (1992). Production of homozygous mutant ES cells with a single targeting construct. *Mol. Cell. Biol.* **12,** 2391–2395.

Orr-Weaver, T., Szostak, J. W., and Rothstein, R. J. (1981). Yeast transformation: A model system for the study of recombination. *Proc. Natl. Acad. Sci. USA* **78,** 6354–6358.

Pevny, L., Simon, M. C., Robertson, E., Klein, W. H., Tsai, S.-F., D'Agati, V., Orkin, S. H., and Constantini, F. (1991). Erythroid differentiation in chimaeric mice blocked by a targeted mutation in the gene for transcription factor GATA-1. *Nature (London)* **349,** 257–260.

Robertson, E. J. (1987). Embryo-derived stem cell lines. *In* "Teratocarcinomas and Embryonic Stem Cells: A Practical Approach" (E. J. Robertson, ed.), pp. 71–112. IRL Press, Oxford.

Rosahl, T. W., Geppert, M., Spillane, D., Herz, J., Hammer, R. E., Malenka, R. C., and Südhof, T. C. (1993). Short-term synaptic plasticity is altered in mice lacking synapsin I. *Cell* **75,** 661–670.

Sambrook, J., Fritsch, E. F., and Maniatis, T. (1989). "Molecular Cloning: A Laboratory Manual." Cold Spring Harbor Laboratory Press, Cold Spring Harbor, New York.

Seabra, M. C., Brown, M. S., Slaughter, C. A., Südhof, T. C., and Goldstein, J. L. (1992). Purification of component A of Rab geranylgeranyl transferase: Possible identity with the choroideremia gene product. *Cell* **70,** 1049–1057.

Sedivy, J. M., and Sharp, P. A. (1989). Positive genetic selection for gene disruption in mammalian cells by homologous recombination. *Proc. Natl. Acad. Sci. USA* **86,** 227–231.

Smithies, O. (1993). Animal models of human genetic diseases. *Trends Genet.* **9,** 112–116.

Smithies, O., Gregg, R. G., Boggs, S. S., Koralewski, M. A., and Kucherlapati, R. S. (1985). Insertion of DNA sequences into the human chromosomal $\beta$-globin locus by homologous recombination. *Nature (London)* **317,** 230–234.

Snouwaert, J. N., Brigman, K. K., Latour, A. M., Malouf, N. N., Boucher, R. C., Smithies, O., and Koller, B. H. (1992). An animal model for cystic fibrosis made by gene targeting. *Science* **257,** 1083–1088.

Soriano, P., Montgomery, C., Geske, R., and Bradley, A. (1991). Targeted disruption of the c-*src* proto-oncogene leads to osteopetrosis in mice. *Cell* **64,** 693–702.

Strickland, D. K., Bhattacharya, P., and Olson, S. (1984). Kinetics of the conformational alterations associated with nucleophilic modification of $\alpha_2$-macroglobulin. *Biochemistry* **23,** 3115–3123.

te Riehle, H., Maandag, E. R., Clarke, A., Hooper, M., and Berns, A. (1990). Consecutive inactivation of both alleles of the pim-1 proto-oncogene by homologous recombination in embryonic stem cells. *Nature (London)* **348,** 649–651.

te Riehle, H., Maandag, E. R., and Berns, A. (1992). Highly efficient gene targeting in embryonic stem cells through homologous recombination with isogenic DNA constructs. *Proc. Natl. Acad. Sci. USA* **89,** 5128–5132.

Thomas, J. E., Soriano, P., and Brugge, J. S. (1991). Phosphorylation of c-*src* on tyrosine 527 by another protein tyrosine kinase. *Science* **254,** 568–570.

Thomas, K. R., and Capecchi, M. R. (1987). Site-directed mutagenesis by gene targeting in mouse embryo-derived stem cells. *Cell* **51,** 503–512.

Willnow, E., and Herz, J. (1994). Genetic deficiency in low density lipoprotein receptor-related protein (LRP) confers cellular resistance to pseudomonas exotoxin A—evidence that LRP is required for uptake and degradation of multiple ligands. *J. Cell Science* **107,** 719–726.

Zheng, H., and Wilson, J. H. (1990). Gene targeting in normal and amplified cell lines. *Nature (London)* **344,** 170–173.

# CHAPTER 16

# Regulation of Protein Activities by Fusion to Steroid Binding Domains

**Tiziana Mattioni, Jean-François Louvion, and Didier Picard**

Département de Biologie Cellulaire
Université de Genève
CH-1211 Genève 4, Switzerland



## I. Introduction

### A. A Novel Approach to Constructing Conditionally Active Proteins

With few exceptions, cell biologists have relied on inducible expression systems when they wanted inducible protein activity. Not surprisingly, this solution is often associated with severe disadvantages such as a long lag phase for protein accumulation. Until a few years ago, only a few alternative techniques, for example microinjection and the use of temperature-sensitive mutants, provided a way to circumvent the difficulties inherent in inducible expression systems. Table I compares the features of some existing inducible systems.

In this chapter we will discuss a novel approach that involves the post-translational regulation of a constitutively expressed fusion protein (Picard *et al.*, 1988; Yamamoto *et al.*, 1988; reviewed by Picard, 1993). This approach overcomes many of the shortcomings of the other systems mentioned in Table I. The method is based on the fact that the hormone binding domain (HBD) of steroid receptors can be used as an autonomous regulatory cassette to subject many heterologous protein functions to hormonal control *in cis*. Fusion proteins are maintained in an inactive state in the absence of hormone and are rapidly activated by addition of the cognate hormone.

## B. A Model for the Regulatory Mechanism

We previously hypothesized that the inactivation of functions within the heterologous moiety of a fusion protein by the unliganded HBD is mediated

**Table I**
**Induction Systems for Studying Proteins *in Vivo*[a]**

| System/approach | Activation of endogenous genes | Lag | Basal level | Number of plasmids introduced | Further comments |
|---|---|---|---|---|---|
| Viral infection (vaccinia, VSV, SV40, adenovirus, etc.) | | Yes | No | 0–1 | Often severe side effects on cellular metabolism; host range restricted |
| Transcriptional systems | | | | | |
| T7, T3 RNA polymerase | No | Yes | Yes | 1–2 | Needs vaccinia superinfection for efficient translation |
| MMTV (GRE) | Yes | Yes | Yes | 1 | Requires endogenous steroid receptor |
| Heat-shock promoter | Yes | Yes | Yes | 1 | Severe side effects on cells |
| Metal-regulated promoter | Yes | Yes | Yes | 1 | Heavy metals are toxic |
| GAL4.ER.VP16 | No | Yes | Yes | 2 | |
| *lac* repressor | No | Yes | Yes | 2 | Inducer IPTG inefficient and expensive |
| *Tet* repressor | No | Yes | Yes | 2 | Repression in presence of tetracycline |
| Microinjection[b] | No | No | No | 0 | Need protein; biochemical analysis difficult |
| Temperature-sensitive mutants[b] | Yes | No | No | 1 | Rarely available |
| **Fusion to steroid binding domain[b]** | **No** | **No** | **No–yes** | **1** | **Five different regulators** |

[a] Examples are restricted to vertebrate systems, mostly tissue culture cells. The comparison, however, also applies to other organisms.

[b] Post-transcriptional systems.

by a complex containing heat-shock protein 90 (HSP90) (Picard *et al.*, 1988; Yamamoto *et al.*, 1988; further discussed by Picard, 1993). HSP90, as well as several other proteins, is associated with the unliganded HBDs of all five vertebrate steroid receptors, that is, the glucocorticoid (GR), mineralocorticoid (MR), androgen (AR), progesterone (PR), and estrogen (ER) receptors (for review, see Pratt, 1990; Smith and Toft, 1993). Hormone binding results biochemically in the release of the HSP90 complex and genetically in protein activation (or derepression). The hormone-reversible protein inactivation function of the HBD may therefore work by a relatively unspecific mechanism involving steric hindrance by the HSP90 complex. This method should be applicable to any protein with at least one essential function that is sensitive to steric hindrance. Hence, this regulatory system should work in the cytosolic and nuclear compartments of any organism that provides the components of the HSP90 complex.

## II. Advantages and Limits

### A. Advantages

The advantages can be summarized easily. (1) HBDs can subject a wide variety of heterologous proteins to hormonal control. (2) Induction (activation) occurs with very rapid kinetics (seconds to minutes) and is easily reversible. (3) Intermediate levels of induction can be obtained at subsaturating concentrations of hormone. (4) Activation of endogenous steroid receptors can be avoided by using a distinct one of the five different available HBDs, each with a distinct ligand specificity. (5) Given a set of five regulatory domains, several proteins can be regulated independently in the same cell. (6) The system has been shown to work in vertebrate cells and in yeast, and is likely to function in cells of a wide range of other organisms including plants and insects (for discussion, see Picard, 1993).

#### 1. Regulation of a Wide Variety of Functions

A long list (Table II) of regulatable heterologous proteins has accumulated since the first report in 1988 showing that the adenovirus protein E1A can be subjected to hormonal control by fusion to an HBD (Picard *et al.*, 1988). Table II shows that the approach is not restricted to transcription factors. Notably, the list includes the RNA-binding protein Rev and tyrosine and serine/threonine kinases such as Abl, Src, STE11, and Raf1. Considering the aforementioned model, one could expect that the HSP90 complex might block any functional domain that requires the access of macromolecular partners (DNA, proteins, RNA, polysaccharides). In contrast, the interaction with small molecules might not be regulatable. Indeed, preliminary findings with β-galactosidase, galactoki-

**Table II**
**Current List of Regulatable Heterologous Proteins**

| Protein X | Regulated as [a] | Reference |
|---|---|---|
| E1A (adenovirus) | Transcription factor | Picard *et al.* (1988) |
| E1A | Oncoprotein | Spitkovsky *et al.* (1994) |
| c-Myc | Oncoprotein | Eilers *et al.* (1989) |
| v-Myb | Transcription factor | Burk and Klempnauer (1991) |
| c-Fos, v-Fos, FosB-L, FosB-S | Oncoprotein, transcription factor | Superti-Furga *et al.* (1991); Schuermann *et al.* (1993) |
| Jun | Transcription factor | H. Beug and M. Busslinger, personal communication. |
| GCN4 | Transcription factor | Fankhauser *et al.* (1994) |
| C/EBP | Transcription factor | Umek *et al.* (1991) |
| v-Rel | Oncoprotein, transcription factor | Boehmelt *et al.* (1992) |
| GATA-1, -2, -3 | Transcription factor, promoter of proliferation (GATA-2) | Briegel *et al.* (1993) |
| GAL4-VP16 | Transcription factor in yeast and in tissue culture cells | Braselmann *et al.* (1993); Louvion *et al.* (1993) |
| GAL4 | Transcription factor in yeast | J.-F. Louvion and D. Picard, unpublished data. |
| MyoD | Transcription factor | Hollenberg *et al.* (1993) |
| p53 | Transcription factor, tumor suppressor | Roemer and Friedmann (1993) |
| E7 (HPV16) | Oncoprotein | J. M. Bishop, personal communication. |
| Rev (HIV) | Transactivation (RNA-binding protein) | Hope *et al.* (1990) |
| c-Abl | Oncoprotein, tyrosine kinase | Jackson *et al.* (1993) |
| Src | Tyrosine kinase | J. M. Bishop, personal communication. |
| erbB1 | Tyrosine kinase | J. M. Bishop, personal communication. |
| STE11 | Serine/threonine kinase in yeast | J.-F. Louvion and D. Picard, unpublished data. |
| Raf1 | Oncoprotein, serine/threonine kinase | Samuels *et al.* (1993) |

[a] Fusion proteins were assayed in vertebrate tissue culture cells unless indicated otherwise.

nase, dihydrofolate reductase, and URA3 show that HBDs cannot inactivate their enzymatic functions in the absence of hormone (D. Picard, unpublished results). The enzymatic activity of dihydrofolate reductase fused to the GR HBD is even slightly reduced on hormone addition (Israel and Kaufman, 1993; D. Picard, unpublished results).

It is generally sufficient to regulate only one essential function of a heterologous protein to establish overall hormonal control. For example, in the case of a transcription factor, its activity will be hormonally regulated irrespective of whether it is only the dimerization, the nuclear localization, the DNA binding, or the transcriptional regulatory function that is directly inactivated by the

HBD. Indeed, there are few cases where the precise activity that is inhibited has been determined (for discussion, see Picard, 1993).

### 2. Combinatorial Regulation with Several Signals

Experimentally, a hormone-reversible protein inactivation function has been demonstrated for the HBDs of GR (Picard *et al.*, 1988), ER (Eilers *et al.*, 1989), MR (Fankhauser *et al.*, 1994), and AR (F. Stewart, personal communication). Our model predicts that the HBD of PR should also work. Hence, in all likelihood, there are up to five different regulatory domains that can all be regulated independently and specifically with their cognate ligands. One can therefore avoid the activation of endogenous steroid receptors, which may be expressed in a particular vertebrate cell line, by choosing the appropriate regulatory domain. It should often be possible to regulate several fusion proteins, each with a different HBD, in the same cell.

Whether the ligand binding domains of other members of the nuclear receptor superfamily may be widely applicable for this type of regulation is currently unclear. The HBDs of the thyroid receptor (Hollenberg *et al.*, 1993) and the *Drosophila* ecdysone receptor (Christopherson *et al.*, 1992) have been reported to work for the regulation of fusion proteins. However, we have been unable to confirm this finding for the ecdysone receptor HBD using Fos as a test protein (M. Worek and D. Picard, unpublished results). Since these types of HBDs do not appear to form hormone-reversible stable complexes with HSP90, their regulatory potential may be limited to special cases and may depend on the interaction with other proteins such as the retinoic acid X receptor (RXR).

## B. Limits

The following concerns will be discussed in this chapter: (1) Will the heterologous protein tolerate a fusion? (2) Does the fusion protein have qualitatively altered properties? (3) Is there a basal level of expression?

The approach can only work if the protein of interest tolerates fusion to the rather large (about 300 amino acids) regulatory domain. It is often difficult to predict whether N- or C-terminal or even internal additions are compatible with the activity of the heterologous moiety. The number of successful examples (see Table II) is indeed quite surprising, confirming that many proteins consist of fairly independent modules.

A priori it is difficult to exclude the possibility that the activity of a fusion protein may be different than that of the wild-type unfused protein. Remarkably, fusion proteins have largely been found to have full or only slightly reduced activity after hormone induction (see references in Table II). A major concern is the possibility that the fusion protein may have qualitatively different properties than the wild-type protein. This concern stems from the complication that

HBDs carry more than the inactivation function; they also contain hormone-dependent nuclear localization (Picard and Yamamoto, 1987; Guiochon-Mantel *et al.*, 1989; Ylikomi *et al.*, 1992), dimerization (Kumar and Chambon, 1988; Wrange *et al.*, 1989; Fawell *et al.*, 1990), and transactivation (denoted TAF-2) functions (reviewed by Gronemeyer, 1991). With one exception (see subsequent discussion), significant alterations of activity have not been reported to date and should be avoidable in most cases. Only the HBD of GR contains an autonomously active and hormone-dependent nuclear localization signal (Picard and Yamamoto, 1987; Picard *et al.*, 1990b). The dimerization activity of the HBD appears to be relatively weak, which is consistent with the notion that fusion proteins with Fos (Superti-Furga *et al.*, 1991) and Myc (Eilers *et al.*, 1989) were still able to heterodimerize with their partners Jun and Max, respectively. The transactivation function of the HBD (TAF-2) is a potential problem in fusions with transcription factors. The contribution of TAF-2 has been shown to be considerable with weak transactivators (Schuermann *et al.*, 1993). However, this problem could be avoided altogether by using specific TAF-2 mutants (Danielian *et al.*, 1992; Ince *et al.*, 1993; Wrenn and Katzenellenboger, 1993). Note that HBD functions other than the inactivation function have also been exploited to regulate heterologous proteins. The nuclear localization and dimerization functions have been exploited for the regulation of Rev (Hope *et al.*, 1990,1992) and c-Abl (Jackson *et al.*, 1993), respectively.

Another common concern about inducible systems is the basal level of activity under uninduced conditions. Ultimately the basal level of activity of a fusion protein must be determined experimentally. As for any other inducible system, the basal level that can be tolerated will depend on the biological assay system. Note also that the position of the HBD relative to active sites in the fusion protein will determine whether an activity is inhibited. As previously summarized (Picard, 1993), often only a subset of all functions of a heterologous moiety is regulated. Despite considerable flexibility of the regulatory mechanism, the tightness of regulation must depend on the positioning of heterologous function(s) relative to the HBD or, more specifically, to components of the HBD–HSP90 complex. This issue will be further illustrated in our discussion of E1A fusion proteins.

## III. Designing and Using Fusion Proteins

### A. Choice of Hormone Binding Domain

The choice of HBD is primarily dictated by the biological system rather than by the heterologous protein itself, since the activation of endogenous steroid receptors is usually undesirable. Although the expression pattern of steroid receptors may not be known for a given cell line, the literature on the tissue- and cell-type-specific effects of steroid hormones is vast. GR is almost ubiquitously expressed, at least in mammals, whereas other steroid receptors are much more tissue specific; MR may display the narrowest distribution. Indeed, by choosing between the MR and ER HBDs, cross-talk with endogenous receptors should almost always be avoidable. Free choice applies to other organisms such as yeast, plants, and invertebrates because vertebrate steroids tend to have no or little effect on their physiology.

Although the GR HBD has been used extensively (Picard *et al.*, 1988; Eilers *et al.*, 1989; Hope *et al.*, 1990,1992; Superti-Furga *et al.*, 1991; Umek *et al.*, 1991; Hollenberg *et al.*, 1993; Jackson *et al.*, 1993; Louvion *et al.*, 1993), it should be reserved for special cases. Disadvantages of the GR HBD are (1) the potential cross-activation of endogenous GR and (2) a severe loss of hormone binding affinity by about two orders of magnitude when the HBD is isolated from the intact GR (Rusconi and Yamamoto, 1987). As a consequence, the addition of high concentrations of glucocorticoids may even cross-activate other steroid receptors such as the MR. Moreover, in yeast very high ligand concentrations are needed to induce GR (see, for example, Picard *et al.*, 1990a,c; Wright *et al.*, 1990; Garabedian and Yamamoto, 1992; Wright and Gustafsson, 1992); thus, it may even be difficult to achieve sufficient hormone concentrations for full activation of the GR HBD.

In contrast, mutant HBDs with only moderately reduced hormone binding affinity are extremely useful tools to avoid activation by steroids that may be present in the biological assay system. For example, it is often difficult to avoid partial activation of the wild-type ER HBD by estrogens in tissue culture medium. The point mutation Gly 400 to Val 400 in the human ER reduces affinity about 10-fold (Tora *et al.*, 1989), rendering the HBD less sensitive to estrogenic contaminants in medium. HBD mutants with altered specificities offer the additional advantage that endogenous receptors may not be activated or even repressed. For example, a C-terminal deletion mutant of the human PR has been described that, instead of being activated by progesterone, now responds to RU486, normally a PR antagonist (Vegeto *et al.*, 1992). Moreover, several point mutations in the ER HBD strongly reduce estrogen binding and receptor activation but allow the estrogen antagonist hydroxytamoxifen to act as an agonist (Danielian *et al.*, 1993; Ince *et al.*, 1993; Wrenn and Katzenellenbogen, 1993).

## B. Designing Fusion Proteins and Their Expression Vectors

For maximal tightness of control, it is advisable to place the HBD relatively "close" to an essential function of the heterologous moiety despite all the apparent flexibility of the regulatory mechanism. Spacial proximity is almost certainly critical. Although the spatial arrangement usually remains undetermined, it may often be reasonably approximated by the primary structure. In principle, the HBD can be fused to the N terminus (Eilers *et al.*, 1989) or the C terminus (vast majority of examples) or inserted into the protein (Picard *et al.*, 1988; Braselmann *et al.*, 1993; Hollenberg *et al.*, 1993; Louvion *et al.*,

1993). Fusion of the HBD, a C-terminal domain in wild-type receptors, to the C terminus of another protein is obviously easier to achieve since it only requires one in-frame junction. An additional complication with N-terminal additions may be the possibility for translation initiation at an internal AUG, resulting in the partial or complete deletion of the regulatory HBD from a subset of molecules. This appears to explain the elevated basal level of ER-Myc fusion proteins compared with Myc-ER in the absence of hormone (Eilers *et al.*, 1989; M. Eilers, personal communication).

Vectors for constitutive or even inducible expression *in vivo* can be chosen according to the needs of a given biological assay system. Note that proper hormonal regulation cannot be expected to work in bacteria or cell-free systems. In the latter case, the conditions for hormonal regulation and assay of the heterologous activity (for example, by gel shift) are often incompatible.

## C. Tissue Culture Media

To avoid the activation of fusion proteins by steroid "contaminants" present in culture medium the following precautions may be necessary.

### 1. Media without Phenol Red

Phenol red, added to the majority of mammalian cell culture media as a pH indicator, has been shown to mimic the effect of certain steroids (Berthois *et al.*, 1986; Picard and Yamamoto, 1987). Therefore, it is preferable to use media without phenol red. Such media are now commercially available.

### 2. Removal of Steroids from Serum by Charcoal Treatment

Serum, added as a supplement to the culture medium for many cell lines of mammalian and non-mammalian origin, is a source of steroid contaminants. These contaminants can be eliminated by charcoal treatment:

1. To 100 ml serum, add 2 g acid-washed activated charcoal (e.g., Sigma, St. Louis, MO).
2. Stir for 90 min at 4°C.
3. Remove the charcoal by passage through filter paper (e.g., pleated filter; Schleicher & Schuell, Keene, NH).
4. Sterilize the serum by filtration (0.22-$\mu$m filter).

Note that charcoal treatment of the serum may deplete other important components. Therefore, certain cell lines or cell types may not be able to grow in medium supplemented with charcoal-treated serum. As an alternative, a synthetic serum substitute could be tested [e.g., Nutridoma (Boehringer Mannheim, Indianapolis, IN) or NuSerum (GIBCOBRL, Grand Island, NY)]. These particular serum substitutes are completely steroid free. However, certain cells may fail to grow with any of these medium supplements. In this case, they could be maintained in normal complete medium and only switched to the special medium for the last 24–48 hr of a short-term experiment (e.g., a transient transfection experiment).

### D. Activation and Deinduction

The induction of fusion proteins is achieved by adding specific ligands directly to the medium (Table III). For deinduction hormone is removed by washing the cells several times with medium, phosphate-, or Tris-buffered saline prior to feeding them with fresh complete medium devoid of specific ligand. In long-term experiments, the medium is replaced again 24 hr later.

## IV. Examples

Three representative examples will be discussed. (1) E1A: This adenovirus early protein can be regulated as a transcription factor or an oncoprotein, depending on the relative positioning of E1A domains and the HBD. (2) Abl: As a fusion protein, this nonreceptor tyrosine kinase has two opposite activities: it is an oncoprotein and a growth inhibitor in the presence and absence of hormone, respectively. (3) STE11: The regulation of this serine/threonine kinase of the yeast pheromone signaling pathway illustrates that fusion proteins can also be used in yeast.

**Table III**
**Hormone Binding Domains and Their Cognate Ligands**

| HBD | Amino acid positions[a] | Ligand | Final concentration[b] |
|---|---|---|---|
| GR | 500–540 to 793–795 (rat) | Dexamethasone | 10 $\mu M$ |
| MR | 685 to 981 (rat) | Aldosterone | 10 n$M$ |
| ER | | | |
| Wild-type G400 | 282 to 576–595 (human) | 17$\beta$-Estradiol | 0.1 $\mu M$ |
| Mutant V400 | 282 to 576–595 (human) | 17$\beta$-Estradiol | 0.1 $\mu M$ |
| AR | ~650 to 918 (human) | 5$\alpha$-Dihydrotestosterone | 0.1 $\mu M$ |
| PR | ~670 to 933 (human) | Progesterone | 0.1 $\mu M$ |

[a] For the GR and ER HBDs there is an experimentally determined flexibility with respect to N and C termini. Amino acid positions are characteristic for the species indicated in parentheses, but given the extensive evolutionary conservation the corresponding HBD sequences from other species can be utilized as well.

[b] Ligands and concentrations are appropriate for vertebrate tissue culture cells and should not yield cross-activation of other endogenous steroid receptors. For other organisms, both the type of ligand and the concentration may have to be adapted. In the yeast *Saccharomyces cerevisiae*, 10 $\mu M$ deoxycorticosterone should be used with the GR and MR HBDs. Hormones can be stored as 1000-fold concentrated stock solutions in ethanol at $-20$°C.

## A. E1A

E1A is an adenovirus early protein that can activate the transcription of several adenoviral and cellular genes, apparently by being tethered to DNA indirectly via a partner protein (reviewed by Nevins, 1990). E1A can also immortalize certain types of primary cells and can collaborate with other oncoproteins to transform fibroblast cells (Land *et al.*, 1983; Ruley, 1983; Shenk and Flint, 1991). With E1A as an example (Fig. 1), we will emphasize how fusion proteins are used in short-term ("transient") and long-term ("stable") gene transfer experiments and how different functions of a heterologous fusion protein can be differentially regulated by an HBD.

### 1. Transactivation

Conserved region 3 (CR3) of E1A is necessary and sufficient for transactivation (reviewed by Flint and Shenk, 1989). Therefore, the N-terminal 222 amino acids of E1A, which retain CR3, were fused to the GR HBD. Picard *et al.* (1988) demonstrated that such E1A–HBD fusion proteins enhance transcription of a reporter plasmid in a fully hormone-dependent fashion. Typically, expression plasmid for the effector protein and reporter plasmid are co-transfected into HeLa or CV1 cells by calcium phosphate co-precipitation. Cells are washed

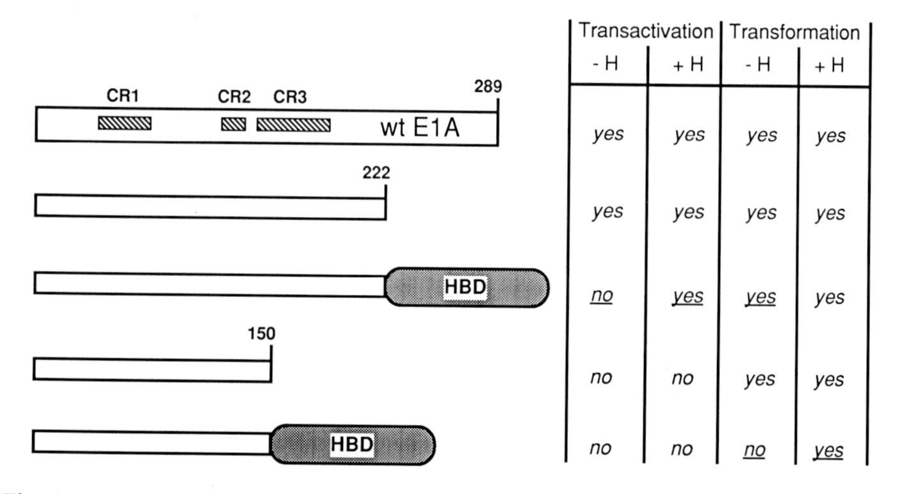


**Fig. 1** Differential regulation of E1A functions by a steroid hormone binding domain (HBD). The transactivation and transformation functions of E1A require conserved regions CR3 and CR1–CR2, respectively. In fusion proteins these regions can be subjected to hormonal control by the HBD depending on the relative positioning of regulatory and functional domains (after Picard *et al.*, 1988; D. Picard, unpublished results). wt, Wild-type; transactivation, transcriptional stimulation of an E1A-reponsive reporter gene; transformation, collaboration with the *ras* oncogene in immortalizing and transforming primary fibroblast cells; H, steroid hormone (added as indicated in Table III).

18 hr after adding DNA, and glucocorticoid hormone is added with fresh medium for another 24 hr before harvesting the cells. In this case, hormone is present during accumulation of the fusion protein. Alternatively, hormone can be added once fusion protein levels reach a maximum since regulation is post-translational. Using this approach, it could very easily be confirmed, by adding cycloheximide during hormonal induction, that protein synthesis is not required for E1A action (D. Picard, unpublished results).

### 2. Hormone-Reversible Transformation

The collaboration of E1A with the *ras* oncogene in the transformation of fibroblasts, which requires CR1 and CR2, can be assayed by co-transfecting expression vectors for an E1A fusion protein and for Ras into primary rat embryo fibroblast cells. Within 2–3 wk, transformed cells will form foci. If they are also immortalized, they can be established as a permanent cell line.

The transformation function turns out to be constitutively active in fusion proteins retaining the first 222 amino acids, including CR3 (Fig. 1). In the case of the "long" E1A fusions, the transformation function is not regulated (D. Picard, unpublished results) whereas the transactivation function is (Picard *et al.*, 1988). However, even the former could be regulated by moving the HBD (from GR or ER) closer to CR1 and CR2 on the linear protein map (D. Picard, unpublished results; Fig. 1). This observation is consistent with the idea that physical proximity of the HBD may be important for tight regulation despite remarkable spatial flexibility.

Such a shortened E1A fusion protein induces foci in primary cells exclusively and reversibly, in the presence of steroid hormone. In fact, we have been able to derive a permanent cell line that grows only in the presence of steroid hormone and becomes arrested in $G_1$ after hormone removal (Spitkovsky *et al.*, 1994).

## B. Abl

The product of the c-*abl* proto-oncogene is a member of the nonreceptor class of protein tyrosine kinases. As a consequence of genetic alterations such as viral transduction or chromosomal translocation events, c-*abl* can acquire the capacity to induce malignant transformation (reviewed by Rosenberg and Witte, 1988; Daley and Ben-Neriah, 1991). In addition, several transforming variants of c-*abl* have been constructed by *in vitro* mutagenesis (Jackson and Baltimore, 1989). Oncogenic transformation is reflected, for example, by altered morphology and growth properties and always correlates with increased Abl tyrosine kinase activity. To facilitate the investigation of the transformation process, regulatable Abl derivatives have been constructed by fusion to the HBD of the ER (Fig. 2A; Jackson *et al.*, 1993). Here we describe the protocols used to generate and to study cell lines containing these fusion proteins.

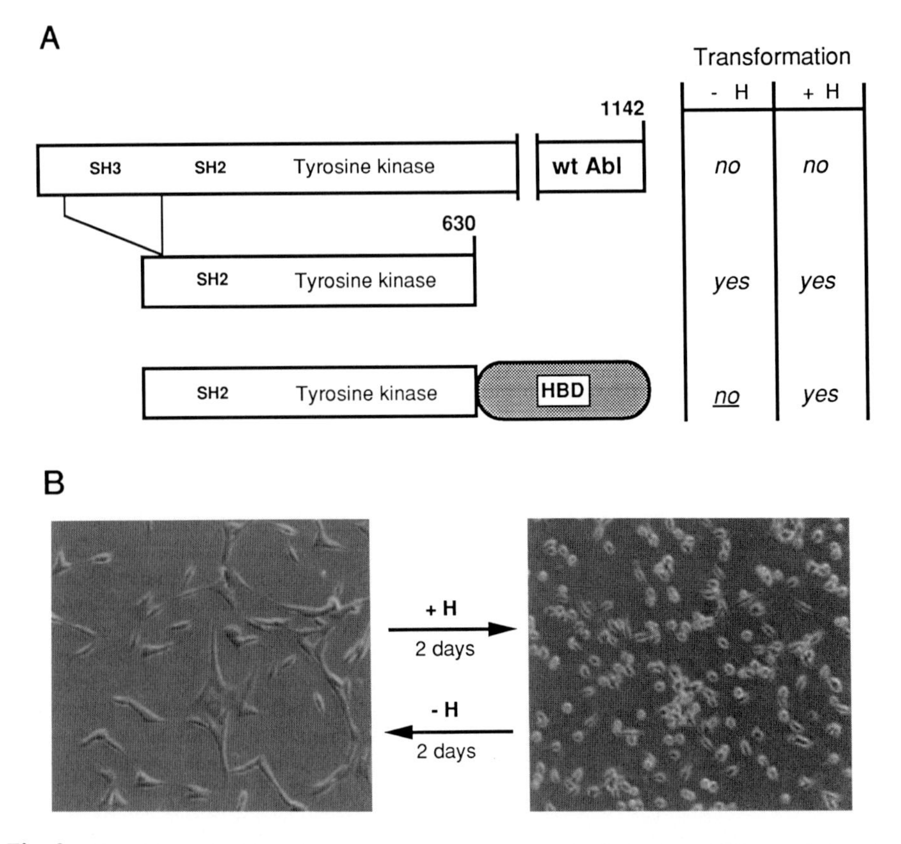


**Fig. 2** The Abl tyrosine kinase becomes hormone dependent by fusion to an HBD. (A) Schematic representations of wild-type (wt) c-Abl type IV, an oncogenic derivative (SH3 deletion mutant of c-Abl), and a fusion derivative. SH, Src homology domains; H, steroid hormone (added as indicated in Table III). Oncogenic transformation was evaluated by focus assay and soft agar assay with NIH 3T3 cells (after Jackson *et al.*, 1993). (B) Reversible hormone-dependent morphological transformation. Phase contrast micrographs of a stable cell line transformed with the Abl–HBD fusion derivative. Cells were cultured in the absence (*left*) or presence (*right*) of hormone. The interconversion in both directions is completed in 2 days (Jackson *et al.*, 1993).

### 1. Generation of Abl-Transformed Cell Lines

To obtain cell lines that are transformed with the Abl fusion protein, there are two alternative protocols. (1) Cells are co-transfected with a Moloney murine leukemia proviral clone (helper virus) and a replication-defective retroviral vector expressing the fusion protein (Jackson and Baltimore, 1989). In co-transfected cells, the helper virus allows replication and packaging of the recombinant virus, which can subsequently infect many neighboring cells with high efficiency. Morphologically transformed cells can be scored based on their ability to form a focus within a monolayer of normal cells (focus assay) or to form a colony in a soft agar assay (Superti-Furga *et al.*, 1991). The soft agar assay is the more stringent assay for transformation because it requires anchorage-independent growth. (2) An expression vector encoding the fusion protein can also be co-transfected with a plasmid containing a selectable marker. The resulting resistant colonies are heterogeneous with respect to their transformed morphology (in the presence of steroid hormone) because of the variability in the copy number of integrated expression plasmid. Therefore, the oncogenic transformation of such colonies must be confirmed by a soft agar assay.

#### *a. Retroviral Infection*

1. Co-transfect $8 \times 10^5$ NIH 3T3 cells in a 10-cm dish with 10 $\mu$g recombinant retroviral DNA (encoding the fusion protein) and 0.5 $\mu$g of the Moloney proviral clone pZAP (Goff *et al.*, 1982) using the calcium phosphate transfection protocol.
2. Maintain the cells in medium containing 0.1 $\mu M$ 17$\beta$-estradiol which is changed every 3 days.
3. Score plates for transformed foci by microscopic inspection.
4. Dislodge isolated transformed foci 15 days after transfection and clone by limiting dilution.
5. Screen independent clones for the hormone-dependence of morphological transformation and, by immunoblotting, for expression of the fusion protein.
6. Further analyze clones by a soft agar assay to test for anchorage-independent growth. The soft agar assay also provides an alternative to cloning by limiting dilution for the isolation of individual colonies.

#### *b. Co-transfection with a Selectable Marker*

1. Co-transfect $8 \times 10^5$ NIH 3T3 cells in a 10-cm dish with 10 $\mu$g expression vector (encoding the fusion protein) and 1 $\mu$g plasmid carrying the neomycin resistance gene using the calcium phosphate transfection protocol.
2. Maintain the cells in medium containing 0.1 $\mu M$ 17$\beta$-estradiol and 0.5 mg/ml G418. This medium is changed every 3 days.
3. Pick independent colonies after about 2 wk and test for anchorage-independent growth by soft agar assay in the presence or absence of hormone. Alternatively, all colonies from a plate can be pooled for the soft agar assay.
4. Isolate single colonies that appear exclusively in hormone-containing soft agar plates and expand.
5. Confirm expression of the fusion protein by immunoblotting.

### 2. Reversibility of Hormone-Dependent Transformation

Regardless of the method of choice for the gene transfer, cell clones are isolated in the transformed state. Morphological transformation is easily recognizable, as shown in Fig. 2B (*right*). Note that Abl-mediated transformation is completely reversible (*left*) (see also Jackson *et al.*, 1993). Reversion to normal morphology is obtained within 2 days by switching off the Abl-HBD fusion protein following the aforementioned "deinduction protocol." It is possible to repeatedly pass from the transformed to the reverted state simply by adding or washing out hormone.

Deinduction experiments reveal yet another facet of Abl: its ability to inhibit cell proliferation. At 48 hr after hormone depletion, not only are cells morphologically reverted, as discussed already, but their growth is inhibited as well. A cytostatic effect of Abl overexpression has been reported by several groups, but until today could only be examined by comparing different Abl derivatives (Jackson and Baltimore, 1989) or the same derivative in different cellular contexts (Renshaw *et al.*, 1992). In contrast, our hormone-regulatable system has the great advantage that the two distinct functions of Abl (transformation versus growth inhibition) can be studied using the same derivative expressed in one cell line. The nature of this inhibitory effect remains to be determined, but fusion proteins can once more be of great help for its elucidation.

## C. STE11

Vertebrate steroid receptors can also function in yeast in a ligand-dependent fashion, suggesting that all other components of this particular signal transduction pathway are conserved (Metzger *et al.*, 1988; Schena and Yamamoto, 1988; Mak *et al.*, 1989; McDonnell *et al.*, 1989; Picard *et al.*, 1990c). Steroids are specific inducers with little or no effect on the yeast cell (Tanaka *et al.*, 1989). Furthermore, we were able to demonstrate that the inactivation function of HBD also works in yeast (Louvion *et al.*, 1993). Fusion of the ER HBD to the artificial transcriptional activator GAL4.VP16 subjects its activity to hormonal control. This prompted us to test whether the HBDs of various steroid receptors could regulate a true yeast protein. We chose STE11, a serine/threonine kinase involved in the pheromone signal transduction pathway (Chaleff and Tatchell, 1985; Rhodes *et al.*, 1990).

In the budding yeast *Saccharomyces cerevisiae*, mating is initiated by the binding of mating type-specific peptides (the a- and $\alpha$-pheromones) to G-coupled receptors. The signal is then propagated through the sequential activation of a series of kinases (STE11, STE7, FUS3, and KSS1), leading finally to the phosphorylation and activation of STE12. STE12 is a transcriptional activator for genes involved in cellular differention and in the specific cell-cycle arrest in late $G_1$ (for review, see Marsh *et al.*, 1991). The HBDs of ER, GR and MR were fused to STE11 (Fig. 3). Typically, yeast plasmids encoding the STE11 derivatives were introduced into an *ste*$11^-$ strain (Rhodes *et al.*, 1990) by electroporation. STE11-induced growth arrest was then assessed by plating the different clones on agar–media containing $\alpha$ factor, the cognate steroid, or both.

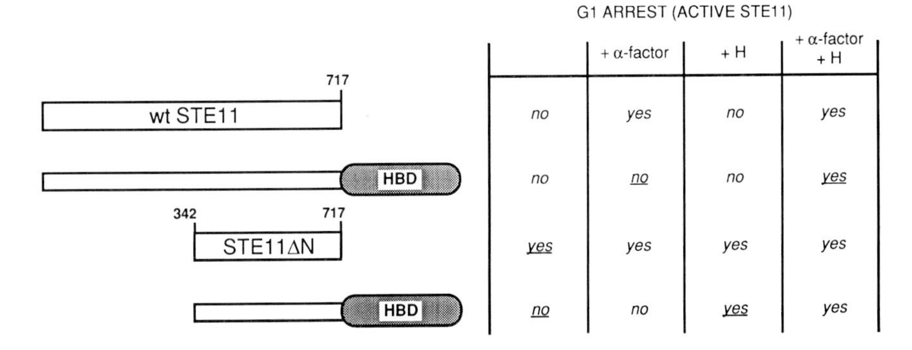


**Fig. 3** An STE11-HBD fusion protein is hormone dependent in budding yeast. Wild-type (wt; amino acids 1–717) and N-terminally truncated (STE11ΔN; amino acids 342–717) STE11 were fused in frame with the HBDs of ER, GR, or MR. Episomal expression vectors for these fusion proteins were introduced into a *ste11*$^-$ strain by the electroporation technique. Individual clones were then streaked onto agar plates and tested for STE11 activity. Active STE11 effects a cell-cycle arrest ($G_1$ arrest). $\alpha$ Factor was added at 5 $\mu M$ and steroid hormones (H) were added as indicated in Table III.

As indicated in Fig. 3, the cells expressing the STE11-HBD fusion proteins are efficiently and specifically growth inhibited only when both $\alpha$ factor and steroid hormone are present, indicating that the unliganded HBD blocks the normal activation pathway. FUS1 is one of the genes activated by the mating response (McCaffrey *et al.*, 1987); therefore, the specificity of the growth arrest was confirmed by measuring the activation of a chromosomally integrated FUS1-*lacZ* reporter gene (Rhodes *et al.*, 1990).

The deletion of the N-terminal regulatory domain of STE11 (derivative STE11ΔN) results in constitutive activation of the kinase (Cairns *et al.*, 1992). Fusion of the HBD to STE11ΔN renders $G_1$ arrest exclusively dependent on steroid hormone (with no requirement for $\alpha$ factor). As a result, this system can even be transferred to STE11$^+$ strains. This example illustrates very nicely how the fusion protein approach can be used to replace a yeast-specific control (such as pheromone regulation) by an HBD-mediated hormonal control.

## Acknowledgments

We are grateful to a large number of colleagues for sharing unpublished results. We thank Katharina Strub for critical reading of the manuscript. Our work was supported by the Swiss National Science Foundation and the Canton de Genève.

## References

Berthois, Y., Katzenellenbogen, J. A., and Katzenellenbogen, B. S. (1986). Phenol red in tissue culture media is a weak estrogen: Implications concerning the study of estrogen-responsive cells in culture. *Proc. Natl. Acad. Sci. USA* **83,** 2496–2500.

Boehmelt, G., Walker, A., Kabrun, N., Mellitzer, G., Beug, H., Zenke, M., and Enrietto, P. J.

(1992). Hormone-regulated v-rel estrogen receptor fusion protein: Reversible induction of cell transformation and cellular gene expression. *EMBO J.* **11,** 4641–4652.

Braselmann, S., Graninger, P., and Busslinger, M. (1993). A selective transcriptional induction system for mammalian cells based on Gal4-estrogen receptor fusion proteins. *Proc. Natl. Acad. Sci. USA* **90,** 1657–1661.

Briegel, K., Lim, K.-C., Planck, C., Beug, H., Engel, J. D., and Zenke, M. (1993). Ectopic expression of a conditional GATA-2/estrogen receptor chimera arrests erythroid differentiation in a hormone-dependent manner. *Genes Dev.* **7,** 1097–1109.

Burk, O., and Klempnauer, K.-H. (1991). Estrogen-dependent alterations in differentiation state of myeloid cells caused by a v-*myb*/estrogen receptor fusion protein. *EMBO J.* **10,** 3713–3719.

Cairns, B. R., Ramer, S. W., and Kornberg, R. D. (1992). Order of action of components in the yeast pheromone response pathway revealed with a dominant allele of the STE11 kinase and the multiple phosphorylation of the STE7 kinase. *Genes Dev.* **6,** 1305–1318.

Chaleff, D. T., and Tatchell, K. (1985). Molecular cloning and characterization of the *STE7* and *STE11* genes of *Saccharomyces cerevisiae*. *Mol. Cell. Biol.* **5,** 1878–1886.

Christopherson, K. S., Mark, M. R., Bajaj, V., and Godowski, P. J. (1992). Ecdysteroid-dependent regulation of genes in mammalian cells by a *Drosophila* ecdysone receptor and chimeric transactivators. *Proc. Natl. Acad. Sci. USA* **89,** 6314–6318.

Daley, G. Q., and Ben-Neriah, Y. (1991). Implicating the Bcr-Abl gene in the pathogenesis of Philadelphia chromosome-positive human leukemia. *Adv. Cancer Res.* **57,** 151–184.

Danielian, P. S., White, R., Lees, J. A., and Parker, M. G. (1992). Identification of a conserved region required for hormone dependent transcriptional activation by steroid hormone receptors. *EMBO J.* **11,** 1025–1033.

Danielian, P. S., White, R., Hoare, S. A., Fawell, S. E., and Parker, M. G. (1993). Identification of residues in the estrogen receptor that confer differential sensitivity to estrogen and hydroxytamoxifen. *Mol. Endocrinol.* **7,** 232–240.

Eilers, M., Picard, D., Yamamoto, K. R., and Bishop, J. M. (1989). Chimeras of the MYC oncoprotein and steroid receptors cause hormone-dependent transformation of cells. *Nature (London)* **340,** 66–68.

Fankhauser, C. P., Briand, P.-A., and Picard, D. (1994). The hormone binding domain of the mineralocorticoid receptor can regulate heterologous activities *in cis*. *Biochem. Biophys. Res. Commun.* **200,** 195–201.

Fawell, S. E., Lees, J. A., White, R., and Parker, M. G. (1990). Characterization and colocalization of steroid binding and dimerization activities in the mouse estrogen receptor. *Cell* **60,** 953–962.

Flint, J., and Shenk, T. (1989). Adenovirus E1A protein paradigm viral transactivator. *Annu. Rev. Genet.* **23,** 141–161.

Garabedian, M. J., and Yamamoto, K. R. (1992). Genetic dissection of the signaling domain of a mammalian steroid receptor in yeast. *Mol. Biol. Cell* **3,** 1245–1257.

Goff, S. P., Tabin, C. J., Wang, J. Y.-J., Weinberg, R., and Baltimore, D. (1982). Transfection of fibroblasts by cloned Abelson murine leukemia virus DNA and recovery of transmissible virus by recombination with helper virus. *J. Virol.* **41,** 271–285.

Gronemeyer, H. (1991). Transcription activation by estrogen and progesterone receptors. *Annu. Rev. Genet.* **25,** 89–123.

Guiochon-Mantel, A., Loosfelt, H., Lescop, P., Sar, S., Atger, M., Perrot-Applanat, M., and Milgrom, E. (1989). Mechanisms of nuclear localization of the progesterone receptor: evidence for interaction between monomers. *Cell* **57,** 1147–1154.

Hollenberg, S. M., Cheng, P. F., and Weintraub, H. (1993). Use of a conditional MyoD transcription factor in studies of MyoD trans-activation and muscle determination. *Proc. Natl. Acad. Sci. USA* **90,** 8028–8032.

Hope, T. J., Huang, X. J., McDonald, D., and Parslow, T. G. (1990). Steroid-receptor fusion of the human immunodeficiency virus type 1 Rev transactivator: Mapping cryptic functions of the arginine-rich motif. *Proc. Natl. Acad. Sci. USA* **87,** 7787–7791.

Hope, T. J., Klein, N. P., Elder, M. E., and Parslow, T. G. (1992). *trans*-Dominant inhibition of human immunodeficiency virus type 1 Rev occurs through formation of inactive protein complexes. *J. Virol.* **66,** 1849–1855.

Ince, B. A., Zhuang, Y., Wrenn, C. K., Shapiro, D. J., and Katzenellenbogen, B. S. (1993). Powerful dominant negative mutants of the human estrogen receptor. *J. Biol. Chem.* **268,** 14026–14032.

Israel, D. I., and Kaufman, R. J. (1993). Dexamethasone negatively regulates the activity of a chimeric dihydrofolate reductase/glucocorticoid receptor protein. *Proc. Natl. Acad. Sci. USA* **90,** 4290–4294.

Jackson, P., and Baltimore, D. (1989). N-terminal mutations activate the leukemogenic potential of the myristoylated form of c-*abl*. *EMBO J.* **8,** 449–456.

Jackson, P., Baltimore, D., and Picard, D. (1993). Hormone-conditional transformation by fusion proteins of c-Abl and its transforming variants. *EMBO J.* **12,** 2809–2819.

Kumar, V., and Chambon, P. (1988). The estrogen receptor binds tightly to its responsive element as a ligand-induced homodimer. *Cell* **55,** 145–156.

Land, H., Parada, L. F., and Weinberg, R. A. (1983). Tumorigenic conversion of primary embryo fibroblasts requires at least two cooperating oncogenes. *Nature (London)* **304,** 596–602.

Louvion, J.-F., Havaux-Copf, B., and Picard, D. (1993). Fusion of GAL4-VP16 to a steroid binding domain provides a tool for gratuitous induction of galactose-responsive genes in yeast. *Gene* **131,** 129–134.

Mak, P., McDonnell, D. P., Weigel, N. L., Schrader, W. T., and O'Malley, B. W. (1989). Expression of functional chicken oviduct progesterone receptors in yeast (*Saccharomyces cerevisiae*). *J. Biol. Chem.* **264,** 21613–21618.

Marsh, L., Neiman, A. M., and Herskowitz, I. (1991). Signal transduction during pheromone response in yeast. *Annu. Rev. Cell Biol.* **7,** 699–728.

McCaffrey, G., Clay, F. J., Kelsay, K., and Sprague, G. F., Jr. (1987). Identification and regulation of a gene required for cell fusion during mating of the yeast *Saccharomyces cerevisiae*. *Mol. Cell. Biol.* **7,** 2680–2690.

McDonnell, D. P., Pike, J. W., Drutz, D. J., Butt, T. R., and O'Malley, B. W. (1989). Reconstitution of the vitamin D-responsive osteocalcin transcription unit in *Saccharomyces cerevisiae*. *Mol. Cell. Biol.* **9,** 3517–3523.

Metzger, D., White, J. H., and Chambon, P. (1988). The human estrogen receptor functions in yeast. *Nature (London)* **334,** 31–36.

Nevins, J. R. (1990). Adenovirus E1A-dependent trans-activation of transcription. *Sem. Cancer Biol.* **1,** 59–68.

Picard, D. (1993). Steroid-binding domains for regulating the functions of heterologous proteins in *cis*. *Trends Cell Biol.* **3,** 278–280.

Picard, D., and Yamamoto, K. R. (1987). Two signals mediate hormone-dependent nuclear localization of the glucocorticoid receptor. *EMBO J.* **6,** 3333–3340.

Picard, D., Salser, S. J., and Yamamoto, K. R. (1988). A movable and regulable inactivation function within the steroid binding domain of the glucocorticoid receptor. *Cell* **54,** 1073–1080.

Picard, D., Khursheed, B., Garabedian, M. J., Fortin, M. G., Lindquist, S., and Yamamoto, K. R. (1990a). Reduced levels of hsp90 compromise steroid receptor action *in vivo*. *Nature (London)* **348,** 166–168.

Picard, D., Kumar, V., Chambon, P., and Yamamoto, K. R. (1990b). Signal transduction by steroid hormones: Nuclear localization is differentially regulated in estrogen and glucocorticoid receptors. *Cell Reg.* **1,** 291–299.

Picard, D., Schena, M., and Yamamoto, K. R. (1990c). An inducible expression vector for both fission and budding yeasts. *Gene* **86,** 257–261.

Pratt, W. B. (1990). Interaction of hsp90 with steroid receptors: Organizing some diverse observations and presenting the newest concepts. *Mol. Cell. Endocrinol.* **74,** C69–76.

Renshaw, M. W., Kipreos, E. T., Albrecht, M. R., and Wang, J. Y. J. (1992). Oncogenic v-Abl tyrosine kinase can inhibit or stimulate growth, depending on the cell context. *EMBO J.* **11,** 3941–3951.

Rhodes, N., Connell, L., and Errede, B. (1990). STE11 is a protein kinase required for cell-type-specific transcription and signal transduction in yeast. *Genes Dev.* **4,** 1862–1874.

Roemer, K., and Friedmann, T. (1993). Modulation of cell proliferation and gene expression by a p53–estrogen receptor hybrid protein. *Proc. Natl. Acad. Sci. USA* **90,** 9252–9256.

Rosenberg, N., and Witte, O. N. (1988). The viral and cellular forms of the Abelson (abl) oncogene. *Adv. Virus Res.* **35,** 39–81.

Ruley, H. E. (1983). Adenovirus early region 1A enables viral and cellular transforming genes to transform primary cells in culture. *Nature (London)* **304,** 602–606.

Rusconi, S., and Yamamoto, K. R. (1987). Functional dissection of the hormone and DNA binding activities of the glucocorticoid receptor. *EMBO J.* **6,** 1309–1315.

Samuels, M. L., Weber, M. J., Bishop, J. M., and McMahon, M. (1993). Conditional transformation of cells and rapid activation of the mitogen-activated protein kinase cascade by an estradiol-dependent human raf-1 protein kinase. *Mol. Cell. Biol.* **13,** 6241–6252.

Schena, M., and Yamamoto, K. R. (1988). Mammalian glucocorticoid receptor derivatives enhance transcription in yeast. *Science* **241,** 965–967.

Schuermann, M., Hennig, G., and Müller, R. (1993). Transcriptional activation and transformation by chimaeric Fos-estrogen receptor proteins: Altered properties as a consequence of gene fusion. *Oncogene* **8,** 2781–2790.

Shenk, T., and Flint, J. (1991). Transcriptional and transforming activities of the adenovirus E1A proteins. *Adv. Cancer Res.* **57,** 47–85.

Smith, D. F., and Toft, D. O. (1993). Steroid receptors and their associated proteins. *Mol. Endocrinol.* **7,** 4–11.

Spitkovsky, D., Steiner, P., Lukas, J., Lees, E., Pagano, M., Schulze, A., Joswig, S., Picard, D., Tommasino, M., Eilers, M., and Jansen-Dürr, P. (1994). Modulation of cyclin gene expression by adenovirus E1A in a cell line with E1A-dependent conditional proliferation. *J. Virol* **68,** 2206–2214.

Superti-Furga, G., Bergers, G., Picard, D., and Busslinger, M. (1991). Hormone-dependent transcriptional regulation and cellular transformation by Fos-steroid receptor fusion proteins. *Proc. Natl. Acad. Sci. USA* **88,** 5114–5118.

Tanaka, S., Hasegawa, S., Hishinuma, F., and Kurata, S. (1989). Estrogen can regulate the cell cycle in the early G1 phase of yeast by increasing the amount of adenylate cyclase mRNA. *Cell* **57,** 675–681.

Tora, L., Mullick, A., Metzger, D., Ponglikitmongkol, M., Park, I., and Chambon, P. (1989). The cloned human estrogen receptor contains a mutation which alters its hormone binding properties. *EMBO J.* **8,** 1981–1986.

Umek, R. M., Friedman, A. D., and McKnight, S. L. (1991). CCAAT-enhancer binding protein: A component of a differentiation switch. *Science* **251,** 288–292.

Vegeto, E., Allan, G. F., Schrader, W. T., Tsai, M.-J., McDonnell, D. P., and O'Malley, B. W. (1992). The mechanism of RU486 antagonism is dependent on the conformation of the carboxy-terminal tail of the human progesterone receptor. *Cell* **69,** 703–713.

Wrange, Ö., Eriksson, P., and Perlmann, T. (1989). The purified activated glucocorticoid receptor is a homodimer. *J. Biol. Chem.* **264,** 5253–5259.

Wrenn, C. K., and Katzenellenbogen, B. S. (1993). Structure-function analysis of the hormone binding domain of the human estrogen receptor by region-specific mutagenesis and phenotypic screening in yeast. *J. Biol. Chem.* **268,** 24089–24098.

Wright, A. P., and Gustafsson, J.-Å. (1992). Glucocorticoid-specific gene activation by the intact human glucocorticoid receptor expressed in yeast. Glucocorticoid specificity depends on low level receptor expression. *J. Biol. Chem.* **267,** 11191–11195.

Wright, A. P., Carlstedt-Duke, J., and Gustafsson, J.-Å. (1990). Ligand-specific transactivation of gene expression by a derivative of the human glucocorticoid receptor expressed in yeast. *J. Biol. Chem.* **265,** 14763–14769.

Yamamoto, K. R., Godowski, P. J., and Picard, D. (1988). Ligand regulated nonspecific inactivation of receptor function: A versatile mechanism for signal transduction. *Cold Spring Harbor Symp. Quant. Biol.* **53,** 803–811.

Ylikomi, T., Bocquel, M. T., Berry, M., Gronemeyer, H., and Chambon, P. (1992). Cooperation of proto-signals for nuclear accumulation of estrogen and progesterone receptors. *EMBO J.* **11,** 3681–3694.

# CHAPTER 17

# Gene Gun Transfection of Animal Cells and Genetic Immunization

**Stephen A. Johnston* and De-chu Tang†**

*Departments of Internal Medicine and Biochemistry
†Simmons Cancer Center
and Department of Internal Medicine
University of Texas Southwestern Medical School
Dallas, Texas 75235



## I. Introduction

A central component of the revolution in molecular biology has been the ability to introduce genes into cells by transfection or transduction. The three different modes of introducing a transgene are (1) infectious viral vectors such as retrovirus or adenovirus, (2) chemical methods such as $Ca_3(PO_4)_2$, polyethylene glycol (PEG), DEAE–dextran, or liposomes, and (3) physical methods such as gene guns, microinjection, muscle injection, or electroporation. Each one of these methods has inherent advantages and limitations, so no universal method of gene delivery has emerged.

The ideal gene delivery method would allow introduction of exogenous DNA into any cell type. It would do so at high efficiency and with little or no damage

METHODS IN CELL BIOLOGY, VOL. 43

to the target cells. It would be simple and safe, and would allow control over the amount of DNA introduced. Importantly for both research and clinical applications, it would permit introduction of DNA into the cells of living animals, that is, *in situ* gene delivery. Infectious agents can often introduce the transgene into nearly 100% of the target cells. Although retrovirus can only infect dividing cells, adenovirus has been shown to be effective on both dividing and nondividing cells (Breakefield, 1993; Mulligan, 1993). Viral vectors also have the advantage of being adaptable to *in situ* transduction protocols, and offer some control over the amount of DNA that enters each cell. Chemical methods are not as species sensitive as infectious agents, and have fewer complications and dangers in their use. However, the efficiency of transfection can vary considerably with different targets, and there is little control over the amount of DNA introduced per cell. DEAE–dextran, PEG, and $Ca_3(PO_4)_2$ methods do not adapt well to *in situ* transfection, but liposome-type methods do (Nicolau *et al.,* 1983; Plautz *et al.,* 1993). Microinjection is only applicable to certain cell types, and only when a few transgenic events are required. On the other hand, electroporation is effective with a large number of different cell types. Neither of these physical methods has proven useful for *in situ* transfection. Muscle injection of DNA with a needle is only applicable to *in situ* transfection (Wolff *et al.,* 1990).

One of the newer additions to the arsenal for gene delivery is the gene gun. This technology has a number of names including "biolistic method" (the one chosen by its inventor, John Sanford), "particle bombardment," and "gene gun." The basics of the technique are that DNA is coated onto 1 to 5-$\mu$m heavy metal particles (usually gold or tungsten) that are accelerated to sufficient velocity to penetrate the target cells. Gene gun transfection was first demonstrated on plant cells, and has become the most important technology for plant transformation (Sanford, 1990). Its most notable application was in producing the first transgenic corn. Gene gun transfection remains the major method of producing transgenic maize, but it is now also used to transform all other important crop plants. This technique has also been used to transform microbes, including fungi and bacteria (Klein *et al.,* 1992; Johnston and Tang, 1993).

The focus of this chapter is the application of the gene gun to transfection of and gene expression in animal cells, particularly in the new technique of genetic immunization. However, researchers have far less experience in applying the gene gun to animals than to plants and microbes. Three applications in plants and microbes might anticipate future use in animal cells. First, for both plants and microbes, the gene gun has proven capable of transforming recalcitrant species. As mentioned before, the most notable example was the creation of the first transgenic maize. A fungal pathogen of rice, *Magnoportha griseia* (Klein *et al.,* 1992), and one of humans, *Crytococcus neoformans* (Toffaletti *et al.,* 1993), were transformed for the first time with the gene gun, as was the prokaryote *Bacillus megaterium* (Sanford *et al.,* 1991). Second, the first organelle transformations were accomplished with the gene gun. The chloroplasts of *Chlamydomonas* and tobacco have been transformed (Boynton *et al.,*

1988; Daniell, 1993), as have the mitochondria of yeast (Johnston *et al.*, 1988). The transformation of the mitochondria of animal cells remains unaccomplished. Third, the gene gun has been used to study *cis*-acting transcription signals and *trans*-acting transcription factors by shooting the test constructs directly into the intact tissue of interest. This method has proven to be a simple and effective way to assess genetic factors in an authentic context, avoiding the problems of tissue culture artifacts or the inability to culture tissues of interest (reviewed in Klein *et al.*, 1992).

In this chapter we will review the basic elements of the gene gun and its operation. Its application to transfecting cultured cells, tissue explants, and tissues *in situ* will be outlined with attention to the potentials and limitations of each process. Finally, the new application of the gene gun to genetic immunization and genetic vaccines will be summarized.

## II. Details of Operation

### A. Various Gene Guns

#### 1. Helium Gun

The original gene gun as invented by Sanford and co-workers (Sanford *et al.*, 1987) involved placing the DNA-coated microprojectiles in an aqueous slurry onto a small plastic bullet. This bullet was placed into a 22-caliber barrel in front of a gunpowder cartridge. When the cartridge was fired, it propelled the plastic bullet down the barrel until it was stopped by a solid plate. The plate had a small hole in the center that allowed the microprojectiles to continue their trajectory into the target cells. The target cells or tissues were placed in a chamber for the bombardment, with a vacuum drawn in the chamber so air would not slow the microprojectiles. This design was effective for plant and microbial transformation and was the first system to be commercially available. Several home-made versions of this type are now being used.

When the Sanford and Johnston laboratories initiated collaboration on a device for *in situ* introduction of DNA into animals, it was clear that the gunpowder device was not adequate. It led to too much tissue damage, the velocity of the projectiles was not readily regulatable, and it was somewhat dangerous. Our goal was to design a unit that was hand-held, gentle, and more controllable. We envisioned being able to hold the unit against any surgically accessible organ or tissue in the animal or human and shooting the cells in a defined area several layers deep.

The currently available unit that meets these requirements uses helium for propulsion (Fig. 1). A small amount of high pressure helium gas is restrained by a pierceable Kapton disc. Another Kapton disc, the macroprojectile, is positioned approximately 1 cm in front of the restraining membranes. The DNA-coated microprojectiles are placed dry onto the front of this macroprojectile.

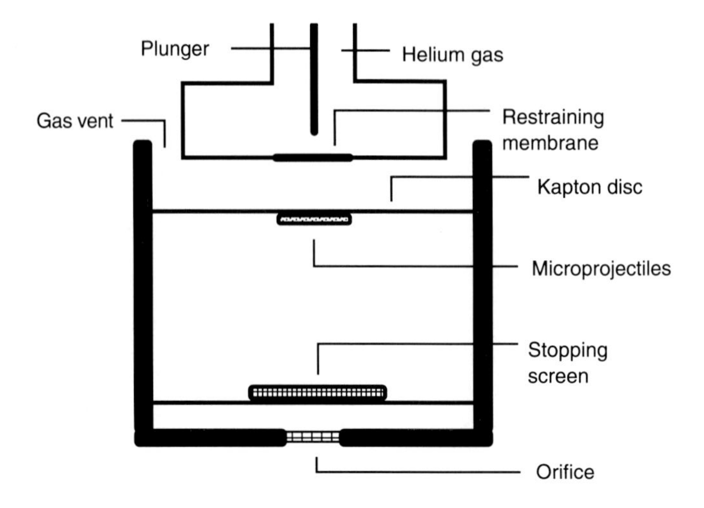


**Fig. 1** Schematic diagram of the helium-driven biolistic device. High-pressure helium gas in the chamber is used as the power source. When the membrane restraining the helium gas is ruptured and the disc is propelled toward the target tissue by the resulting shock wave, DNA-coated microprojectiles are launched to penetrate the tissue as the disc hits the screen. Further details are given in the text.

Approximately 1 cm in front of the macroprojectile is a fixed stopping screen. The target tissue is placed just in front of the stopping screen. When the high pressure helium gas is released by piercing the restraining membranes, a supersonic shock wave is produced. This shockwave, traveling at 3–4 times the speed of sound, hits the macroprojectile in front of the gas front. The macroprojectile is then launched against the stopping screen and the microprojectiles continue on into the target cells. An important technical feature is that the macroprojectile seals off the chamber when it hits the stopping screen and deflects the trailing gas front, protecting the target tissue. All the workings of the device are contained in a cylindrical chamber about the size of a soft drink can (4.5 × 11.5 cm). A vacuum can be applied over the tissue between it and the membranes. The canister is attached to a gas line for the high pressure helium, to a vacuum line if needed, and to an electric line to a battery pack to operate the lance that pierces the restraining membrane. Ironically, although this instrument was originally designed for animal applications, we found that it was a marked improvement over the gunpowder unit when fitted on the in-chamber device. Almost all in-chamber devices have now been retrofitted for the helium design, and this is the design currently commercially available from Bio-Rad (PDS 1000/He; Hercules, CA). The PDS 1000/He design involves allowing the restraining membrane to break spontaneously at the desired gas pressure, rather than lancing it as in the hand-held device. The hand-held unit described is not yet commercially available, but apparently Bio-Rad plans to produce one. The plans for construction of the helium device have been published elsewhere (Sanford *et al.*, 1991).

The in-chamber helium device is basically the same as the gunpowder unit, but with a different propellant. The flight path of the macroprojectile can be varied to some extent. The target cells can be placed from 2 to 14 cm from the stopping screen. Varying the distance the microprojectiles travel to the cells affects the transfection efficiency in at least three ways. First, especially for small particles, there is a large decrease in velocity with distance travelled. Although velocity is not an important problem with cultured cells, it is with tissue explants. Second, placing the target cells closer to the stopping screen increases the exposure to the gas blast effects and to secondary shockwaves. This is an important consideration with delicate cells. Third, the area of bombardment increases with distance from the stopping screen. The bombarded area covers a 100-mm culture plate when it is placed ~8 cm from the screen. For bombardment of most cultured cells, a vacuum [~25 mm Hg (~0.03 atm) or about 29 inches on conventional gauges] is applied to the chamber. The velocity of the microprojectiles, particularly the smaller ones, falls off in a nonlinear fashion with vacuums greater than 25 mm Hg (see Klein *et al.*, 1992, for details). With one exception (a macrophage cell line), none of the cultured cell lines tested has been sensitive to vacuum. Heiser (1994) reported that some cell lines had limited sensitivity to a vacuum of 29 inches, but none to 25 inches Hg.

### 2. Other Gene Guns

Several other forms of the gene gun have been devised. The most successful alternative is that designed by the scientists at Agracetus (Madison, WI) (Yang *et al.*, 1990). McCabe and co-workers use a gene gun that is basically the same as the helium devise, except that a large electric discharge is used to vaporize a water droplet to create the impelling shockwave. A hand-held version of this device has been constructed. This ''Accell'' unit has been used to transform agriculturally important crops, as well as for animal cell transfection. This design is not commercially available, but is being used by several laboratories in collaborative projects with Agracetus. The Accell unit and the helium device seem to have similar capabilities (S. A. Johnston and D. McCabe, unpublished results). In addition, several simpler versions of the gene gun have been reported. One interesting variation uses entrainment to focus the microprojectiles on a very small area (as small as 0.15 mm). Whether these alternative designs will prove useful remains to be seen. Our laboratory and others are currently working on improved gun designs, particularly for genetic vaccination (see subsequent discussion).

## B. Microprojectiles

### 1. Physical Characteristics

The most efficient transfection is accomplished with microprojectiles of high density. Gold and tungsten are commonly used because they are readily available and will coat with DNA. Numerous other materials have been tested including platinum, iron, irridium, uranium, and glass (J. C. Sanford and S. A. Johnston, unpublished results). Gold microprojectiles of varying sizes can be obtained from Bio-Rad, Alpha (Ward Hill, MA), and several other suppliers. Lots can vary considerably in their size distributions and physical characteristics. The most important features are the tendency to clump and the ability to bind DNA. The DNA-binding character can be assessed to some extent *in vitro*. The lot of microprojectiles is prepared with DNA and an aliquot loaded in an agarose gel well. The gel is run as normal and the DNA resolved. With good lots of microprojectiles, the DNA will resolve in a band at its correct size position. Poor lots will not release the DNA, the DNA will be degraded, or it will release the DNA slowly, resulting in a smear from the well. The final assessment of quality must still be done *in vivo*.

## 2. Preparation of Microprojectiles

The methods for coating the microprojectiles with DNA have been extensively studied. The basic protocol follows.

1. The microprojectiles are washed in water and ethanol and are stored at 60 mg/ml in water at −20°C.
2. The coated microprojectiles are prepared by mixing sequentially 25 μl microprojectiles in a slurry, 2.5 μl plasmid DNA (1 mg/ml), 25 μl $CaCl_2$ (2.5 *M*), and 5 μl spermidine (1 *M*).
3. After a 15-min incubation at room temperature, the microprojectiles are pelleted by brief centrifugation, and are washed with 70% ethanol.
4. The DNA-coated microprojectiles are resuspended in 100% ethanol and are spread into thin films on Kapton membranes. The microprojectiles are allowed to dry in a desiccator before use.

The details of the procedure have been published elsewhere (Armaleo *et al.*, 1990). Although it may be advantageous to vary the load of microprojectiles and the amount of DNA per load for specific applications, generally each shot uses 500 μg microprojectiles and 0.1–2 μg DNA. More DNA can lead to aggregation of the microprojectiles and cellular damage. Other methods of preparation may have particular uses. For example, to coat RNA or proteins onto the microprojectiles we have lyophilized the RNA from ethanol or the protein from water onto the microprojectiles. The microprojectiles were then pulverized through a 20-μm mesh and dispersed onto the macroprojectile. Figure 2 shows that the inoculation of *in vitro* transcribed and capped mRNA molecules, when coupled with the tobacco mosaic virus Ω element as the leader sequence, is more effective in producing proteins in inoculated live animals than the inoculation of RNA messages without the cap and/or the Ω leader sequence. Large DNA molecules or viruses can also be coated onto microprojectiles.

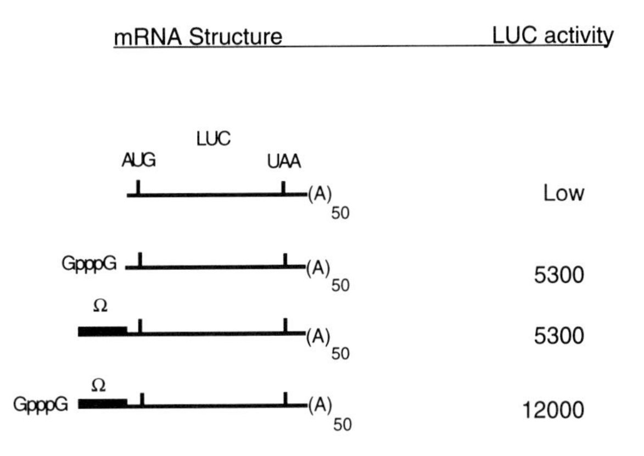


**Fig. 2** Inoculation of mRNA molecules into live animals with the gene gun. Firefly luciferase mRNA molecules were synthesized *in vitro* with phage T7 RNA polymerase; 5–15 μg mRNA molecules were mixed with 500 μg gold microprojectiles (1–3 μm), lyophilized, and pulverized through 20-μm mesh onto Kapton membranes. The mRNA-coated microprojectiles were launched into mouse ears with the hand-held gun in an upward orientation. The tobacco mosaic virus leader element Ω which enhances translation in plant and animal cells, the cap (GpppG), and the poly(A) tail consisting of 50 A residues [$(A)_{50}$] have been described (Gallie *et al.*, 1991).

Researchers have determined that ~20 biologically active 6-kb plasmids can be carried on a 0.5-μm microprojectile (Armaleo *et al.*, 1990). The commonly used microprojectile for animal cells in culture is 1–1.6 μm (for *in situ,* 1–5 μm), so potentially hundreds of plasmids are carried on each microprojectile. However, the coating of individual microprojectiles is quite uneven. A significant area for improvement in gene gun technology is in particle quality. Ideally, one would like to have a range of discrete sizes of microprojectiles that coat DNA evenly, with high density, and in such a way that the DNA is all biologically active when it enters the cell. We are currently far from this goal, but if realized it would certainly extend the use of the technology.

## III. Applications

### A. In-Chamber Gun

#### 1. Transfecting Cells in Culture

The in-chamber helium gun can be used to transfect tissue culture cells.

1. For adherent cells in a dish, the plate is tipped on edge and the medium is removed. This leaves a thin film of medium over the cells.

2. The cells are placed into the chamber and a vacuum of 25–29 inches Hg drawn.

3. The plate can be placed at various distances from the stopping screen, depending on the coverage desired and the sensitivity of the cells to the shockwave.

4. After bombardment, the medium is reapplied to the cells.

We have noted no toxicity of the gold particles but some cells are sensitive to tungsten. As little as 100 ng DNA can be used for transfecting a 60-mm plate. Nonadherent cells can be collected on filter paper or attached artificially, for example, with Cell-Tak (Collaborative Biomedical Products, Bedford, MA). We have noted a nonlinear increase in transfection frequency with increasing confluence on the plate. This may be due to physical support between cells during the bombardment. If cells are particularly sensitive to the bombardment, they can be overlayed with helium. The dispersal of clumped microprojectiles can be improved by placing a 100-$\mu$m mesh immediately below the stopping screen.

Many different cell types have now been transfected with the gene gun including NIH 3T3 cells (Zelenin *et al.*, 1989), CHO cells, T-lymphocyte cell lines (Fitzpatrick-McElligott, 1992), primary myotubes (Williams *et al.*, 1991), primary cardiac myocytes (S. A. Johnston, unpublished data), COS-7 cells, preadipocytes (BMS-2), macrophages (J774) (Heiser, 1994), and a large number of carcinoma lines (Yang *et al.*, 1990; Tang *et al.*, 1994). Using histochemical staining, the range in transient transfection has been between 0.5 and 7% of the cells on the plate. Stable transformants of a number of lines have also been produced with a range in frequency between $1 \times 10^{-4}$ and $7 \times 10^{-4}$.

There are three possible advantages to using the gene gun for transfecting cell lines. First, the gun does seem capable of transfecting cells that are difficult or impossible to transfect by other means. For example, primary cultures of chick myotubes were only transfected by the gun. Calcium phosphate precipitation and lipofection only transfected residual myoblasts in the culture, whereas the gun was capable of transfecting up to 12% of the myotubes (Williams *et al.*, 1991). Heiser (1994) also reported the successful transfection of cell lines that were impossible or very difficult to transfect by other means. Second, transfection with the gun uses very little DNA or RNA compared with other methods. Third, there are some indications that more co-transfectants are generated using the gun (Heiser, 1994). Although not extensively studied, the fate of the DNA (e.g., single versus tandem integrations, methylation patterns) may be different with the gun.

The gene gun can also be applied to tissue explants. Ductal segments of rat mammary gland (Yang *et al.*, 1990), as well as liver and kidney explants, have been transiently transfected. We have transformed limb bud explants of newts that were then grafted onto the whole animal (D. Lo, L. Pecorino, and S. A. Johnston, unpublished results). Neuronal tissue in ferret brain slices has also been successfully transfected (D. Lo, personal communication). *Drosophila* embryos have been transiently transfected with the gun, as have fertilized fish eggs (Zelenin *et al.*, 1991).

When bombarding organs or tissue explants, concerns about the balance between velocity and damage become more important than with cultured cells. Intact tissue is much harder to penetrate. Whereas most tissue culture cells are easily penetrated with 1-$\mu$m particles using 600–1000 psi helium pressure, it may be necessary to use larger particles and higher pressures for intact organs and tissues. One can also decrease the flight path of the microprojectiles; however, this can lead to cell damage if the target is placed too close to the stopping screen. It may be necessary to add modifications such as the helium overlay or the dispersing mesh described earlier.

## B. Hand-Held Gun

An adaptor can be placed in the in-chamber gun to allow bombardment of live animals (Johnston *et al.*, 1991). This adaptation has been used to bombard the skin of mice and the limb buds of newts. However, this system is difficult to apply to larger animals or to internal organs. On the other hand, the hand-held unit described earlier has been used to bombard liver, skin, intestine, spleen, muscle, and neurons primarily in mice but also in rabbits, newts, rats, and monkeys.

### 1. Transfecting Skin

We have had the most experience bombarding skin and liver. For skin, the results are fairly reproducible.

1. The mouse is first anesthetized and the hair in the target area is cut away.
2. A depilatory (Nair or Neet) is then applied to clear away residual hair and to remove the dead skin. Figure 3 shows the inoculation of DNA into mouse ear skin.

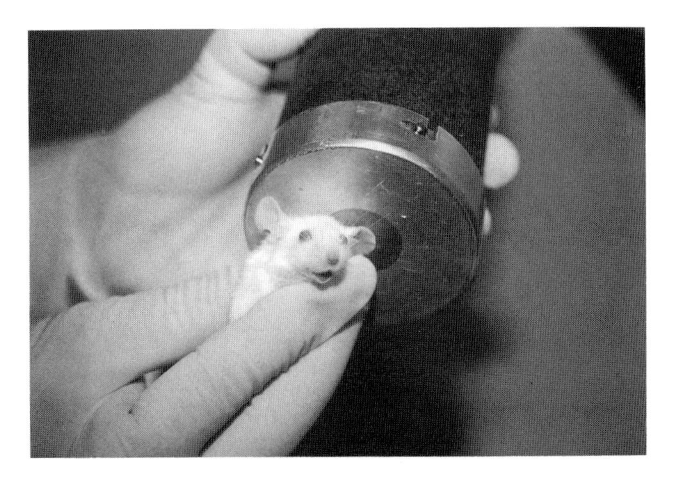

**Fig. 3** Transfection of mouse ear skin cells *in situ* with the hand-held gun.

3. Typically a vacuum of 20 inches Hg is applied to the skin and 1200 psi He is used.

4. When areas of the skin greater than ~0.6 cm are being bombarded, a wide mesh nylon screen is placed directly on the skin to support it in the vacuum.

Under these conditions, a significant number of the particles will penetrate into the dermis whereas the majority come to rest in the epidermis. Little if any damage to the skin occurs from the bombardment or subsequent inflammation. Cells through which the microprojectiles pass show no evidence of damage, nor is there any evidence from histochemical staining that cells on the pathway of the microprojectile are transfected. Interestingly, almost none of the transfected cells have the particle in the nucleus. Apparently, delivering the DNA to the cytoplasm is sufficient. We have noted in yeast and cultured mammalian cells that the DNA on the microprojectile relocates quickly to the nucleus, suggesting an active transport system.

The expression of reporter genes shot into the skin peaks in 18–36 hr and begins a marked decline after day 3. Expression is still detectable at least 3 wk later, but at very low levels. When skin bombarded with a nuclear-localized $\beta$-galactosidase gene is stained 1 day after treatment, most of the $\beta$-galactosidase-expressing cells are in the epidermis (Fig. 4). The epidermis in mouse is turned

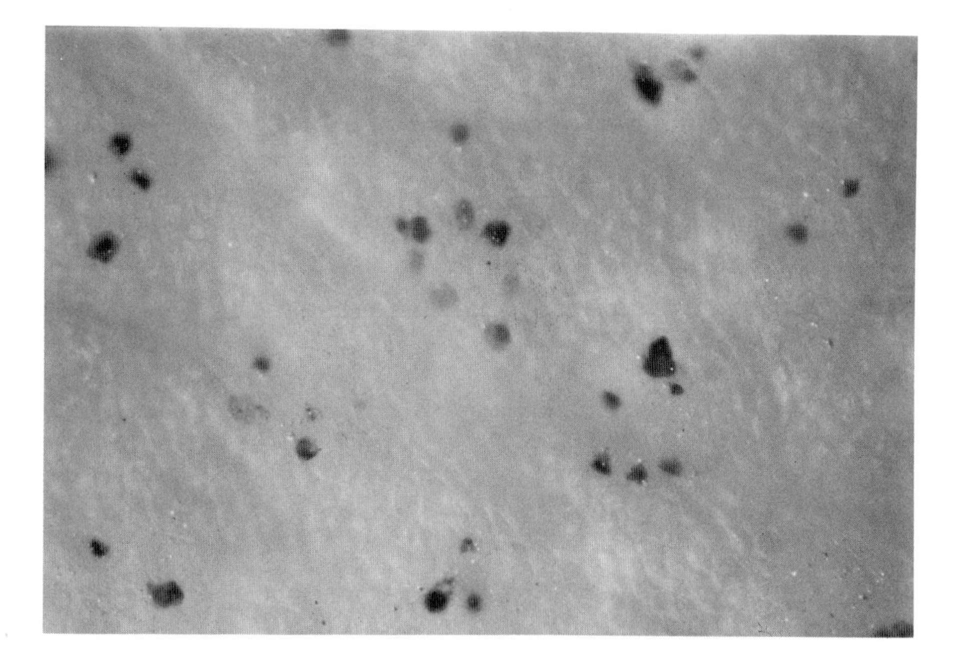

**Fig. 4** Visualization of transfected skin cells in a section of a bombarded mouse ear 1 day after inoculating a nuclear-localized $\beta$-galactosidase gene with the hand-held gun. The whole ear was fixed in 4% paraformaldehyde, stained with X-gal, and examined under a dissection microscope (40 ×).

over in 3–5 days, so by day 3 there are no particles or expression in the epidermis, probably explaining the precipitous decline in expression after day 3. The continued expression after day 3 may come from transfection of mitotically inactive dermal cells or the occasional stable transformation of stem cells. Wounding of the target area leads to lower reporter gene expression. As long as the tissue is not wounded, the same area can be bombarded as many as 3 times with a linear increase in reporter expression. We have tested several promoters for relative expression level in the skin, using luciferase as the reporter. We find that the cytomegalovirus (CMV) early promoter, the human $\beta$-actin promoter, and the Rous sarcoma virus (RSV) long terminal repeat (LTR) give comparable levels. The simian virus 40 (SV40) promoter and the Moloney murine sarcoma virus LTR are lower. The CMV promoter usually gives the highest expression level in the various tissues we have tested.

### 2. Transfecting Liver

1. To bombard the liver in an anesthetized mouse, a small incision (1 cm) is made in the muscle wall of the abdominal cavity just below the ribcage.
2. One lobe is exposed for treatment.
3. A 4-$mm^2$ area is bombarded at 1500 psi with either no vacuum or 5 inches Hg. Any vacuum that can be applied leads to better penetration, but it is often hard to apply the vacuum without tissue damage to soft organs.
4. After bombardment the lobe is repositioned and the incision sutured.

We have noted no problems with this treatment. The level of expression of the reporter plasmid is ~10-fold lower than that seen when skin is bombarded. The expression level also falls off after several days, although it can be detected for at least 3 wk. If human growth hormone gene is introduced, the hormone can sometimes be detected in the serum. There is high variability in expression level from different treatments, which would make it difficult to use liver bombardment to test different promoter constructs, for example, as has been successfully done in plants. We believe the primary limitation is variable penetration when a vacuum cannot be used. This problem may be solved with newer designs of the gun. The penetration problem could be at least partially solved by using particles of greater mass, since the liver hepatocytes tolerate penetration by microprojectiles of 2–6 $\mu$m. Particles of this size that coat well with DNA are not currently available.

## IV. Genetic Immunization

Good vaccines are immunogens that can induce high titer and long-lasting immunity for targeted intervention against diseases with the minimum number of inoculations. Genetic immunization is an approach that can elicit immune responses against specific proteins by expressing genes encoding the proteins in an animal's own cells. The substantial antigen amplification and immune stimulation resulting from prolonged antigen presentation *in vivo* would be expected to induce a solid immunity against the antigen. Genetic immunization simplifies the vaccination protocol and shortens the time required to produce antibodies against particular proteins because the often difficult steps of protein purification and combination with adjuvant, both routinely required for vaccine development, are eliminated. Since genetic immunization does not require the isolation of proteins, it is especially valuable for proteins that may lose conformational epitopes when extracted and purified biochemically. Genetic immunization is a variation on previous strategies in which the gene encoding the antigen was delivered with bacterial or viral vectors. Genetic immunization, in contrast, is with naked DNA. Using the hand-held gun, we have inoculated gold microprojectiles coated with plasmids expressing human proteins into the ear skin of mice, and have demonstrated humoral immune responses to several antigens (Tang *et al.*, 1992). Most animals tested produced antibodies against the human proteins within a few weeks of inoculation. The primary response could be augmented by subsequent DNA boosts on the same bombarded area multiple times without doing any harm to the animal. This is the first time that plasmid DNA has been used to elicit immune responses. Monoclonal antibodies could also be produced by creating hybridoma cells from plasmid-bombarded animals (Barry *et al.*, 1994). When compared with previous modes of antigen gene expression (e.g., vaccination by bacterial or viral vectors), the delivery of pure plasmid DNA avoids the risk of inoculating any infectious agents and the problem of eliciting an immune response to bacterial or viral vectors (Lanzavecchia, 1993). The shed-off of transfected skin cells as a normal process further strengthens the safety of plasmid-mediated vaccination. Heat stability of plasmid DNA is another advantage because it eliminates the requirement of refrigeration.

Genetic immunization may be a new approach to vaccine (genetic vaccine) development. This area of research is being actively explored by several laboratories. This technology may also have research utility in the following situations. First, if it is difficult to purify the protein of interest but the gene is available, one can still produce antibody. Second, the array of monoclonal antibodies available may be different from those produced by conventional methods. Third, the character and level of the cytotoxic T lymphocyte response is likely to be different than the one elicited by protein inoculation.

## V. Summary

Gene gun technology at this point has the most utility in animal protein expression as a back-up technology. In other words, when other conventional systems fail, it will generally work. Most notable is its usefulness for hard-to-transfect cells or in some particular *in situ* applications. Improvements in the gun itself and in the microprojectiles present the potential for this technology to expand in utility. The one area in which it now appears to be the method of choice is genetic immunization.

## References


Armaleo, D., Ye, G.-N., Klein, T. M., Shark, K. B., Sanford, J. C., and Johnston, S. A. (1990). *Curr. Genet.* **17,** 97–103.

Barry, M. A., Barry, M. E., and Johnston, S. A. (1994). *BioTechniques* **16,** 616–620.

Boynton, J. E., Gillham, N. W., Harris, E. H., Hosler, J. P., Johnson, A. M., Jones, A. R., Randolph-Anderson, B. L., Robertson, D., Klein, T. M., Shank, K. B., and Sanford, J. C. (1988). *Science* **240,** 1534–1538.

Breakefield, X. O. (1993). *Nature Genet.* **3,** 187–189.

Daniell, H. (1993). *Meth. Enzymol.* **217,** 536–556.

Fitzpatrick-McElligott, S. (1992). *BioTechnology* **10,** 1036–1040.

Gallie, D. R., Feder, J. N., Schimke, R. T., and Walbot, V. (1991). *Mol. Gen. Genet.* **228,** 258–264.

Heiser, W. C. (1994). *Anal. Biochem.* (*in press*).

Johnston, S. A., and Tang, D. (1993). *Genet. Eng.* **15,** 225–236.

Johnston, S. A., Anziano, P. Q., Shank, K., Sanford, J. C., and Butow, R. A. (1988). *Science* **240,** 1538–1541.

Johnston, S. A., Williams, R. S., Riedy, M., DeVit, M. J., Sanford, J. C., and McElligott, S. (1991). *In Vitro Cell. Dev. Biol.* **27,** 11–14.

Klein, T. M., Arentzen, R., Lewis, P. A., and Fitzpatrick-McElligott, S. (1992). *BioTechnology* **10,** 286–291.

Lanzavecchia, A. (1993). *Science* **260,** 937–944.

Mulligan, R. C. (1993). *Science* **260,** 926–932.

Nicolau, C., Pape, A. L., Soriano, P., Fargette, F., and Juhel, M.-F. (1983). *Proc. Natl. Acad. Sci. USA* **80,** 1068–1072.

Plautz, G. E., Yang, Z., Wu, B., Gao, X., Huang, L., and Nabel, G. J. (1993). *Proc. Natl. Acad. Sci. USA* **90,** 4645–4649.

Sanford, J. C. (1990). *Physiol. Plant.* **79,** 206–209.

Sanford, J. C., Klein, T. M., Wolf, E. D., and Allen, N. (1987). *Particle Sci. Technol.* **5,** 27–37.

Sanford, J. C., DeVit, M. J., Russell, J. A., Smith, F. D., Harpending, P. R., Roy, M. K., and Johnston, S. A. (1991). *Technique* **3,** 3–16.

Tang, D., DeVit, M., and Johnston, S. A. (1992). *Nature* (*London*) **356,** 152–154.

Tang, D., Johnston, S. A., and Carbone, D. P. (1994). *Cancer Gene Ther.* (*in press*).

Toffaletti, D. L., Rude, T. H., Johnston, S. A., Durack, D. T., and Perfect, J. R. (1993). *J. Bacteriol.* **175,** 1405–1411.

Williams, R. S., Johnston, S. A., Riedy, M., DeVit, M. J., McElligott, S. G., and Sanford, J. C. (1991). *Proc. Natl. Acad. Sci. USA* **88,** 2726–2730.

Wolff, J. A., Malone, R. W., Williams, P., Chong, W., Acsadi, G., Jani, A., and Felgner, P. L. (1990). *Science* **247,** 1465–1468.

Yang, N.-S., Burkholder, J., Roberts, B., Martinell, B., and McCabe, D. (1990). *Proc. Natl. Acad. Sci. USA* **87,** 9568–9572.

Zelenin, A. V., Titomirov, A. V., and Kolesnikov, V. A. (1989). *FEBS Lett.* **244,** 65–67.

Zelenin, A. V., Alimov, A. A., Barmintzev, V. A., Beniumov, A. O., Zelenina, I. A., Krasnov, A. M., and Kolesnikov, V. A. (1991). *FEBS Lett.* **287,** 118–120.

# INDEX

## H

I



K



L



M



N

T



V

**Y**



**Z**

# VOLUMES IN SERIES

**Founding Series Editor**
**DAVID M. PRESCOTT**

**Volume 1 (1964)**
**Methods in Cell Physiology**
*Edited by David M. Prescott*

**Volume 2 (1966)**
**Methods in Cell Physiology**
*Edited by David M. Prescott*

**Volume 3 (1968)**
**Methods in Cell Physiology**
*Edited by David M. Prescott*

**Volume 4 (1970)**
**Methods in Cell Physiology**
*Edited by David M. Prescott*

**Volume 5 (1972)**
**Methods in Cell Physiology**
*Edited by David M. Prescott*

**Volume 6 (1973)**
**Methods in Cell Physiology**
*Edited by David M. Prescott*

**Volume 7 (1973)**
**Methods in Cell Biology**
*Edited by David M. Prescott*

**Volume 8 (1974)**
**Methods in Cell Biology**
*Edited by David M. Prescott*

**Volume 9 (1975)**
**Methods in Cell Biology**
*Edited by David M. Prescott*

**Volume 10 (1975)**
**Methods in Cell Biology**
*Edited by David M. Prescott*

**Volume 11 (1975)**
**Yeast Cells**
*Edited by David M. Prescott*

**Volume 12 (1975)**
**Yeast Cells**
*Edited by David M. Prescott*

**Volume 13 (1976)**
**Methods in Cell Biology**
*Edited by David M. Prescott*

**Volume 14 (1976)**
**Methods in Cell Biology**
*Edited by David M. Prescott*

**Volume 15 (1977)**
**Methods in Cell Biology**
*Edited by David M. Prescott*

**Volume 16 (1977)**
**Chromatin and Chromosomal Protein Research I**
*Edited by Gary Stein, Janet Stein, and Lewis J. Kleinsmith*

**Volume 17 (1978)**
**Chromatin and Chromosomal Protein Research II**
*Edited by Gary Stein, Janet Stein, and Lewis J. Kleinsmith*

**Volume 18 (1978)**
**Chromatin and Chromosomal Protein Research III**
*Edited by Gary Stein, Janet Stein, and Lewis J. Kleinsmith*

**Volume 19 (1978)**
**Chromatin and Chromosomal Protein Research IV**
*Edited by Gary Stein, Janet Stein, and Lewis J. Kleinsmith*

**Volume 20 (1978)**
**Methods in Cell Biology**
*Edited by David M. Prescott*

## Advisory Board Chairman
## KEITH R. PORTER

**Volume 21A (1980)**
**Normal Human Tissue and Cell Culture, Part A: Respiratory, Cardiovascular, and Integumentary Systems**
*Edited by Curtis C. Harris, Benjamin F. Trump, and Gary D. Stoner*

## Series Editor
## LESLIE WILSON

Volume 31 (1989)
**Vesicular Transport, Part A**
*Edited by Alan M. Tartakoff*

Volume 32 (1989)
**Vesicular Transport, Part B**
*Edited by Alan M. Tartakoff*

Volume 33 (1990)
**Flow Cytometry**
*Edited by Zbigniew Darzynkiewicz and Harry A. Crissman*

Volume 34 (1991)
**Vectorial Transport of Proteins into and across Membranes**
*Edited by Alan M. Tartakoff*

Selected from Volumes 31, 32, and 34 (1991)
**Laboratory Methods for Vesicular and Vectorial Transport**
*Edited by Alan M. Tartakoff*

Volume 35 (1991)
**Functional Organization of the Nucleus: A Laboratory Guide**
*Edited by Barbara A. Hamkalo and Sarah C. R. Elgin*

Volume 36 (1991)
***Xenopus laevis:* Practical Uses in Cell and Molecular Biology**
*Edited by Brian K. Kay and H. Benjamin Peng*

## Series Editors
## LESLIE WILSON AND PAUL MATSUDAIRA

Volume 37 (1993)
**Antibodies in Cell Biology**
*Edited by David J. Asai*

Volume 38 (1993)
**Cell Biological Applications of Confocal Microscopy**
*Edited by Brian Matsumoto*

Volume 39 (1993)
**Motility Assays for Motor Proteins**
*Edited by Jonathan M. Scholey*

Volume 40 (1994)
**A Practical Guide to the Study of Calcium in Living Cells**
*Edited by Richard Nuccitelli*

**Volume 41 (1994)**
**Flow Cytometry, Second Edition, Part A**
*Edited by Zbigniew Darzynkiewicz, J. Paul Robinson, and Harry A. Crissman*

**Volume 42 (1994)**
**Flow Cytometry, Second Edition, Part B**
*Edited by Zbigniew Darzynkiewicz, J. Paul Robinson, and Harry A. Crissman*

**Volume 43 (1994)**
**Protein Expression in Animal Cells**
*Edited by Michael G. Roth*